HOME VEGETABLE GARDENING

FIRST EDITION

By Alan Walters
Southern Illinois University – Carbondale

cognella
academic publishing

Bassim Hamadeh, CEO and Publisher
Michael Simpson, Vice President of Acquisitions
Jamie Giganti, Managing Editor
Jess Busch, Senior Graphic Designer
Angela Kozlowski, Acquisitions Editor
Michelle Piehl, Project Editor
Alexa Lucido, Licensing Associate
Mandy Licata, Interior Designer

First published in the United States of America in 2015 by Cognella, Inc.

ISBN: 978-1-5165-5295-5

cognella
academic publishing
www.cognella.com 800-200-3908

CONTENTS

CHAPTER NINE

CHAPTER TEN

CHAPTER ELEVEN

CHAPTER TWELVE

CHAPTER THIRTEEN

CHAPTER FOURTEEN

CHAPTER FIFTEEN

CHAPTER SIXTEEN

PREFACE

Gardening is in our blood, so to speak. Home gardening is an important part of our culture and for many of us who grew up in rural areas, it was a yearly ritual of planting in the spring and spending a lot of our time in the summer and fall working in the garden. Although gardening is labor-intensive, sweaty work, the benefit received from having fresh vegetables was well worth the effort. To me, it brings back such great memories of my youth and the people in my family who has since passed who taught me many things.

However, not only those of us who grew up in rural areas have interests in gardening. Many gardeners now live in highly urbanized areas and are now growing their own vegetables. They realize that those bought from local supermarkets do not compare in taste, quality, or freshness with those grown in a home garden. So, many are now trying their hand at container vegetable gardening to raise fresh vegetables on any sunlit space that is available, such as balconies or patios. A vegetable garden can also be used to supplement a family's food budget, as well as a source of those vegetables that can't easily be found in the supermarket, such as various Chinese vegetables, kohlrabi, culinary or medicinal herbs, or various heirloom vegetables. Flowers and various other herbaceous and woody ornamental plants also play important roles in gardening by increasing the aesthetics around or in the home, and many gardeners combine both food and beauty.

Home gardening can be seen all around us, as all we have to do is observe our surroundings a little more. As you drive along rural highways during the summer months, take notice that many homes have small flower, fruit, or vegetable gardens adjacent to them. When you drive in urbanized areas, observe the number of city residents who try to grow flowers and vegetables in confined areas with many of these plants grown in some type of container. Almost anyone can grow these plants and it can be easily done. It is possible to enliven our living spaces with flowers, fruits, or vegetables and all one needs is a little sunlight and motivation to do so. Home gardening is often hot, sweaty work, but it is rewarding in the end—and gets easier with experience.

This book had its beginnings as my set of lecture notes that I developed for the Home Gardening class that I teach at Southern Illinois University in Carbondale. For several years, I have made these notes available for the students who took this class. These notes have now been modified even more and developed into a book. My goal is that this text will not only provide a useful source for students, but also to others who are interested in gaining a better understanding of vegetables and the methods used to produce these crops in home gardens.

Dr. Alan Walters is Professor of Vegetable Science at Southern Illinois University in Carbondale, Illinois.

INTRODUCTION

The interest in home gardening seems to increase every year and this is occurring for several reasons. In the past, many would garden as a hobby, to have fresh vegetables available for the kitchen, or to save money and reduce the amount spent for food in a grocery store. However, today, consumers are becoming increasingly interested in and aware of developing healthier lifestyles and eating habits, and are becoming more knowledgeable about fresh fruits and vegetables. Both gardeners and consumers are concerned about their nutrition and health, and are becoming more interested in eating fresh fruits and vegetables that do not have toxic pesticide residues. Some consumers perceive organically produced vegetables as being safer primarily from reduced pesticide residues. Many consumers are willing to pay significantly more for organic produce, as they associate these foods as being healthier, more nutritious, better tasting, safer, and produced through more environmentally friendly methods than those produced through conventional agricultural practices. However, organic produce is sometimes two or three times the cost of that produced conventionally with pesticides. Thus, a major reason that many have their own garden is to produce organic foods for their family. This allows them to possibly avoid the high costs associated with organic produce by growing their own. This gives gardeners and consumers assurance that they, their children, or other family members are eating products that are entirely natural and free of pesticides.

From an environmental cost standpoint, the carbon emissions associated with the transport of food from farm to supermarket can be substantial. Greenhouse gases, which include the carbon dioxide and other gases released when fossil fuels are burned for energy, are an environmental concern. These gases absorb and trap the heat in the air and re-radiate that heat downward, contributing to the trend in global warming. The carrots or lettuce that you put in your grocery cart at a supermarket travels an average distance of about 1,600 miles to reach you. Thus, growing your own produce has environmental benefits, as the less your food travels, the less energy is being used, and less carbon dioxide gas emissions are being created in getting the food to you.

Growing your own produce in a home garden is its own reward. Although many home gardeners gain the satisfaction of watching their food grow, they also get to enjoy the taste of fresh fruits and vegetables once crops are ready to harvest. Whether a gardener is growing a few herbs on the windowsill of a studio apartment or putting in rows of veggies in an acre-sized backyard, cultivating your own food is a great way to eat healthy, cut down on your carbon footprint, reduce pesticide-residue consumption, and encourage a produce-heavy diet, full of easy-to-grow vegetable crops, like lettuce and tomatoes, and herbs. Anyone can grow a lot of tasty produce in a small space if they are careful about picking the right crops and providing the proper growing conditions. A good rule of thumb is that about a half-pound of vegetables can be grown in a 1 ft x 1 ft space.

Chapter One

HOME GARDEN PREPARATION

Garden preparation is the key to successful gardening. What we do before we even plant a seed in the soil, water or fertilize a plant is very important to how successful a garden will be. It doesn't matter how you garden outside, whether in containers or in the ground. All of these factors that I am going to discuss next are very important for a gardner to think about prior to establishing a home garden.

SITE SELECTION

Some home gardeners do not select the proper site in which to locate their garden and have problems that can be observed year after year due to the location of the garden. So, site selection is an important step in developing a successful garden. There are many things that you should consider before selecting and developing a site for garden use. A few things that the new home gardener should consider that will influence their gardening success are the climate in the area, sunlight exposure, the soil on the site, the water supply, and the accessibility to the garden from the house.

CLIMATE

Climate is weather conditions over time for a particular region. The climate has a direct effect on which plants can be grown in a given area and the best planting dates to achieve optimum results. For example, southern Illinois is considered to be in the lower Midwest and is in Zone 6b according to the USDA Plant Hardiness Map (Fig. 1), which divides the U.S. into eleven plant hardiness zones based on average minimum temperatures. The minimum temperatures for Zone 6b is –5° to 0°F. It is easy to find the region of the U.S. in which you have an interest; see Fig. 1.1.

There are four critical components of climate that we need to be concerned about before we decide on a planting a home garden. These are

1) **Temperature**
2) **Precipitation**
3) **Solar Radiation**
4) **Air Movement and Composition (hot or cold air)**

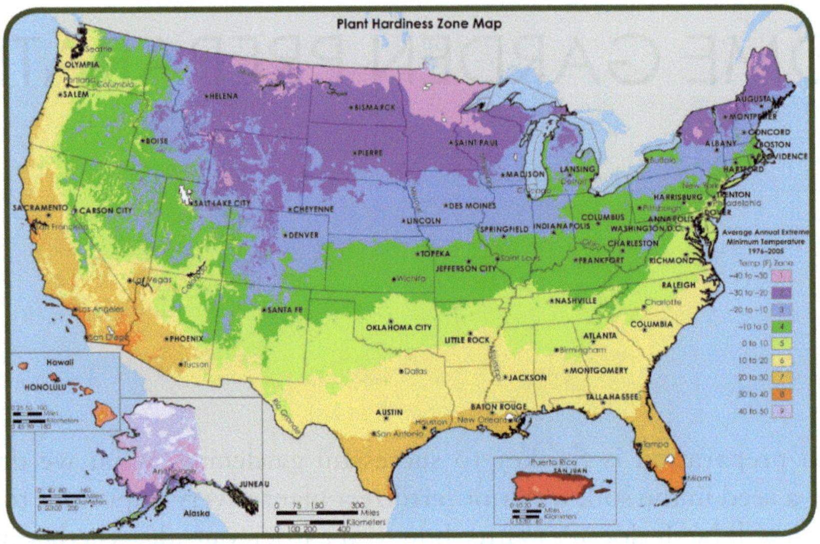

Figure 1.1 USDA Plant Hardiness Map.

The first component is **temperature,** which really determines the length of growing season and is affected by:

Latitude. We generally associate that it will be 1°F cooler for every 200 miles from equator, and this temperature decrease with increasing latitude is due to a decrease in the as the angle of sunlight reaching a surface.

Altitude or elevation. There is also an average decrease of 3° to 4°F for every 1,000 feet of an increase in elevation.

Large bodies of water. The proximity to large bodies of water affects the temperature for a given area; bodies of water such as oceans, rivers, and large lakes reduce temperature extremes and give a land area cooler summers and warmer winters.

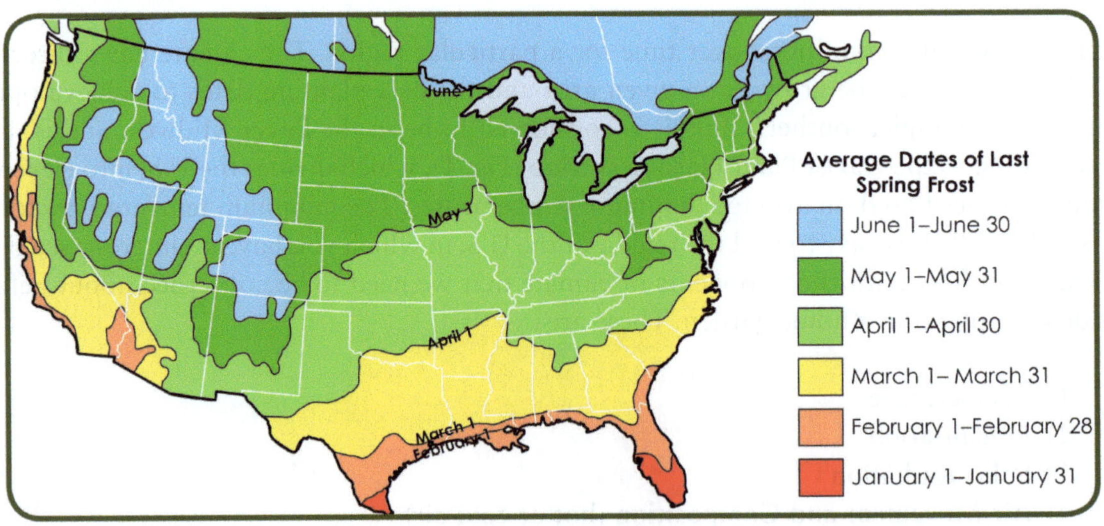

Figure 1.2 The average dates of the last spring frost for specific regions of the U.S.

Temperature definitely affects the length of the frost-free growing season, which is the average number of days between the last 32°F temperature in the spring (last frost date) and the first 32°F temperature in the fall (first frost date). The length of this frost-free period determines the length of the growing season in which we can effectively grow plants outdoors. The season length is the most important factor that determines what—the types of plants—you can grow in a given region. Many areas of the lower Midwest, such as southern Illinois, has about a 180-day frost-free growing season, which is plenty of time to grow most annual vegetable crops. Compare this to the middle South (which extends from Richmond, Virginia, westward to Abilene, Texas, and includes most large southern cities—this encompasses USDA zones 6 to 8 which have about a 200-day growing season. Remember that the frost-free growing season is determined primarily by temperature. Most planting directions for annual plants are based on the average last-frost date. The last-frost date for an area is the last day in the spring that you might have a killing frost (Fig. 1.2). Table 1.1 lists the last- and first-frost dates for selected cities in the U.S.

Table 1.1 Length of growing season and average dates of last (spring) and first (fall) frost for selected cities in the continental United States and Alaska

City	Growing Season (days)	Spring (last frost date)	Fall (first frost date)
Akron, OH	192	18 April	28 October
Albany, NY	153	2 May	3 October
Amarillo, TX	185	18 April	20 October
Athens, GA	227	24 March	7 November
Baltimore, MD	200	11 April	29 October
Bismarck, ND	129	14 May	21 September
Boise, ID	147	10 May	6 October
Burlington, VT	147	8 May	3 October
Carbondale, IL	189	10 April	15 October
Carlsbad, NM	215	31 March	2 November
Cedar Rapids, IA	163	25 April	6 October
Charleston, SC	260	9 March	25 November
Chicago, IL	187	20 April	24 October
Cincinnati, OH	192	13 April	23 October
Columbia, SC	214	1 April	1 November
Concord, NH	124	20 May	21 September
Dallas, TX	243	18 March	17 November
Denver, CO	157	30 April	4 October
Duluth, MN	124	21 May	23 September
Eureka, CA	323	27 January	16 December
Fayetteville, NC	222	28 March	5 November
Flagstaff, AZ	106	9 June	22 September
Green Bay, WI	150	6 May	4 October

Continued

Hartford, CT	166	26 April	9 October
Helena, MT	121	19 May	18 September
Houston, TX	281	28 February	6 December
Indianapolis, IN	181	17 April	16 October
Janesville, WI	164	28 April	10 October
Jefferson City, MO	188	13 April	18 October
Juneau, AK	148	8 May	4 October
Kingston, RI	147	8 May	3 October
Lansing, MI	145	10 May	3 October
Las Vegas, NV	283	16 February	27 November
Lexington, KY	192	15 April	25 October
Los Alamos, NM	149	11 May	8 October
Memphis, TN	235	22 March	13 November
Miami, FL	*	*	*
Mobile, AL	273	28 February	29 November
Monroe, LA	256	3 March	15 November
Nashville, TN	204	6 April	28 October
New Orleans, LA	300	12 February	11 December
Newark, NJ	217	3 April	7 November
Norfolk, VA	247	20 March	23 November
North Platte, NE	137	9 May	24 September
Parkersburg, WV	183	21 April	22 October
Pine Bluff, AR	240	16 March	12 November
Phoenix, AZ	*	*	*
Portland, ME	156	2 May	6 October
Portland, OR	236	23 March	15 November
Prescott, AZ	163	7 May	18 October
Rapid City, SD	140	9 May	27 September
Richmond, VA	206	6 April	30 October
Sacramento, CA	296	10 February	4 December
San Antonio, TX	270	28 February	25 November
San Francisco, CA	*	*	*
Savannah, GA	268	1 March	25 November
Seattle, WA	251	10 March	17 November
South Bend, IN	175	26 April	19 October
Spokane, WA	153	2 May	3 October
Springfield, IL	182	13 April	13 October
Syracuse, NY	168	28 April	13 October
Tampa, FL	338	28 January	3 January
Topeka, KS	174	19 April	11 October
Tucson, AZ	324	19 January	18 December
Tulsa, OK	225	27 March	7 November

Vicksburg, MS	240	20 March	16 November
Williamsport, PA	168	30 April	15 October
Wilmington, DE	202	10 April	30 October
Worcester, MA	170	26 April	14 October

* Frosts do not occur every year.

Dates given are normal averages for a light freeze or frost (32°F). Local weather and topography will cause considerable variations. There is a 50% possibility of frost occurring after the spring dates and before the fall dates provided. The classification of freezing temperatures is usually based on their effect on plants, with the following commonly accepted categories:

Light freeze: 29°F to 32°F—tender plants killed, with little destructive effect on other vegetation (see Figs. 1.3 and 1.4).

Moderate freeze: 25°F to 28°F—widely destructive effect on most vegetation, with heavy damage to fruit blossoms and tender and semi-hardy plants.

Severe freeze: 24°F and colder—cold-injury damage will occur to most plants.

Another component of climate is precipitation. **Precipitation**, or the amount of rainfall, determines if the climate is wet or dry or is somewhere in between. If temperatures fall within a given range for plant growth, then water is probably the most important factor in allowing plants to grow. Water stress results in growth suppression and eventual plant death. Water is an essential component of plant life and the amount of water available for plant growth is directly related to the amount of precipitation for a given area.

Solar Radiation (or amount of sunlight) is another factor that influences climate and is determined by the amount of cloud cover/fog, which will reduce the available sunlight that reaches plants. Solar radiation also determines day lengths for specific regions, which plays an important role for the growth and development of some vegetables. Most vegetables prefer full

Figure 1.3 A late October frost on tomatoes. Notice ice crystals on leaves (A) and frost damage that will develop later in the day as blackening necrotic areas on tomato foliage (B).

sun, but some can tolerate shade conditions, including beets, cabbage, carrots, chives, kale, leeks, lettuce, mustard, green onions, parsley, radishes, Swiss chard, and turnips, which are all cool-season crops. All warm-season crops prefer full sun conditions and require at least 6 to 8 hours of direct sunlight every day. Cloud cover plays a major role in gardening, as it can definitely reduce the amount of solar radiation that reaches plants, and is an important consideration in determining the types of plants that can be grown in specific regions and the length of time needed to grow certain crops (Table 1.2).

***Figure 1.4** A late October frost on romaine lettuce. Even with the ice-crystal formation on leaves, this plant will survive, since it is a cool-season plant and can survive a few light frost events.*

Solar radiation also plays an important role in maintaining warm conditions around plants. Since there is no incoming heat at night to warm the soil and plants, they continue to lose heat through radiation and conduction until they are cooler than the surrounding air. The air then passes heat to the soil and plants, and the area in the lower atmosphere cools. Furthermore, if there is no cloud cover present to block the outgoing radiation, the soil, plants, and air temperatures will continue to decrease significantly, especially during cooler periods of the year. This is very important when we are trying to grow plants early and late in the growing season. However, the greenhouse effect of cloud cover can limit this temperature decrease at night (Fig. 1.5). A ***radiation frost*** occurs when a clear sky and calm winds (less than 5 mph) allow an inversion to develop, and temperatures near the surface drop below freezing. The inversion layer varies in thickness from 30 to 200 feet. There are two types of frost. A ***white frost*** results when atmospheric moisture freezes in small crystals on solid surfaces, while a ***black frost*** results when few or no ice crystals form because the air in the lower atmosphere is too dry. The formation of ice crystals depends on the dew point, or frost point, which is the temperature to which air must be cooled to cause atmospheric moisture to condense. The drier the air, the lower the dew point.

What Is an Inversion? During clear nights, the heat from solid objects will continue to radiate out into the atmosphere. This results in a significant temperature drop at the Earth's

***Table 1.2** Comparison of tomato cultivars for maturity based on normal sunny days compared to cloudy conditions*

	Maturity from Transplants	
Tomato Cultivar	**Normal (# days)**	**Cloudy (# days)**
Early Girl	57	100
Better Boy	70	120
Celebrity	70	120

surface. The temperature in the lower parts of the atmosphere inverts, that is, the temperature increases with altitude to the top of the air layer. The term **inversion** comes from atmospheric conditions being inverse to that of normal daytime condition where air temperature decreases with increases in height. The warm air in an inversion is important for some frost protection methods that depend on this source of heat to protect crops.

Frost vs. Freeze. Although the terms *frost* and *freeze* are often interchanged, they describe two distinct phenomena. An *advective*, or *windborne*, freeze occurs when a cold air mass moves into an area bringing freezing temperatures. Wind speeds are usually above 5 mph and clouds may be present. The thickness of the cold air layer can range from about 500 feet to more than 5,000 feet above the Earth's surface. Attempts to protect crops by modifying the environment are very limited under these freeze conditions.

Air Movement and Composition also affects the growth and development of plants. Cold air and warm air move in different ways. *Cold air flows like water* to the lowest possible area, so it is best to place gardens in higher, more elevated areas to avoid damage to plants by the effects of cold-air drainage. Early or late in the season, frost is more likely to develop in these areas to which cold air drains compared to higher areas. *The best method of frost/freeze protection is good site selection.* Microclimate monitoring of a particular site location can be used to determine the suitability of the site before planting. Visualizing the flow of cold air and its possible build-up in low spots or behind cold air dams, such as fences, hedges, wooded areas, is the most effective and quickest method to determine site suitability. If a site has good cold-air drainage, then it is probably a good production site as far as frost/freeze protection is concerned.

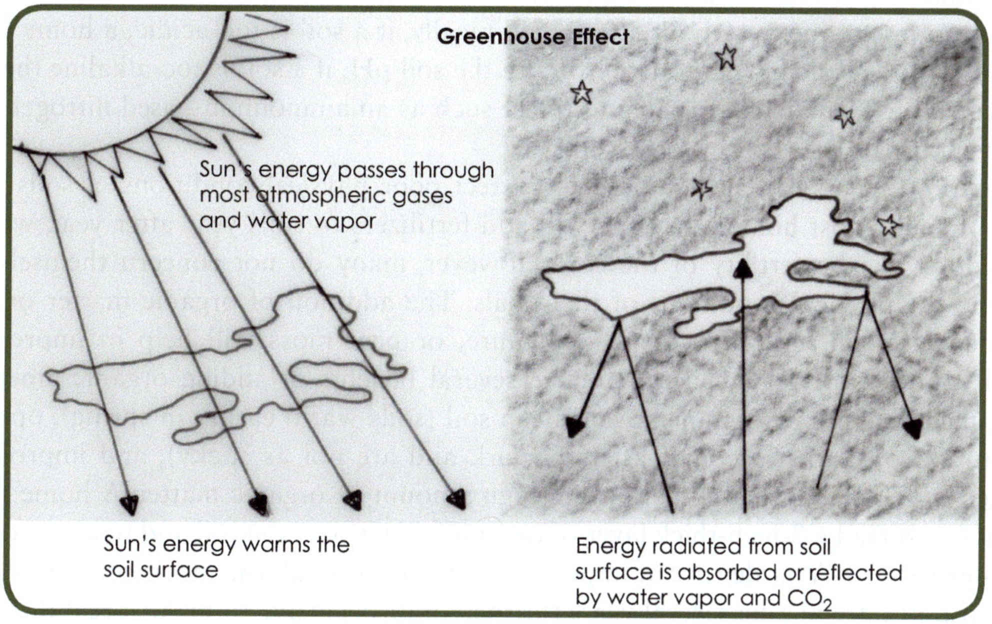

Figure 1.5 Sunlight energy exchange showing the greenhouse effect of cloud cover.

SOIL CONDITIONS

Soil is another factor that the home gardener should consider when selecting the proper site to locate the garden. Many gardeners think of soil as dirt, but soil should be thought of as a living component of the garden. There are several aspects of a soil that one needs to consider before planting a garden including soil drainage, soil texture, and soil acidity.

Soil drainage is the ability of a soil to drain excess water through the soil profile. If drainage is a problem in your garden soil, then the use of raised beds should be considered. Excess water reduces the soil-oxygen levels and inadequate amounts of oxygen in a soil will cause plants to grow poorly or die. Wet soils are also slower to warm in the spring, which can lead to seed germination problems. Vegetables and other plants will not grow properly in waterlogged soils, so the home gardener needs to be aware of the negative effects that poorly drained soils will have on plant growth.

Soil texture is the percentage of sand, silt, and clay in a given soil. Most vegetable crops grow best in a loamy soil consisting of 35% to 45% sand, 35% to 45% silt, and 10% to 25% clay. If a soil has high amounts of sand, then the soil will often dry out very quickly, resulting in wilting of plants. A high-clay soil will hold water better than a sandy soil, but often has other problems, such as water-logging or preventing seedling emergence. Seedling emergence is often prevented when the clay soil dries out and a crust forms on the top of the soil that will not allow the seedling to emerge.

Soil acidity is a common problem in many soils, including those soils that are highly oxidized (or have a reddish color). Vegetable plants have both an optimum pH and suitable ranges in which to grow. Most vegetable crops require a pH of 6.0 to 7.5 for optimum growth and most of the major plant nutrients are readily available in this range. Certain vegetables such as potatoes require a slightly acid soil (4.8 to 5.4 pH). Blueberries also require a slightly acidic soil to be productive (about 5.0 pH). Typically, if a soil is too acidic, a home gardener should apply lime (calcium carbonate) to raise the soil pH; if a soil is too alkaline (has a high pH), then sulfur or an acidic-forming fertilizer such as an ammonium-based nitrogen ($NH4^+$) fertilizer should be used to lower the soil pH.

Soil improvements can also be used to correct poor physical conditions of soils and low fertility of soils. Most home gardeners will add fertilizers to soils year after year, which will aid in improving the fertility of the soil. However, many do not concern themselves with improving the physical conditions of their soils. The addition of organic matter or organic amendments to soils such as compost, manure, or peat moss will help in improving the physical condition of poor soils. There are several benefits of adding organic amendments to soils, including improving air movement in soil (soils warm earlier in spring), opening up fine-textured soils (soils become easier to work and are not as sticky), and improving soil drainage. A well-managed soil will have a high amount of organic matter. A home gardener should try to spread a 2-inch-thick layer of organic matter over the soil and work to a depth of 6 inches every year. A gardener can improve a clay soil through the incorporation of manure or compost, but it will take several years for these improvements to make a real difference in soil quality. However, improvements made to soils will be highly beneficial in a few years and will be worth the additional effort, allowing a gardener to grow certain vegetables that were once thought impossible to grow.

WATER SUPPLY

Before deciding where to locate a garden, a gardener should keep the garden within an acceptable distance to a water supply (Fig. 1.6). The water supply is crucial since plants cannot grow and develop without water. Vegetables and other plants in the garden usually need about 1 to 1½ inches of water on a weekly basis, and natural rainfall will not always supply the necessary water, especially at the exact time that matches plant growth requirements. During times of little or no rainfall, a supplemental source of water is essential for proper plant development. So, it is essential to locate a garden near a water source. Gardeners need to also remember to water their garden in the early morning (especially if sprinkler irrigation is used) because plants will lose less water to evaporation, and plants need to stay dry at night. Plants watered at night tend to stay wet until morning due to the lack of heat build-up in leaves resulting from no solar radiation. Dry plants are less susceptible to disease-causing organisms and most disease-causing organisms need high humidity to develop.

Gardeners also need to be aware of the problem of over-watering. The use of excess water in a garden can stop root growth, and when a soil becomes deprived of air, roots become more susceptible to rot-causing organisms that can eventually cause plant death. The best way to deal with watering is to prepare your soil properly. Clay soils have a high water-holding capacity but will not drain properly. *Most vegetables do best in well-drained soils.* The use of raised beds in a garden can help alleviate this problem if you have high-clay soils. However, under-watering can be worse than over-watering. Water stress in vegetable plants will often cause flowers to abort (e.g., snap beans and tomatoes), whereas lettuce and cucumbers will develop a bitter flavor under low soil-moisture conditions. Furthermore, overall vegetable productivity will decrease without adequate moisture in the soil.

Gardens should be properly watered. The rule of thumb is to water thoroughly twice a week, with deep watering important to get water down to the root zone. It is best to not water a small amount every day, as that tends to produce a shallow root system that is close to the soil surface. This watering method will make plants more vulnerable to drought, especially during periods when no moisture is supplied to the garden.

***Figure 1.6** A water supply, such as this spigot with the attached hose, is essential for a home garden to grow adequately.*

Irrigation Methods

There are several types of irrigation that gardeners can use in a home garden, including sprinkler, furrow, flood, basin, and trickle. Most gardeners will use a hose connected to their water supply and use a sprinkler type of irrigation. Although **sprinkler irrigation** is a common delivery method in home gardens to get adequate amounts of water to plants, it is only about 50% efficient, which means only about 50% of the water applied is used by plants. Furrow and flood irrigation practices are methods that are not commonly used in home gardens. For **furrow irrigation**, shallow trenches between raised beds channel water between the beds to plant roots (Fig. 1.7). The raised beds are generally 1 to 3 feet apart, with the garden placed slightly on a slope to allow gravity to move water from one side of the garden to the other, and more water is required as the area widens between beds. With this irrigation method, it takes time for water to flow from one end of the garden to the other, and water is often wasted through evaporation, as it often sits in the furrow for extended periods of time and areas at the beginning of a row always gets more water than the end. Flood and basin irrigation methods are similar. For **flood irrigation**, portions or even entire garden areas are surrounded by berms to keep water from escaping. The area within the berms is then filled with water and the water is allowed to percolate down into the soil. This irrigation method is commonly used to provide water to a large field or orchard areas (Fig. 1.8). With **basin irrigation**, a donut-shaped depression around a plant in the garden is made, filled with water, and the water is then allowed to percolate down into the soil. **Trickle** (or **drip**) irrigation delivers water to plants at low pressure in plastic pipes or tubes laid under or on the soil surface (Fig. 1.9). Water is applied much more slowly than with other irrigation methods and 90% water-use efficiency rates are often exceeded. Trickle irrigation is the most widely used method for providing water to commercial vegetable plantings (Fig. 1.10).The downside to trickle irrigation is that it is more costly than other methods due to the equipment required.

Figure 1.7 Furrow irrigation being used in a home garden to moisten the soil prior to planting (left) and in a commercial tomato field near Davis, California (right).

GARDEN ACCESSIBILITY

The accessibility of the garden to the gardener is a very important for developing and maintaining a garden. Easy accessibility to a garden is a necessity for the gardener. A vegetable garden that is close to the gardener's home will be easy to reach and more likely be used and maintained. The home gardener will be more likely to harvest vegetables at their peak harvest stages, and thus, take advantage of fresh garden produce. It is also more likely that the gardener will keep up with tasks such as weeding, watering, and insect and disease control if the garden is easily accessible.

Figure 1.8 Flood irrigation being used in a commercial pecan orchard near Fresno, California.

Figure 1.9 Trickle (or drip) irrigation in both a raised bed (left) and in a container to supplement water needs for garden plants.

GARDEN PLANNING

The planning of a garden is important to its success. A plan needs to be developed several months before the planting is done to ensure that there is adequate space in the garden for the vegetables or other plants that you want to grow. Garden planning will allow the gardener to make the best use of the available garden space. Cultivar selection, garden design, and seeding and field planting methods are some of the more important elements in garden planning. Other techniques that a gardener needs to be aware of are the advantages of thinning vegetable seedlings and proper seed collection and seed storage conditions.

Figure 1.10 Trickle (or drip) irrigation used for commercial lettuce production in the Central Valley of California during the summer. Notice the shade cloth above to prevent exposure to intense sunlight.

CULTIVAR SELECTION

The first step in planning a garden is selecting the cultivars of flowers, shrubs, fruits, or vegetables that you want to grow. There are several things that a gardener needs to pay attention to when selecting specific cultivars to grow in the garden. Discussion will be targeted toward vegetables, but the topic areas relate to other horticultural crops as well.

Adaptability to area is determined primarily by the length of frost-free season, and not all vegetables can be grown in all areas (e.g., sweet potatoes require approximately 120 days to mature after transplanting and can't be grown in areas with short growing seasons).

The **Earliness/Maturity** of vegetable varieties needs to be considered. Some vegetable varieties produce a crop quicker than others, and the number of days from planting to maturity can vary considerably from one variety to another. Several varieties of one vegetable with differing maturity dates can be planted at one time to increase the harvest period or one variety can be planted at 3- to 4-week intervals to get a continuous harvest.

Growth habits can differ among vegetable varieties. Some vegetable varieties are dwarfs and do not require as much space in the garden to grow and produce a crop. Also, many dwarf varieties do not require trellis support, while others varieties of the same vegetable do require support (e.g., bush beans versus pole beans, and determinate tomatoes versus indeterminate tomatoes). Make sure you know which growth habit you want or prefer before buying seed or transplants.

Vegetable varieties will definitely differ in their **productivity**. Some varieties are more productive than others. Typically, hybrids will produce more than open-pollinated varieties. New vegetables varieties are continuously being developed and usually outperform older varieties.

The **quality** of the vegetable product will definitely differ among varieties. Vegetable varieties differ greatly in flavor, texture, keeping ability, and adaptability to processing procedures such as canning and freezing. The final use of the product (either fresh or processed) will help determine the best variety for any given situation. Heirloom vegetables are becoming popular and they often have better flavors than the newer varieties, but remember that they usually have other problems associated with them. For example, heirloom tomatoes have good flavor, color, and size, but have problems with green shoulders and fruit cracking, and most have no disease resistance.

Disease resistance differs among vegetable varieties. Most of the newer vegetable varieties have some disease resistances, whereas older varieties and heirloom varieties usually do not. Growing disease tolerant/resistant varieties is important since fewer pesticides will be needed to control these pathogens in a garden.

SEED SELECTION

When purchasing vegetable seed, remember it is best to purchase new seed from year to year. Also, remember that hybrid vegetable seed cannot be collected from the garden and replanted at a later date due to the genetic makeup of the seed, as they will not come back true to the hybrid plant that was first planted in the garden. When new seeds are purchased, the seed packet should indicate the vegetable, the variety, the percent germination, the date of germination testing, and the year packeted. It is best to order and buy seed during the late fall or winter months prior to spring seeding, so that there will be no delay in starting your garden. There are many seed companies from which to purchase seed depending on your needs, including Burpee (burpee.com), Johnny's Selected Seed (johnnyseeds.com), Jung Quality Seed (www.jungseed.com), Parks Seed (parkseed.com), Southern Exposure Seed Exchange (www.southernexposure.com), Territorial Seed Company (www.territorialseed.com), Tomato Growers Supply Company (www.tomatogrowers.com), and Totally Tomatoes (www.totallytomato.com), to name a few.

Also, consider if the particular vegetable variety is an **All-America Selection** (AAS). What is an All-America Selection? AAS is an independent, non-profit organization that began 80 years ago to conduct confidential and impartial trials of new, never-before-sold flower and vegetable seed varieties throughout North America. Entries are trialed each year in more than 50 locations throughout the U.S. and Canada. These new varieties are grown side by side for comparison with currently available varieties to evaluate garden performance, earliness, flower or fruit size, or any other characteristic important to the home gardener.

The trial judges are horticulture professionals at universities, public gardens, extension offices, seed companies, breeding companies, retailers, and commercial growers. Only the best trial performers that show superiority are declared AAS winners. When you purchase an AAS winner, you can be assured you have a variety that has been evaluated extensively and has out-performed other similar varieties in the marketplace. Many All-American Selection Gardens can be found throughout the U.S., such as the one in Fresno, California (Fig. 1.11)

GARDEN DESIGN

The first rule in home vegetable garden design is to keep it simple. A gardener should consider several factors when designing a garden and these are discussed next.

Garden size is an important component of a garden design. Most first-time gardeners try to do too much and the garden typically becomes overrun with insects, diseases, and weeds. This often discourages potential gardeners from gardening again. Most can initially handle a 20' × 30' or 600 ft^2 garden space. This size garden will be large enough for one person to work alone and be successful, but this amount of space will require one person to work at least ½ hr/day and most likely longer during the growing season (due to constant harvesting, weeding, monitoring and possibly spraying for insect/diseases, etc.). If the garden size is much larger, the gardener is likely to be overwhelmed with work activities that need to be dealt with in a garden, which often leads to frustration and a lost incentive for future gardening activities. However, if the gardener wants to work less, then the size of the garden should be reduced to about 300 ft^2, which would still require about 15 minutes of work per day to adequately maintain.

Gardeners should try to **plan for a continuous harvest** of vegetables. One of the most frustrating problems for gardeners is when everything ripens at once, and after that intense

Figure 1.11 All American Selections (AAS) Display Garden at the Garden of the Sun in Fresno, California.

harvest period, there are no more vegetables left in the garden. It isn't easy to plan for a continuous harvest, but it's possible to have fresh vegetables year round with succession plantings. With succession plantings, you don't use all of your garden space initially, but save some space for later plantings.

Gardeners need to also develop a **Garden Calendar** so they can determine when to plant various crops. Some vegetables can be planted all at once and will produce throughout the growing season, while others need to be planted at defined intervals to provide a continuous harvest. For example, the harvest window is very small for sweet corn (usually about 2 weeks), so it is necessary to plant every 2 to 3 weeks at 4 or 5 different planting dates spaced a few weeks apart to get a continuous harvest of sweet corn throughout the season.

A map of the garden should be developed before the gardener starts planting vegetables or other plants in the garden. A gardener needs to have an idea of the garden layout and this can be easily accomplished by drawing a map of the garden. This will allow the gardener to have a feel for what is growing and the areas of the garden in which they will be grown.

It is also important for the gardener to **keep records** of where vegetables or other plants are grown in the garden from year to year. This is important since this will allow gardeners to have an understanding as to where various vegetables or other plants were grown in the garden each year. A good gardener knows where previous vegetables or flowers were grown in a garden for several years.

It is necessary to **rotate vegetables in garden** to prevent disease and insect build-up in the garden, as well as to build soil nutrient levels. Experienced gardeners rotate vegetables in different families to different areas of the garden from year to year. So, it is important to realize that vegetables must be rotated among families and not just between different vegetables. For example, it is best not to plant an area in tomatoes that had been used to grow peppers the previous year (tomatoes and peppers are in the same family: solanaceae or nightshade family). Some of the major vegetable crop families include the **Amaryllidaceae, Apiaceae, Asteraceae, Brassicaceae, Chenopodiaceae, Cucurbitaceae, Fabaceae, Liliaceae, Poaceae**, and **Solanaceae**. It is best to grow vegetables on a 3-year or longer rotation schedule in the garden.

Crop rotation is a disease-control strategy that can be implemented and incorporated into an integrated pest-management plan for home garden vegetable production. The length of rotation varies depending on the nature of the disease-causing agent (pathogen) that needs to be controlled. Some soil-inhabiting fungi (such as *Fusarium* and *Verticillium*) form special structures that allow them to persist in the soil for many years, while other organisms (such as many bacterial pathogens and foliage-infecting fungi) can persist only as long as the infected crop residue remains intact in the garden.

Table 1.3 provides some guidelines for planning a successful crop rotation program. Some factors may shorten the interval. These include using resistant or tolerant varieties, or lengthening the interval, such as extremely high levels of disease incidence in the garden. While these crop-rotation periods will not guarantee that diseases will not occur, they will greatly reduce the appearance of certain diseases.

Table 1.3 Rotation periods required to control certain vegetable diseases.

Vegetable	Disease	Rotation Period Needed Without a Susceptible Crop to Control Disease
Asparagus	Fusarium root & crown rot	8 years
Beans	Root rots	3 to 4 years
	Anthracnose	2 years
	Bacterial blight	2 years
Beets	Cercospora leaf spot	3 years
Cabbage and related Brassicas	Fusarium yellows	Not effective
	Blackleg (Phoma)	3 to 4 years
	Black Rot	2 to 3 years
	Rhizoctonia damping-off, wirestem and head rot	3 years
Carrots	Leaf blights	2 years
Corn	Smut	Not really effective
	Leaf blights	1 year
Cucumber	Scab & leaf spots	2 years
Eggplant and other solanaceous crops	Verticillium wilt	4 years
Lettuce	Bottom rot & drop	3 years
Muskmelon, cucumber, pumpkin, squash, watermelon, and other cucurbits.	Leaf spots	1 to 2 years
	Anthracnose	2 years
	Fusarium wilt	5 years or longer, not really effective
	Gummy Stem Blight	2 years
Onion	Leaf blights	2 years
Peas	Root rots	3 to 4 years
	Fusarium wilt	5 years
Pepper and other solanaceous crops	Bacterial spot	2 years
	Phytophthora blight	2 years
	Anthracnose	2 years
Pumpkin	Black rot	2 years
Spinach	Downy mildew	2 years
	White rust	2 years
Sweet Potato	Scurf	3 years
	Pox	3 years
Tomato and other solanaceous crops	Bacterial canker	3 years
	Bacterial spot	2 years
	Bacterial speck	2 years
	Early blight	2 years
	Anthracnose	2 years

White potato and other solanaceous crops	Verticillium wilt	2 to 3 years
	Rhizoctonia canker	2 to 3 years
	Pink rot and Pythium	3 to 4 years
	Scab	3 to 4 years

In addition, some vegetables have higher nutrient requirements than others and will deplete nutrient levels in the soil as they grow and produce a crop. Those having higher amounts of nutrients required for growth can be rotated with vegetables having fewer nutrient requirements to aid in building soil nutrient levels. Although building and maintaining soil nutrients is important, the rotation of different vegetable families is important primarily for disease/insect control in the garden. Plants in the same family are usually attacked by the same diseases/insects, and by rotating, the gardener is preventing pathogen/insect build-up for vegetables grown in the garden.

DIRECT SEEDING VERSUS TRANSPLANTING

Most vegetables are started from seed and a garden will not grow and develop unless seeds germinate to form plants. Seeds are an important part of the vegetable gardening process. It is important that the current year's seed be used, as older seed will definitely have lower germination rates, which can cause poor plant stands, especially if they are directly seeded into the garden. In many cases, older seed will germinate at less than 50%, which can lead to inadequate plant populations and less productivity. It is best to select and purchase seeds that were produced for the current gardening year. Remember that seed packets are stamped with a statement that indicates the year for which they were packed.

Seed germination is critical to maximize vegetable crop productivity. Without high germination rates, gardeners will waste their time and resources in trying to produce sufficient plant populations in the garden to maximize vegetable crop productivity. The **three most important critical factors required for seed** germination are temperature, moisture, and oxygen. In some cases, light makes up the fourth critical factor. However, **for most vegetable crops, light has no effect on seed germination.** At planting, the medium or garden soil should contain enough moisture to initiate germination, and an easy test is to squeeze soil through fingers to determine if moisture is present. In dry soils, water thoroughly before planting to about 4 to 5 inches deep. This will ensure that enough moisture is present for germination to occur. However, do not overwater the soil, especially if the soil is not well drained, as resulting anaerobic conditions will prevent seed germination.

Directly Seeding Vegetables in the Garden

Most vegetables are usually directly seeded into garden soil, since many vegetable plants cannot always survive the so-called transplant shock when transplanted into a garden. Although seedlings of many vegetable plants can easily be germinated and grown initially indoors and later transplanted successfully into a garden, it is generally easiest to just directly seed these into the soil. However, other vegetables such as sweet corn and beans are not really feasible to transplant into the vegetable garden in high numbers. Regardless of the practicality, many gardeners will start plants indoors and transplant vegetables into the garden to get a jump-start on the growing season. A rule of thumb is to plant seed no deeper than 3 to 4 times their greatest diameter.

Poor seedling emergence is often observed in gardens and may not always be due to poor germination rates of seeds. Gardeners need to be aware of **soil crusting**, which is a problem that hinders seedling emergence. Soil crusting is the formation of a hard, crusty layer at the soil surface, and this crust often prevents seedlings from emerging through the soil, especially if it occurs shortly after seeding. Soil crusting can occur after a heavy rain or after excessive amounts of sprinkler irrigation is applied to a garden. Soils high in clay are more prone to developing a crust on the soil surface than are other soils. Soil crusting can be alleviated by: incorporating 2 to 4 inches of organic matter (or compost) into the soil, covering seeds with organic matter (or compost) rather than soil at planting, and seeding into moist soils to avoid the use of sprinkler irrigation at planting.

The size of vegetable seeds plays a role in determining how to plant seed. Vegetable seeds come in many different sizes and smaller seeds (such as lettuce) are often difficult to distribute evenly when planting. These smaller seeds can be mixed with sand or soil, and then planted to get an even distribution of the seed in garden soil.

There are two seed planting methods used primarily by home gardeners; these are the **Hill planting** and **Row planting** methods. Gardeners need to have an idea of which method to use when planting vegetables.

Hill Planting

Many vegetables can be planted by the hill planting method. Hilling is placing several seeds or a plant in one spot at definite intervals in the garden, which are often called mounds. Cucurbit vegetables such as watermelons, pumpkins, squash, cantaloupes, and cucumbers are often planted by this method. The use of the hill method for these heat-loving vegetables allows the soil to warm up more quickly in the spring, which will improve germination and plant growth. Plants grown in the hill system take up a lot of room in the garden. An advantage of this system is that poor soil can be improved by simply adding compost when making the mounds.

Native Americans planted corn in mounds, with squash and beans growing between the corn plants. This practice is known as "The Three Sisters" and tends to benefit all three crops. The corn shades the squash and beans, while the squash discourages squirrels and raccoons from feeding on the corn, and the beans fix nitrogen in the soil, which makes more nitrogen available to improve corn growth.

The mounds used for hill planting are typically 8 to 10 inches high and between 2 and 8 feet wide, depending on the size of the plant at maturity. Vegetables, such as pumpkins, melons, and winter squash grow on sprawling vines, and mounds for these crops should be spaced between 4 and 8 feet apart. Most cucurbit vegetables are planted at a depth of 1 inch into the soil with 4 to 6 seeds planted into each hill. Once seeds germinate, the hills are thinned, leaving only one or two plants per hill, depending upon the vegetable and the amount of space it needs to grow. Plants are thinned when the plants reach about 4 inches in height and have two or three leaves. Sweet corn is spaced 12 inches apart and is often planted in circles. Remember it is always best to plant more seed than the desired number of plants and then thin to the number that you want per hill. Planting extra seed allows for poor germination and loss of seedlings to disease and/or insects. Also remember not to plant too deep because the seed may germinate but will never make it to the soil surface.

Row Planting

Row planting is the method that most gardeners will use. Row planting is planting seeds more or less evenly down a row. Beans, peas, and sweet corn are examples of vegetables that are commonly planted using this method. After germination, plants are thinned to give the desired spacing. It is best to plant more seeds than needed to allow for poor germination and disease and/or insect loss, and don't plant seeds too deep, as the seedlings may never emerge from the soil.

Thinning Vegetable Seedlings

Once vegetable seedlings emerge, gardeners need to understand that thinning is required to obtain the best yields and highest-quality vegetables. Thinning or pulling up vegetable seedlings is a tough job for many gardeners who see a beautiful healthy plant growing that has potential to provide fresh products from the garden. But thinning must be done to obtain the highest vegetable productivity and best quality that is possible from a garden. Thinning provides a better environment for the remaining seedlings by giving them the proper space for adequate growth and development.

There are several reasons for Thinning Vegetable Seedlings:

- Reduces competition among the seedlings for sunlight, soil nutrients, and water;
- Reduces early disease problems by providing better air circulation around the plants;
- Provides conditions (more growing space/plant) that are better for growing healthier vegetables to get the best possible vegetable product weight, size, and shape; and
- Properly spaced plants will result in higher productivity.

It is best to thin when plants have one or two pairs of true leaves, and when the soil is slightly moist and pliable, as this allows seedlings to be pulled up without damaging adjacent seedlings. If the soil is excessively dry, water the garden site until the soil is moist before seedlings are thinned. Also, thinning in excessively wet soils is not recommended, as it will lead to adjacent seedling damage or removal of too many seedlings, as they tend to clump together as they are pulled. The late afternoon or evening is the best time to thin since the remaining plants have the cooler, dark night to recover after being disturbed. Plants can also be effectively thinned on cloudy days, as minimal damage will occur to remaining plants due to reduced sunlight levels. Thinning is useful for all vegetables including lettuce, beets, radishes, carrots, spinach, and sweet corn, to name a few.

Figure 1.12 Pelleted lettuce seed being planted indoors into 200-cell plastic trays containing a sterilized soil mix.

Starting Seeds Indoors for Transplanting into the Garden

Seed must be started several weeks prior to transplanting in the garden. The seed of vegetables, such as tomato, cabbage, lettuce, broccoli, and cauliflower, should be started about 6 weeks prior to placing in the garden, while pepper seed should be started about 8 weeks before planting outdoors. For cucurbits, such as cucumbers or squash, seed of these vegetables should be started about 3 to 4 weeks prior to transplanting outside. Avoid starting vegetable plants too soon indoors, as plants often become root-bound or leggy when they must wait several days or weeks after they reach optimum size before planting outside. Plant seeds into a sterilized soil mix to prevent soil-borne disease from developing on young seedlings (Fig. 1.12). The optimum germination temperatures and the number of days required for vegetable seed to germinate in optimum soil conditions are provided in Table 1.4.

Table 1.4 Optimum germination temperatures for vegetable seed and days required for germination to occur.

Vegetable Crop	Optimum Soil Temperature for Germination	Days to Germinate at Optimum Soil Temperature
Bean, lima	85°F	7 to 10 days
Bean, snap	75 to 80°F	7 days
Beet	75°F	7 to 14 days
Broccoli	65 to 75°F	5 to 10 days

Brussels sprout	68 to 75°F	5 to 10 days
Cabbage	68 to 75°F	5 to 10 days
Cantaloupe (melon)	80 to 85°F	5 to 10 days
Carrot	75°F	12 to 15 days
Cauliflower	65 to 75°F	5 to 10 days
Celery	70 to 75°F	10 to 14 days
Collard	70 to 75°F	5 to 10 days
Corn	75 to 85°F	7 to 10 days
Cucumber	70 to 85°F	7 to 10 days
Eggplant	75 to 85°F	10 to 12 days
Endive	70 to 75°F	10 to 14 days
Kale	70 to 75°F	5 to 10 days
Kohlrabi	70 to 75°F	5 to 10 days
Lettuce	65 to 70°F	7 to 10 days
Mustard	70°F	5 to 10 days
Okra	80 to 85°F	7 to 14 days
Onion	65 to 75°F	10 to 14 days
Parsnip	70°F	14 to 21 days
Pea	65 to 70°F	7 to 14 days
Pepper	78 to 85°F	10 to 14 days
Pumpkin	70 to 75°F	7 to 10 days
Radish	65 to 70°F	5 to 7 days
Rutabaga	65 to 70°F	7 to 15 days
Spinach	70°F	7 to 14 days
Squash, Summer	75 to 85°F	7 to 14 days
Squash, Winter	75 to 80°F	7 to 14 days
Swiss Chard	70 to 75°F	7 to 14 days
Tomato	75 to 80°F	7 to 14 days
Turnip	65 to 70°F	7 to 14 days
Watermelon	75 to 85°F	7 to 14 days

Although most vegetable seeds prefer germination temperatures around 70°F, notice that the optimum germination temperature of several warm-season vegetables, such as melon, okra, and pepper, is much higher. A waterproof heating pad that enhances seed germination and plant growth, commonly known as a propagation mat, may be required to supply bottom heat to obtain optimum germination rates.

Adequate light is necessary for proper seedling growth and development, and should be provided to seedlings once they emerge from the soil. The amount of light that seedlings receive on a windowsill is often inefficient and may lead to leggy growth, where plants are stretching to reach available sunlight. It may be worthwhile to purchase a grow light and place it over developing seedlings. This will result in more compact and healthier seedlings to transplant

Figure 1.13 Cantaloupe seedlings at cotyledon stage growing in greenhouse. These will be ready to plant outside in about another 2 weeks once they reach the 2- to 3-leaf stage.

into the garden. Maintain adequate moisture to seedlings by watering, but make sure that the seedlings are not sitting in water for long periods of time and allow the soil to dry out a bit before watering again. Also, do not overwater, especially when seeds are germinating, as this can lead to damping-off or other fungal diseases that result in seedling death. In order to reduce transplant shock, seedlings should be gradually conditioned to the outdoor environment for about a week prior to planting into the garden. It is best to begin with just a few hours or so of exposure to the outdoor climate during the first few days, and then gradually increase the time until the transplants are left outdoors all day and night.

Seed Storage

Many gardeners like to keep seed from year to year. They keep not only their unused seed from previous years, but also collect seed of heirloom and other open-pollinated varieties from their gardens. Seed stored under ideal conditions will maintain high germination rates for several years. Depending on the plant, seed properly stored can be kept for 4 to 5 years when harvested off plants, or several years past what's listed on the seed package. However, remember seed germination rates decline with age even when properly stored.

When storing seeds, be aware of the factors that affect their viability. The **two most important factors in seed storage are moisture content of seed and storage temperature.** The moisture content of the seed plays an important role in the potential longevity of seed viability. If the moisture content of the seed is low, the seed will keep well under a wide range of temperatures. Furthermore, it is also important to remember that seed storage temperature plays an important role in their longevity. The moisture content in seed is not as much of a critical factor leading to seed deterioration if the storage temperature is low, compared to when temperatures are high. Lower storage temperatures result in a slower metabolism and increase the period that a seed is viable. **The storage conditions are almost ideal if the seed is both dry and cold.** The general recommendation is to store seeds in seed packets in a refrigerator at about 40°F. Open containers are better for seed storage than are closed containers.

Transplanting

As discussed previously, some plants must be seeded directly into garden soil, as they cannot always effectively handle the shock of transplanting in garden soil. These include root crops such as beets, turnips, and carrots, as well as some cucurbit vegetables, although some cucurbits, such as melons are often transplanted into a garden (Fig. 1.13). But other **crops such as tomatoes, peppers, and eggplants prefer to be set out as transplants.** Depending upon

***Figure 1.14** Sterilized soil mixes like this one can easily be found and purchased at local garden centers for starting vegetable plants indoors.*

the time of the year, some vegetables can be directly seeded into the garden soil or started as transplants, including cabbage and other crucifers, lettuce, onions, and watermelons.

Transplants offer several advantages over direct seeding:

More healthy plants. Plants started as transplants avoid many common hazards to young, developing seedlings such as birds, insects, and seedling diseases.

Earlier vegetables. An earlier harvest will result from use of transplants in a garden. Transplants allow a gardener to get a 2- to 3-week or more head start on the growing season compared to similar vegetables that are directly seeded into the garden soil. This enables plants to grow before the danger of frost is over and before the garden soil is easily tilled. When a vegetable transplant is placed into a garden soil, it is already several weeks ahead of the same vegetable that is direct-seeded into the soil. A vegetable that is direct-seeded must go through the germination process and then develop several leaves, which will take several weeks; by this time, a transplanted vegetable may already be ready to harvest or developing flowers or fruit.

For spring-grown cruciferous vegetables, gardeners will typically use transplants to get a 4- to 5-week jump on the season. This often allows them to avoid growing these cool-season crops during the heat of the summer in which they will often bolt and lose quality.

Transplants can be purchased at local garden centers or started indoors from seed that the gardener has available or has purchased. What a gardener chooses to do often depends on whether specific varieties are available at the garden centers, and in many cases, they aren't. If transplants are bought, remember the best choices are not necessarily the largest and tallest available. Tomato plants in flower are not a good choice since flowering in a small container often puts the plant under stress. High-quality transplants should be stocky with a healthy appearance, be of medium size, have a good green color (not pale green or yellow), be free of insects/diseases, and have a good root ball. It is best to check to make sure that the plant is not root-bound in the container in which it is growing. This can easily be done by pulling the plant out of the container and looking at the root mass growing in the medium. If there are excessive numbers of roots encircling the medium in which the roots are growing and the medium can't be seen, then it is root-bound. Also remember, the number of days to harvest for tomatoes, peppers, and many other commonly transplanted vegetables listed in seed catalogs are often the days after transplanting and not the days after seeding.

To Grow Successful Transplants:

Use a disease-free growing medium. Do not use garden soil, as it often contains soil-borne plant pathogens, but use sterilized soilless mixes that can be purchased at local garden centers (Figs. 1.14 and 1.15). Many gardeners prefer to grow transplants in ready-to-use peat blocks, jiffy pellets, or cubes, which are also a good choice.

Provide adequate warmth and moisture. Heat and water are both needed for seed germination. Many gardeners use propagation mats or heating cables underneath containers or flats to provide seeds and seedlings the proper temperature for germination and growth.

***Figure 1.15** Giant Italian parsley plugs ready for transplanting that has been grown in a sterilized, soilless potting mix.*

Provide adequate light. Sunlight or an artificial source of light (such as a grow lamp) is needed for proper plant growth once seeds have germinated. Under low-light conditions, transplants often become leggy. They become leggy since inadequate light is provided and they are stretching their stems to reach as much light as possible. Remember that light quality, intensity, and duration are all important characteristics that influence plant growth. Thus, low light quality, intensity, and/or duration will often result in leggy transplants. In most cases, once seedlings emerge, they need full sunlight for 8 to 10 hours/day.

Give plants a hardening-off period. Hardening-off toughens plant tissues so that they can withstand the outdoor environment. The hardening-off period allows plants to adjust and adapt to outdoor conditions compared to the indoor or greenhouse environment in which they are growing. A period of adjustment is needed to prepare the indoor plant to outdoor growing conditions. Seedlings need to be gradually exposed to lower temperatures, less water, and greater amounts of sunlight. It is best not to plant young fragile seedlings directly into the open garden directly from the indoor environment. If you grow your own transplants, place the seedlings outside in a protected structure, such as a coldframe, about 2 weeks before transplanting into the garden. If a coldframe isn't available, place plants in a protected area, such as a covered porch or balcony; and if a frost is likely, the plants should be brought back inside at night unless they are adequately protected.

Home gardeners need to be aware of several techniques that can be used to improve transplant survival when planting these plants into a garden. It is best to: 1) transplant on a cloudy day or in the evening (far better than doing it on a hot, bright sunny day) when plants will lose less water; 2) before and during transplanting, thoroughly water soil in which plants will be growing, as this excess water will help roots adjust to their new environment; 3) transplants in the containers in which they are growing should be watered to field capacity before transplanting into the soil; 4) place the transplant slightly deeper than it grew in its container (some plants like tomatoes are an exception, as they will develop roots all along their stems and can be planted deep enough so that only 2 to 3 leaves are left exposed); 5) firmly and gently place the soil around the base of the transplant to make sure that roots are in good contact with the soil; 6) apply a starter fertilizer solution at transplanting to help the plants overcome transplant shock and to ensure proper fertility during the initial growth period (a basic soluble complete fertilizer works well, although there are starter fertilizer solutions available at garden centers, and these have a greater amount of phosphorus to help with root establishment); and, 7) protect young transplants from heat, cold, and damaging winds during the first week after transplanting. Protection structures, such as boxes, baskets, cut milk jugs, and flower pots, can be placed over the transplants when unfavorable weather conditions are likely. However, do not leave the protectors over the plants longer than necessary. Oftentimes, temperatures increase significantly during the daylight hours and transplants can get injured from the excessive heat if the protectors are not removed. The removal of protection structures allows proper ventilation for plants during the daytime. These few techniques should allow better transplant survival and more initial vegetable plant growth in the garden.

Chapter Two

COMPOST, SOIL MODIFICATION, AND MULCHING FOR THE HOME GARDEN

The development and maintenance of a high-quality garden soil is a key component to have a highly productive garden. Thus, composting, soil modification, and mulching all play important roles in garden soil improvement.

COMPOSTING

Gardeners have utilized compost for centuries. Composting is a way of recycling, through a thermophilic process, organic waste materials from the kitchen, yard, or garden that would ordinarily be discarded. Composting is an efficient method of breaking down organic materials into a product that is beneficial to soil and plants (Fig. 2.1). Composting is a natural biological process where bacteria, fungi, and other organisms decompose organic materials such as leaves, grass clippings, and food wastes. The end product is called compost. While composting occurs naturally, the process can be accelerated and improved by gardener intervention. When organic materials are placed into a compost pile with soil, water, and certain other materials, they will decay into a rich humus that can be used in the garden to improve soil structure. Although compost will slowly release a small amount of plant nutrients, it really shouldn't be thought of as a substitute for fertilizer. It should really be used by a home gardener as a soil-building material instead of a fertilizer source.

Figure 2.1 Compost that is ready to use to improve a garden soil.

Vermicomposting is another type of composting method that uses worms to break down organic materials/wastes into a usable compost that can be used to improve the garden soil. However, compared to regular thermophilic composting, vermicomposting is a non-thermophilic way to break down organic matter.

Vermicomposts are finely divided, peat-like materials that have high porosity, aeration, drainage, water-holding capacity, and microbial activity. As stated before, these composts result from a non-thermophilic biodegradation and stabilization of organic materials through interactions of earthworms and microorganisms. A wide range of organic residues, including sewage sludge, animal wastes, crop residues, kitchen scraps, and industrial refuse are increasingly being converted

Figure 2.2 Earthworms being added to a fresh batch of kitchen scraps and paper in a self-made worm bin. This is an easy way to become more sustainable in your home. Don't throw those kitchen scraps into the garbage that will end up in a landfill somewhere, but just place into a small container and add earthworms. In just a short period of time you will have compost ready to use in your garden. Now that is recycling!

Figure 2.3 A large-scale vermicompost facility with raised beds to hold organic materials that are broken down by earthworms. Finished vermicompost is on the bottom of the beds and drops into some type of recepta-cle underneath the beds.

by earthworms to form vermicomposts. Earthworms can be used by home home gardeners to breakdown fresh kitchen scraps to vermicompost (Fig. 2.2). The earthworms break down the organic residues, which stimulate greater microbial activity, increase nutrient mineralization rates, and rapidly convert the wastes into a humus-like substance that has a finer structure than ordinary composts, while possessing greater and more diverse microbial populations. Some say that vermicompost is better than regular compost due to the increased microbes and plant growth stimulants it contains. Since vermicomposts are in such high demand by home gardeners and commercial vegetable growers, large-scale vermicomposting facilities are being built to provide large amounts of this type of compost (Fig. 2.3).

Composting is an important process in a home garden, and adding yard and garden wastes directly to the garden soil without first composting has some undesirable effects. For example, if large amounts of fallen leaves are incorporated into the soil without first being composted, the bacterial microbes that aid in organic matter decomposition will compete with garden plant roots for soil nitrogen. This competition can result in a nitrogen deficiency in plants grown in the garden, resulting in poor plant growth.

Factors that affect rate of decomposition

The decomposition of organic matter in a thermophilic compost pile is dependent on maintaining microbial activity. Any factor that will slow or stop microbial growth will also hinder the composting process. Several factors affect the rate of decomposition, and the breakdown of organic materials will occur more efficiently if the following factors are used to the fullest advantage.

Aeration. Oxygen is required for microbes to efficiently decompose organic wastes. However, some decomposition will occur in the absence of oxygen by anaerobic bacteria, but the process is slow and a foul odor often develops under anaerobic conditions. This can be avoided by turning the compost pile once or twice a month to provide the necessary oxygen to microbes and this will significantly increase the composting process. A compost pile that is not mixed on a regular basis will take 3 to 5 times longer to decompose and develop into usable garden compost compared to one that is turned more frequently. A compost pile that is mixed regularly will also reach higher internal temperatures that help to destroy weed seed and plant pathogens that may have been added to the pile.

Moisture. Moisture is essential for microbial activity and a dry compost pile will not decompose as quickly as one with sufficient moisture. The compost pile may need to be watered once in a while to maintain enough moisture for efficient decomposition to occur. Moisture should be added especially if rainfall is limited, and enough water should be added to completely moisten the pile. Overwatering the compost pile should be avoided, as excess water can lead to anaerobic conditions, which will definitely slow down the decomposition process.

Particle Size. The smaller the particle size of organic wastes, the faster the decomposition process will occur and the quicker compost will be ready to use. Smaller particles have a greater surface area that allows for a quicker microbial breakdown. Many gardeners will use a shredder to reduce particle size before putting organic material, such as branches, sticks, or pine cones, into a compost pile. Another way to reduce the size of organic material, such as fallen leaves, is to run over with a lawn mower. In addition to speeding up the decomposition process, shredding will initially reduce the volume of the compost pile.

Table 2.1 Approximate carbon to nitrogen (C:N) ratios in various organic materials.

Material	C:N ratio
Poultry manure	10 to 20:1
Vegetable wastes	12 to 20:1
Coffee grounds	20:1
Cow manure	20:1
Grass clippings	12 to 25:1
Legume plant material	10 to 20:1
Horse manure	25 to 40:1
Tree leaves	30 to 80:1
Straw	40 to 100:1
Paper	150 to 200:1
Wood chips, sawdust	200 to 500:1
Wood	700:1

Fertilizer and Lime. The addition of a nitrogen fertilizer will speed up the decomposition process in a compost pile. Microbial activity is affected by the carbon-to-nitrogen ratio (C:N) of the organic waste. Microbes require a certain amount of nitrogen for their own metabolism and growth, and a shortage of nitrogen will slow down the composting process considerably. The C:N ratio in plant material is highly variable, ranging from 10:1 to 30:1 in legumes and farm manure to as high as 100:1 to 500:1 in straw or sawdust residues (Table 2-1). Materials high in carbon relative to nitrogen, such as straw or sawdust, will decompose very slowly unless additional nitrogen fertilizer is added to the compost pile. Tree leaves are widely used in compost piles and are higher in nitrogen than straw or sawdust but would still benefit from the addition of nitrogen fertilizer or other materials that are high in nitrogen. Grass clippings are high in nitrogen and when mixed properly with leaves, the decomposition process will speed up compared to only having leaves in a compost pile. Animal manures, poultry litter, or blood meal can be used as organic sources of nitrogen. A fertilizer with a moderate to high nitrogen analysis (10% to 30%) should be used. Most fertilizers found in garden centers or farm stores are within this nitrogen analysis range.

Lime is not necessary for the decomposition of most yard wastes, since the pH or finished compost is usually alkaline (pH of 7.1 to 7.5) without the addition of lime. However, if large amounts of acidic organic materials such as pine needles, pine bark, or vegetable and fruit wastes are composted, additional lime may be necessary and is often recommended.

Materials to Compost

There are many organic materials suitable for composting:

Yard wastes. Leaves, grass clippings, straw, and softwood plant trimmings/prunings are yard wastes that can be easily composted. Leaves are oftentimes the dominant organic waste in many compost piles. Remember that branches and twigs greater than ¼" in diameter should be put through a shredder/chipper before composting to shorten the decomposition process.

Kitchen wastes. Many organic wastes come from the kitchen, like vegetable scraps and coffee grounds. These can also be composted and are widely used in compost piles.

Sawdust. Sawdust can be added to compost piles in moderate amounts and usually only if additional nitrogen is applied. Approximately 1 lb of actual nitrogen is required to aid in the breakdown for about 10 lbs of dry sawdust.

Paper waste. Newspaper or other forms of paper waste is often composted. The nitrogen content of newspaper is low and will consequently slow down the rate of decomposition in a compost pile. If paper is composted, it is best to shred first before placing in a compost pile, and paper should not be more than 10% of the total weight of the material in the compost pile.

Organic materials high in nitrogen. Animal manures, poultry litter, and other materials such as blood meal and bone meal can be composted to add nitrogen and other nutrients to the compost pile. For example, about 100 lbs of poultry litter provides 2 lbs of actual N.

Materials Not to Compost

There are also many waste materials that should not be included in compost piles.

Human or pet feces should not be added to the pile because they can transmit diseases.

Plants treated with herbicides or pesticides should not be used for composting. They will add chemicals to the compost pile that can be toxic to other plants grown in the garden, as well as to the gardener.

Diseased plants or weeds with seeds should not be added to the compost pile as they can create problems later if the compost is to be used in the garden. Most disease-causing organisms and weed seeds are destroyed during the thermophilic composting process because temperatures in the center of the pile often reach 150° to 160°F. However, in most compost piles, it is impossible to mix the contents sufficiently to bring all wastes to the center. So, in most cases, not all pathogenic plant organisms or weed seeds will be killed, since some portions of the compost pile will never reach the 150° to 160°F range that is required to effectively kill these organisms and weed seeds.

Materials that break down slowly such as citrus rinds, corn cobs, sticks, pine cones, and walnut, pecan, and almond shells, should not be placed in compost piles unless shredded in some manner to increase the surface area.

Charcoal resists decay and should not be added since it will not decompose quickly.

Automotive petroleum products (such as car oils) should be taken to recycling centers, and never put into a compost pile.

Location of Compost Pile

The compost pile should be located in an out-of-the-way place in your landscape but close enough to be convenient for the gardener. It must be located in full sun conditions, and on a well-drained site. If the compost pile is well ventilated and has adequate water drainage, it should not emit a foul odor. Compost piles develop foul odors if anaerobic conditions persist in the pile. When animal manures are initially added to the pile, some initial odor should be expected, but will soon dissipate as the pile matures.

COMPOSTING STRUCTURES

Several composting structures can be used to form compost and the structure chosen is usually based upon individual gardeners' needs. Certain structures are more appropriate for those gardeners who have lots of organic materials that require disposal and some are more suited to those who have only small amounts. Composting structures include:

Plastic Garbage Bags. This is one of the easiest ways to make compost, since minimal amounts of maintenance are required. It is best to use large plastic garbage bags (greater than 20-gallon sized) and fill with plant wastes, fertilizers [one tablespoon of a complete granular fertilizer such as 12-12-12 (N-P-K)], lime (about one cup per plastic bag, as this helps to counteract the extra acidity caused by anaerobic composting), and a quart or so of water. The garbage bags should be closed and set in an out-of-the-way place for 6 months to a year, such as a basement or heated garage to allow for better decomposition during the winter months. Although garbage bags require no turning or additional water after closing, the advantage of using bags is that it requires little maintenance. However, due to reduced levels of O_2 in the closed bag, the decomposition process is slow. Thus, the decomposition rate can be sped up if bags are opened and mixed periodically.

Barrel or Drum Composter. This compost structure, shown in Fig. 2.4A, will generate compost in a relatively short period of time and provides an easy mechanism for turning compost. Barrel or drum composters can be purchased or made at home from old barrels or drums, and commonly vary in size to hold just a few gallons of organic materials or up to 50 to 75 gallons of organic waste. Small holes (at least ½" in diameter) are needed in the barrel to allow air circulation and drainage of excess water to prevent water-logging of organic materials/compost. In preparation for breakdown of organic materials, the barrel or drum composting structure should be filled about one-half to three-fourths full with organic wastes. About one cup of a complete granular fertilizer [such as 12-12-12 (N-P-K)] should be added to the mixture with the addition of sufficient water until the organic materials are moist, but not overly wet. Generally, the opening in which organic wastes are placed can be removed or opened after turning to allow for better air penetration into the organic matter.

Figure 2.4 Barrel or Drum Composter (A) and Bin Composting (B).

Bin-Type Composters. For large amounts of organic so wastes, bin-type structures are the most practical since these structures allow the gardener to compost large amounts of organic materials (Fig. 2.4B). A circular wire bin is relatively inexpensive to make and can compost large quantities of organic materials. The bin-type composting structure can be made by using small-panel woven wire fencing that is held together with chain snaps. The circular bin should be about 3 to 5 feet in diameter and at least four feet high. With this bin, it is easy to turn the compost pile by unsnapping the wire, moving the wire bin cylinder a few feet over, and turning the compost back into the bin at its new location. A permanent bin holding unit (Fig. 2.5A) is also simple to make and is often made of wood or wire, or the combination of both. This permanent type of bin composter works best with four posts pounded into the ground at each corner to provide additional support.

Heap Composting. No structure is required for heap composting (Fig. 2.5B). The pile should be at least 3 feet in height and 3 feet in width, and its length will vary depending upon the amount of organic material being composted. This is really the least expensive way to produce compost since no physical structure is needed. Heaps may not appear as neat and aesthetically

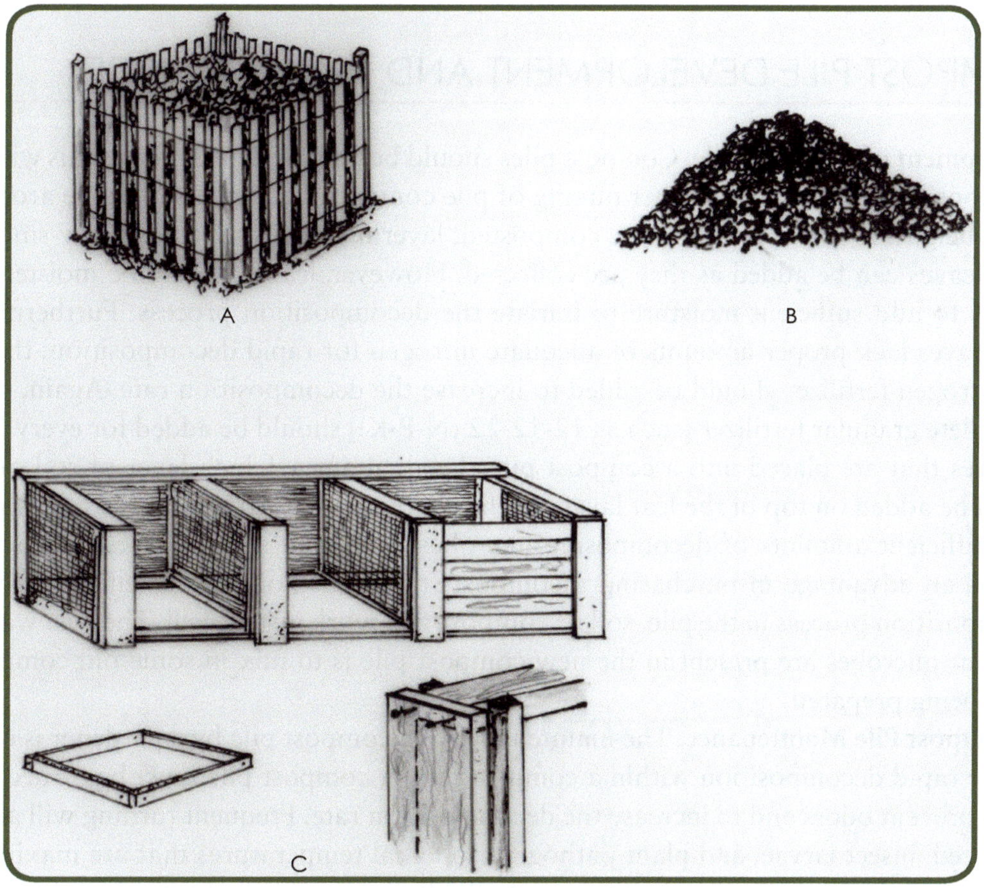

Figure 2.5 Permanent Bin Composter (A), Heap Composting (B), and Three-Chambered Bin Composter (C).

pleasing to the eye as using other bin structures for composting, but it will do the job just as well as the others. Decomposition of organic materials in the heap will occur at a faster rate if the heap is turned on a regular basis.

Three-Chambered Bin. This type of structure (Fig. 2.5C) can hold a considerable amount of organic materials and is designed to provide good air circulation by frequent turning of the organic materials, which produces compost at a quick rate. The three-chambered bin composter works on the idea of an assembly line, as a gardener will have three batches of compost in varying stages of decomposition. The organic materials are started in the first bin and allowed to heat up. After several days, it is turned into the middle bin, while a new batch of organic material is started again in the first bin. Finally, the material in the middle bin is turned into the last bin as finished or nearly finished compost that is ready to use in a garden. When making a three-chambered bin, it is best to use rot-resistant wood such as redwood, salt-treated wood, or wood treated with an environmentally safe preservative or a combination of treated wood and metal posts. Unless the wood is treated or rot-resistant, it will decompose within a few years. Each bin should be at least four to five feet in length, width, and height to contain enough volume to compost organic materials properly. The use of removable slats in the front offers complete access for turning the organic contents.

COMPOST PILE DEVELOPMENT AND MAINTENANCE

Development of Compost Pile. Compost piles should be prepared in layers, as this will facilitate decomposition by ensuring proper mixing of pile contents. Each pile should be around 3 to 5 feet in height. If only tree leaves are composted, layering may not be necessary since recently fallen leaves can be added as they are collected. However, leaves should be moistened if they are dry to add sufficient moisture to initiate the decomposition process. Furthermore, since dead leaves lack proper amounts of adequate nitrogen for rapid decomposition, the addition of a nitrogen fertilizer should be added to increase the decomposition rate. Again, one cup of a complete granular fertilizer [such as 12-12-12 (N-P-K)] should be added for every 20 gallons of leaves that are placed into a compost pile. Then, about a 1 inch layer of soil or compost should be added on top of the leaf layer for the purpose of ensuring that the pile is inoculated with sufficient amounts of decomposing microbes. Although some studies have shown that there is an advantage in purchasing a compost starter or inoculum to enhance the organic decomposition process in the pile, soil or compost will work just as well. The best way to make sure that microbes are present in the new compost pile is to mix in some old compost as the pile is being prepared.

Compost Pile Maintenance. The maintenance of a compost pile by a gardener is essential to achieve rapid decomposition within a compost pile. A compost pile must be turned occasionally to prevent odors and to increase the decomposition rate. Frequent turning will also expose weed seed, insect larvae, and plant pathogens to lethal temperatures that are maximized near the center of the pile. A compost pile with a foul odor is most likely due to overwatering (or too much water in the compost pile) that has created anaerobic conditions. A properly mixed and turned compost pile will not develop a foul odor, and an actively decomposing pile will reach temperatures of 150° to 160°F in the center portion of the pile. Proper water management is also important to the development and maintenance of a compost pile, and in most areas of the eastern and mid-western U.S., natural rainfall will provide adequate moisture for compost pile maintenance.

Generally, a well-managed compost pile will be ready under warm conditions in 2 to 4 months. However, a compost pile started during the late fall will not be ready by spring due to the lower temperatures that result over the winter months, which slows biological activity and thus reduces the decomposition process significantly.

SOIL MODIFICATION

Most home gardeners need to improve the soil conditions in their garden in some manner to provide soil conditions that will allow optimum plant growth and development. However, many home gardeners do not modify their soils, but they really should to maximize biological diversity and improve soil structure. When the same gardening area is used year after year without modifying the soil through organic matter or compost additions, the organic matter in the soil will soon be depleted. Most vegetable plants grow best in well-drained soils that have high organic matter content. The addition of organic matter to a garden soil provides several benefits to the soil, which will allow for better plant growth over the long term.

Soils in gardens can be improved by adding organic matter (e.g., compost), which will build soil structure over time. The addition of organic matter to soils will improve soil drainage, air movement, and moisture-holding capacity, as well as soil nutrient levels, and nutrient-holding capacity of soils.

The addition of organic materials, such as compost, to a garden soil will have several obvious influences on the soil. Compost (or other organic matter sources) will change the soil color over a period of several years (usually from brown to black), and influence soil physical properties (such as increasing water-holding capacity of soil, increasing cation adsorption capacity in soil, and increase supply and availability of nutrients in soil). The utilization of cover crops in a vegetable garden will benefit the soil in many ways and will be discussed in further detail in the organic garden section (Chapter 10). Cover crops that are planted in the autumn and tilled into the soil during the spring or planted in the spring and tilled into the soil during the summer or autumn months are termed "green manure crops." Organic matter and nutrients are being placed back into the soil from a crop that did not mature, as it was still green and actively growing when it was tilled into the soil. Thus, green manure crops are important for building the soil structure through organic matter additions.

MULCHING

A mulch is an organic or inorganic material used to cover the surface of the garden soil and is not worked into the soil. It is used to protect plant roots from heat or cold extremes, suppress weed populations, prevent moisture loss, reduce soil erosion, and to keep fruit cleaner by not touching the soil. Mulches definitely make a garden more attractive, but will also improve plant productivity in gardens. A garden mulch will change the environment where plants are growing, resulting in better plant growth and higher yields. However, if a mulch is used improperly in a garden, it will lower yields or result in plant death.

There are many types of mulching materials, but they can be divided into two general categories: organic (natural) and inorganic (synthetic). Organic mulches include materials such as chopped leaves, bark chips, compost, ground corncobs, chopped vegetable material, hay, manure, grass clippings, peat moss, pine needles, sawdust, wood shavings, and straw. Widely used inorganic mulches include plastics, landscape fabrics, cardboard, and gravel or stone. **Organic mulches** tend to reduce soil temperatures and keep the soil cool, whereas **inorganic mulches** increase soil temperatures. For example, black plastic warms the soil and radiates heat during the night, keeping heat-loving warm-season vegetables, such as eggplant and tomatoes, in optimal growing conditions. Both types will discourage weed growth, but organic mulches also improve the soil as they decompose. Inorganic mulches don't break down and enrich the soil, but under certain circumstances they are the better choice.

Figure 2.6 Organic mulches used in gardens: grass mulch placed around Chinese cabbage (top) and chopped wood mulch used around Chocolate mint (bottom).

Organic Mulches

Organic mulches are commonly used by home gardeners (Fig. 2.6) and provide the gardener several advantages when used in the garden, compared to bare soil conditions. Organic mulches are most useful in conserving soil moisture and reducing soil temperatures in the garden by 8° to 10°F during the summer months, and for this reason, they should not be used too early in the spring. If organic mulches are applied to cold late winter/early spring garden soils, plant maturity will oftentimes be delayed (especially if a warm-season vegetable is being grown) due to the soil warming more slowly. After the soil has sufficiently warmed in the late spring/early summer to about 70°F, it is then the best time to apply an organic mulch over the garden soil. An organic mulch should be applied to a depth of 2 to 4 inches around well-established plants. Remember to make sure that there is adequate moisture in the soil before applying the mulch. However, home gardeners need to realize that organic mulches with high C:N ratios, such as sawdust or wood shavings, will "tie up" soil nitrogen as they decompose. The use of these organic materials can result in soil-nitrogen deficiencies. However, additional nitrogen fertilizer can be applied to these organic mulches to prevent this situation from occurring. To get a quicker decomposition of the high C:N ratio organic materials, just add a small amount of a granular nitrogen-containing fertilizer when the organic mulch in applied to the soil.

Selected Organic Mulching Materials

Bags of wood chips and bark mulch can usually be found at local garden centers to mulch your flower garden and shrub borders. These are usually quite expensive for the small amount of material that is contained within each bag. Some wholesale landscaping companies also make available large amounts of these organic materials at a much lower price. Also, there may be community yard waste collection sites close that offer chipped yard debris or composted grass clippings and fallen leaves to residents for free or for a small fee. Although wood chips and bark mulch work well as a mulch for garden pathways, these materials are not the best choice for vegetable and annual flower beds, since a gardener will be digging into these beds every year and the chips will most likely get in the way.

Grass clippings are a readily available mulch in most home gardens, although we must remember to return at least some of the grass clippings directly back to the lawn as a natural fertilizer. But, it is all right to occasionally collect grass clippings from lawns to use as mulch, and the nitrogen-rich clippings are an especially good choice for mulching flower and vegetable gardens.

Shredded fallen leaves create a nutrient-rich mulch for the home gardener. Many consider raking to be a chore, but leaves can provide an outstanding mulching material. Although a leaf-shredding machine works great, a lawn mower with a bagger will work just as well for collecting leaves and cutting into the perfect size for mulching. Shredded-leaf mulch can be placed anywhere in the garden, but it looks very attractive in flower beds and shrub borders. Shredded leaves make a perfect mulch for a woodland or shade garden.

Compost is fine to use as a mulching material, if a gardener has extra amounts for this purpose. Although compost will definitely improve the soil and make garden plants grow better, remember that when any kind of mulch becomes dry, it can lead to root death, which definitely affects plant growth. So gardeners who use compost as a mulch will usually spread a thin layer around plants and top it with another type of mulch, such as shredded leaves or bark. This allows the compost to stay moist and biologically active, which will provide the greatest benefits for plants in the garden.

Straw and hay are other great mulches for the home garden. Straw or hay mulch has most of the benefits of the other mulches discussed, including maintaining soil moisture, reducing weeds, and adding organic matter to the soil once it decomposes. However, make sure the straw, and especially the hay, is weed- and seed-free, to prevent bringing more weeds and problems into a garden. Remember not to place hay or straw all the way up to the stems of vegetables or the trunks of fruit trees, as this will promote slug and rodent damage.

Weed control with organic mulches

Organic mulches are a great way to combat weeds in a garden. To maximize weed control, place the mulch on soil that is somewhat weed-free, and then place a thick enough layer to prevent new weeds from coming through the mulch layer. A 4- to 6-inch layer of mulch is often used, as this tends to completely suppress weed growth, although a 2- to 3-inch layer is usually sufficient in shady spots where weeds tend to be less troublesome compared to full-sun conditions.

Inorganic Mulches for the Garden

Inorganic mulches are sometimes used by home gardeners but not as often as organic mulches. When discussing inorganic mulches, most gardeners think of plastic mulch. Plastic mulches are springtime mulches, as they help to warm the soil, which permits early planting. These mulches also promote more rapid growth, earlier harvests, moisture retention and weed control. Remember that plastic mulches (clear or black) are easier to use if applied prior to planting, and these are widely utilized in the commercial vegetable industry, with black plastic the one most commonly used (Fig. 2.7).

Warm-season vegetables like cucumbers, melons, squash, tomatoes, peppers, and eggplant will grow and develop better and have higher productivity when grown on black plastic mulch compared to bare soil. Transplants or seeds of warm-season vegetables can be placed into the soil through plastic mulch by cutting holes with a sharpened bulb planter, knife, or other sharp implement. Although plastic mulch is frequently used with warm-season vegetables, it can be used to promote early growth of cool-season vegetables like cabbage, broccoli, and cauliflower. Plastic mulch is generally not used with vegetables that are closely spaced in the rows, like carrots or onions.

Prepare the soil completely before applying a plastic mulch: remove all weeds and debris, break up all large clods, and incorporate all pre-plant fertilizers. It is best if the soil can be raked in some manner to prepare a smooth, level seedbed. Soil moisture should be high before these plastics are placed over the soil, so that the seed planted into the plastic will have enough moisture to germinate. Plastic that is 3 to 4 feet wide is best for covering a standard garden row. Try to select a time to apply plastic mulch when there is little or no wind. The strips of plastic are applied by placing them over the pre-fertilized row to be planted. It is best to bury one end of the plastic and unroll it down the row, and try to get the plastic as straight as possible and in contact with the soil surface. It is also very important to bury and secure the edges of plastic with soil, bricks, or rocks to prevent winds from blowing away or ripping the plastic. Small slits are then made in the plastic with a knife (or other sharp implement) for placing seed or transplants into the soil.

Figure 2.7 Black plastic mulch used in a commercial watermelon field.

Since soils lose less moisture from evaporation with plastic mulches, the garden will not require as much water for irrigation purposes. Supplemental water becomes necessary in prolonged dry periods and in most situations, drip irrigation or soaker hoses are laid on the surface of the rows before covering with plastic to provide a water source for plants in the garden.

Clear plastic results in a soil increase of 8° to 10°F compared to bare soil. Although clear plastic warms the soil more rapidly and to higher temperatures than black, weed seed will germinate and grow under the clear plastic, which will eventually provide competition for the crop being grown. Because clear plastic does not exclude light, weeds are stimulated to grow under the mulch and can become problematic. So, clear mulches are not generally recommended for home gardens. Black plastic mulch results in a soil temperature increase of 5° to 6°F early in the growing season. Moreover, light will not penetrate black plastic, and thus, weeds cannot grow under the plastic without light.

Plastic mulch is highly recommended for warm-season vegetable crops that produce fruit on the ground, such as melons, cucumbers, peppers, and tomatoes. Many other vegetables including broccoli, cauliflower, sweet corn, pumpkin, and winter squash can also be established as transplants into plastic mulch if earliness and a reduction in environmental stress (especially moisture) are warranted. The benefits of inorganic mulches include earlier crop production, higher yield potential, cleaner and higher quality produce, better water- and fertilizer-use efficiency, weed suppression, reduced soil and wind erosion, and opportunity for efficient double or triple cropping. At the end of the season, remove the plastic from the soil since it will not decompose in the soil like organic mulches.

If plastic is to be used on a small garden, it is best to use larger sheets of black plastic to cover the entire area rather than covering individual rows. However, remember that this has the disadvantage of excluding rain and sprinkler irrigation from the soil surface. Also, wet plastic is oftentimes very slippery and working in the garden when there is significant amounts of moisture on the plastic can be quite hazardous to a gardener.

Although plastic warms the soil in spring, this temperature increase in summer can have disastrous results on a garden. Due to the increased temperatures, excess heat can build up under the plastic and high soil temperatures can injure plant roots and reduce yields. However, rather than removing the plastic and losing the additional weed control provided by the plastic, cover the plastic with straw or similar organic material to shade it where the crop foliage does not provide good cover. The black plastic mulch can be sprayed with a white latex paint after the mulch is installed or after an early crop to reduce the build-up of excess heat under the mulch. This will make it useful for summer and fall vegetables. The light-colored surface reflects much of the heat, while the other benefits of the mulch remain.

One of the best time-saving measures that a gardener can do in a garden is to apply a mulch of some sort to cover the soil. This goes for every garden area, not only a vegetable garden or flower bed, but also to other areas, like pathways. Mulched gardens are closer to being weed-free and require less water than gardens that are not mulched. A gardener's choice for using either an organic or inorganic mulch depends on the season and the purpose of the mulch. Organic mulches are best in conserving moisture and reducing soil temperatures during the summer months, and they can also help to reduce soil crusting. Inorganic mulches are best utilized to increase soil temperatures early in the growing season when the soil is cool. Increasing soil temperatures early in the season will increase seed germination and plant growth when air and soil temperatures are typically too low for seeds to germinate or plants to grow adequately.

Chapter Three

GARDEN SOIL FERTILITY PRACTICES

Fertilizers are essential to maximize plant productivity in home gardens, and some type of fertilizer must be applied to all garden soils at some point. Fertilizers provide plants with the essential nutrients that are required for their growth and development, which may be lacking in certain soil types. However, a gardener should first have an idea about the amounts of essential nutrients that are present in their soil prior to applying fertilizers. This is accomplished by a soil test or analysis, which is an important step in starting and maintaining a garden. A soil analysis can help a gardener first decide which plants will grow best under the present soil conditions. It can also help a gardener decide which nutrients and the amounts that should be added to a soil. Soil analyses usually measure the pH of the soil, organic matter content, and the levels of important nutrients like, nitrogen, phosphorus and potassium.

Fertilizers provide the essential mineral elements necessary for plant growth. Although about 90% of the plant weight is made up of water, the remaining plant dry weight is made up primarily of seventeen elements that are required for plant growth. It is first important to understand some common fertilizer terms.

ESSENTIAL NUTRIENTS

There are seventeen elements that are considered essential for proper plant growth and development. They are C-carbon, H-hydrogen, O-oxygen, N-nitrogen, P-phosphorus, K-potassium, Ca-calcium, Mg-magnesium, S-sulfur, Fe-iron, Mn-manganese, Zn-zinc, Cu-copper, B-boron, Mo-molybdenum, Cl-chlorine, and Ni-nickel.

Carbon (C), **hydrogen** (H), **and oxygen** (O) make up 90% of the plant dry weight. These elements are obtained from air or water and are not included in fertilizers.

Macronutrients are those essential nutrients that plants require in large amounts. They are N, P, and K, which are the primary macronutrients, and Ca, Mg, and S, which are the secondary macronutrients.

Micronutrients are those essential nutrients that plants require in small amounts, and include Fe, Mn, Zn, Cu, B, Mo, Cl, and Ni.

It is important to understand that different plants may show different sensitivities to the presence or absence of these nutrients, especially the micronutrients, but they are all required for specific plant functions. Some plants require some of the micronutrients in greater amounts than others. For example, higher amounts of boron are required in cole crops, such as cabbage, broccoli, and cauliflower, compared to most other plants. Thus, a fertilizer containing boron is often applied to the soil in which these crops are grown to ensure that there are adequate amounts of this mineral element for plant growth and development.

FERTILIZERS

Commercial fertilizers come in a wide variety of formulations and nutrient analyses. A commercial fertilizer may contain one or all of the essential elements but the percent of each will be listed on the fertilizer label. Micronutrients may or may not be included in a specific fertilizer formulation. All commercial fertilizers are labeled by the percentages they contain of nitrogen (N), phosphorus (P_2O_5), and potassium (K_2O), although there are many different (N-P-K) formulations, such as 5-10-5, 5-10-10, 12-12-12, and 9-45-15. The percent N is the first number, while the second number is the percentage of P_2O_5, and third number is the percentage of K_2O. These primary macronutrients, N-P-K, are placed on a fertilizer bag by percentage of weight. For instance, if a bag was 10-10-10 and the bag weighed 100 pounds, there would be 10 pounds of N, 10 pounds of P_2O_5, and 10 pounds of K_2O in the bag. The other 70 pounds consist of fillers and carriers. These make the fertilizer easier to spread and prevent plants from getting burned from fertilizer applications. **Complete fertilizers** contain the three primary macronutrients, N, P, and K, but in a variety of proportions, while **incomplete fertilizers** contain less than all three of the primary macronutrients (e.g., 15-0-15 contains 15% N, no P, and 15% K_2O).

It is important for home gardeners to understand simple calculations that often need to be made to apply a certain amount of fertilizer over a given area of the garden. For example, to apply 0.5 lb of actual nitrogen using a 10-10-10 fertilizer over a defined area of a garden soil, divide 0.5 by 10% (or 0.1), and this results in 5 lbs. Thus, 5 lbs of this fertilizer would need to be applied to achieve this fertilizer rate. Organic fertilizers such as animal manures or poultry litters have much lower N levels and should be applied over the surface area of a garden and tilled into the soil before planting. Since organic fertilizers have lower N levels than inorganic fertilizers, later side dressings are usually required for adequate plant growth. However, most home vegetable gardeners tend to use dry inorganic fertilizers due to their higher amounts of nutrients and ease of use.

Of all the essential nutrients, nitrogen is required by plants in the largest quantity and is oftentimes the limiting factor in crop productivity. In plant tissue, nitrogen content ranges from 1% to 6%, and proper management of nitrogen is important since it is required in high amounts by crop plants and easily lost from the soil system by volatilization (loss of gaseous ammonia to the atmosphere), run-off, leaching, or denitrification (conversion of nitrate to atmospheric forms of nitrogen).

Fertilizer Price Comparison. Nitrogen is a major nutrient and is the most likely nutrient to become deficient in garden soils. Thus, nitrogen content is often utilized to compare fertilizer prices. Although dry fertilizers are almost always less expensive and a better buy per pound compared to liquid products, the prices of organic fertilizers will generally be more than the conventional petroleum-based fertilizers.

Various fertilizer products can be compared using this simple equation:

1. Multiply the retail price (let's say $9.95) by 100, which equals 995.
2. Multiply the weight of the package (let's say 10 lbs) by the percentage of nitrogen that is contained in the product (let's say 7-1-2 for N-P-K), which equals 70.
3. Divide the number based on price by the number based on weight and percent nitrogen to get the cost per lb of the nitrogen in the fertilizer; for our example 995/70 = $14.21 is the cost per lb for the nitrogen.

If you are not careful, you can pay outrageous amounts of money (even hundreds to thousands times more) for the nitrogen and other nutrients that your garden needs. So, it is important to have an idea of how to calculate the cost of the nitrogen that you are buying to apply to your garden.

There are many different types of fertilizers on the market. Fertilizers can be readily found that are packaged for certain uses or types of plants. There are fertilizers formulated specifically for roses, for tomatoes, or acid-loving plants like azaleas. Gardeners often notice that different fertilizer brands have different nutrient ratios for the same type of plant such as the rose. Other fertilizers indicate that they will increase the flowering period or produce more flowers, although there is no guarantee that it will occur. There are lots of factors that need to be considered when choosing a fertilizer for a home garden, including whether it is organic or synthetic, and if the gardener is looking to apply either a liquid or granular formulation.

Organic fertilizers contain substantial amounts of organic matter derived from vegetable or animal products. They are made from natural substances without any chemical additives. Many organic fertilizers improve the soil structure, although most tend to be a relatively slow release fertilizer. Most organic fertilizers do not cause as rapid growth as do synthetic fertilizer sources, since much of the nitrogen is tied up in organic forms and is not readily available in the forms (nitrate or ammonium ions) required for plant uptake. Therefore, the non-available forms of this nutrient must be mineralized to the form that is available for plant uptake. Mineralization is the conversion of organic nitrogen to ammonium, while nitrification (another nitrogen soil transformation) is the conversion of ammonium to nitrate. The mineralization of organic fertilizers is generally slow and so there can be problems with the supply of nitrogen, especially when the demand for nitrogen in plants is high. The supply of nitrogen from organic products depends on the speed of their mineralization, which is influenced by not only the composition and formulation of their raw material, but also available moisture and temperature conditions.

Organic fertilizers can be purchased from gardening stores but can also be made from building a compost pile at home from a wide range of substances such as food scraps, grass clippings, and decomposing plants. Organic fertilizers are widely used in some gardens and can be highly effective at improving the growth of vegetables, flowers, lawns, and trees and shrubs.

Manure is also a type of organic fertilizer, and many times it can found for purchase already composted (Fig. 3.1). Animal manures are covered in more detail in the organic gardening section (Chapter 10). Remember that different animal manures differ in the nutrient content and should be considered prior to purchase.

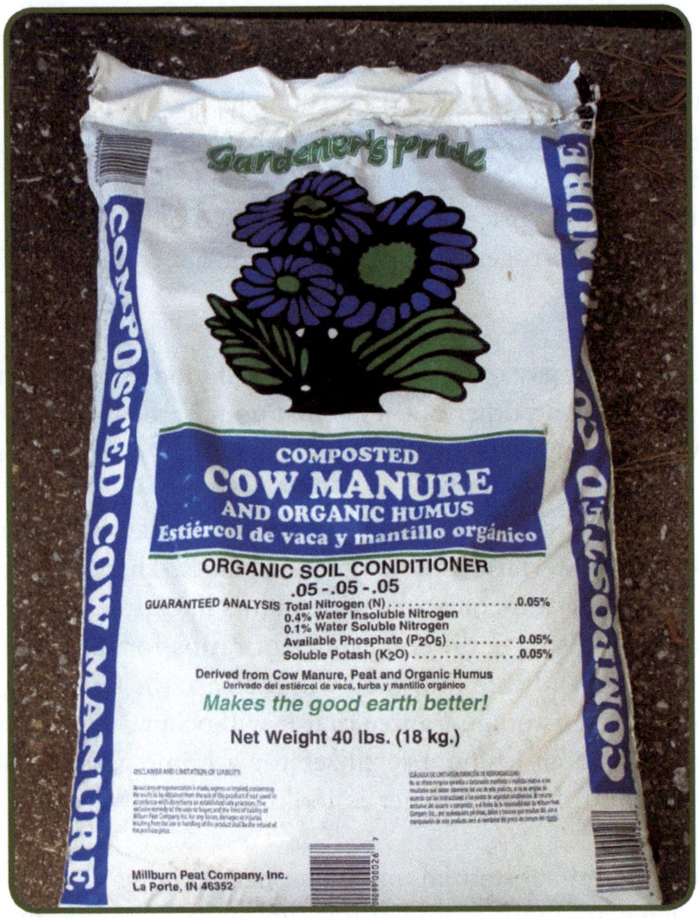

***Figure 3.1** A bag of composted cow manure fertilizer. Notice the low amounts of nutrients present in the material. A good soil conditioner, but it provides few nutrients.*

Fish emulsion is a common organic fertilizer that is derived from gutted fish waste. It provides an excellent source of phosphorous and potassium with a few other nutrients that are highly effective in helping vegetable plants grow. Fish emulsion is easily soluble, which gives it a distinct advantage over many other dry organic fertilizers that break down more slowly.

Synthetic fertilizers can greatly increase crop productivity, since they have nutrients in a concentrated form that is readily available for plant uptake. There are high energy costs associated with their development, since fossil fuels are required in the preparation of the fertilizer. These are common granular type fertilizers formulations that many home gardeners will spread over their garden soil. Although synthetic fertilizers are most often used in home gardens, it is important that they are not overused since excess nutrients can leach into the groundwater supplies. When excessive nitrates leach into the water supplies, it can greatly affect surrounding ecosystems. A large percentage of our rivers and lakes are polluted because of agricultural runoff. High concentrations of plant nutrients often result in large algal blooms, and these algal blooms consume lots of oxygen and block the sunlight, causing other organisms in the eosystem to die. Regions of water with low concentrations of oxygen are called "hypoxia regions," and often referred to as "dead zones." Hypoxia regions can be found in the Gulf of Mexico and the Mississippi River due to all the excess nutrients that are flowing into these areas from agricultural production.

Another problem with synthetic fertilizers is that they feed the plants, not the soil. This results in a gradual decrease in organic matter content of the soil. Once the organic matter is consumed in a soil, the soil will become more compacted and tillage operations become more difficult. The lack of organic matter also leads to a reduction in soil organisms, making it more susceptible to insect or disease infestations.

Although organic fertilizers are natural slow-release fertilizers, there are many different synthetic fertilizers that are also formulated as slow release. These fertilizers provide plants an ongoing nutrient source over several months without the gardener having to make additional applications. These slow-release fertilizers are released for different lengths of time depending on the specific type of slow-release formulation. Nutrient-release rates depend on the fertilizer coating and its thickness, as well as the temperature and level of moisture in the environment in which they are applied. Slow-release fertilizers are often used on plants grown in containers. However, a major disadvantage is they tend to be much more expensive compared to other quick-release fertilizers.

Fertilizer Application Method. The method of how a particular fertilizer is going to be applied must be considered by the gardener. There are basically five ways to apply fertilizers (broadcast, band, sidedress, liquid-soil, or liquid-foliar), although some are more commonly used than others.

Broadcast application. With a broadcast application, the fertilizer is spread over an area of the garden and mixed into the soil before planting. The fertilizer should be spread uniformly over the soil surface and then worked into soil with a garden rake or to a deeper depth with a tiller.

Band application. Oftentimes with a broadcast application, many of the nutrients are absorbed by weeds or lost through other means, and not used for crop growth. Band applications are effective at placing the nutrients close to the seed or transplant, so that there is little waste. For a band application, fertilize in narrow bands in furrows about 3 to 4 inches from where the seed or transplant is placed in the soil and 2 to 3 inches deeper than the planting depth.

Sidedress application. A sidedress application refers to a fertilizer application that is placed near the base of a growing plant several weeks after the initial seeding or transplanting. This application is usually needed to replenish the nutrients that have been depleted from the soil to allow plants to continue to grow at an optimum rate without suffering from nutrient deficiencies. The fertilizer is normally scattered on both sides of the row about 6 to 8 inches from the crop plants. During midsummer, it is important to rake certain fertilizers (those high in ammonium) into soil to prevent ammonia (NH_3) from damaging plants on very hot days. Also, by watering after fertilizing, it will help to dissolve the fertilizer, and thus placing into soil solution which allow plants to more readily take up the nutrients.

Liquid fertilizers are concentrated chemicals formulated to encourage plant growth by either applying to the soil or foliage. These fertilizers can be mixed with water and applied through a garden hose directly onto the plant or on the soil surrounding the plants.

Various formulations of liquid fertilizers can be purchased that dissolve in water and are then applied to the soil. Most of these have a high amount of nitrogen and/or other nutrients. However, the nutrients are available to plants for only a short period of time since they tend to be in the readily available form and will leach from the soil. Nitrogen tends to encourage vegetative growth, while phosphorus encourages root and bud growth (for fruit and flowers). Liquid soil–applied fertilizers tend to work best for container plants because they react more quickly than granular or organic fertilizers, and can be applied easily and quickly. A liquid-fertilizer application can be done either by attaching a mixing unit to the end of your garden hose or by mixing the correct amount of the fertilizer in a watering can or pail and then watering the container plant with the mixture.

There are also foliar fertilizers that can be purchased for feeding plants through their foliage. Plants have the ability to absorb a small amount of nutrients from dilute solutions sprayed onto their leaves. Again, there are multiple fertilizer formulations that contain specific nutrients that can be applied to plant foliage. Foliar feeding has generally been used as a means of supplying supplemental doses of micro- and macronutrients to plants. Since the amount of micronutrients needed by plants is very small, these can be supplied effectively as foliar sprays, especially if nutrients are chelated for better absorption. Foliar application of macronutrients can help plants recover from temporary stress due to moisture problems, pests, or disease. Sprays containing calcium can also be used to prevent blossom-end rot in tomatoes and other fruits, or tip burn in lettuce and cabbage, which often occurs during dry periods. Although plants can absorb macronutrients through the leaves, it is not possible to supply all their needs in this manner, as it is best if these nutrients are taken up by the roots. Thus, foliar fertilization should not be considered as a substitute for a soil-fertility program.

It is important for home gardeners to check fertilizer rates for the different vegetables and other plants grown in a home garden because they vary from low fertilizer feeders, such as sweet potatoes, to those that require high amounts of fertilizers, such as tomatoes.

Liming

The addition of lime is essential in acidic soils for increasing pH for vegetable growth. The limestone available to most vegetable gardeners is calcic (or Calcium Carbonate-$CaCO_3$) or dolomitic [$CaMg(CO_3)_2$]. Dolomitic limestone is often used since it provides a substantial amount of magnesium (Mg^{+2}), as well as calcium (Ca^{+2}) for plant growth. Vegetable gardeners also should be aware that there are different grades of lime to apply, from fine to coarse granules. With fine granules, the Ca^{+2} cations will not last long in soil; this grade is used to give quick results. Coarser grades will release Ca^{+2} cations over a longer period of time and will persist for a long amount of time in the soil until the granules totally degrade. It is probably best to use a mixture of fine and coarse grades, which has the benefits of both fine and coarse materials. Lime is usually applied in spring before planting to avoid losses due to leaching from excessive rains over winter months. Most vegetables grow best in slightly acid soils (pH ~6.5), and excess liming can make the soil too alkaline, which reduces the availability of some nutrients and reduces plant growth.

Wood ashes. Many home gardeners like to apply wood ashes to their garden soils. Wood ashes will supply potassium as well as some other nutrients to the soil. However, wood ashes will also increase the soil alkalinity, and gardeners need to be aware that excessive amounts of wood ashes should not be applied to neutral or alkaline soils. It is recommended to not use wood ashes when the soil pH is above 7.0, as it will reduce the availability of many nutrients in the soil.

Fertilizers are essential to maximize plant growth in a home garden. A fertilizer source, whether organic or inorganic (or synthetic), must be added to a garden soil to achieve adequate plant growth. Remember that synthetic fertilizers provide a quicker plant response in nutrient-deficient situations compared to organic fertilizers, since most of the nutrients in organic fertilizers must be mineralized for plant use.

Chapter Four

VEGETABLE PROPAGATION METHODS FOR THE HOME GARDEN

Many types of vegetables are grown in home gardens. Before we discuss any specific vegetable, it is important to first have an understanding of the propagation techniques for vegetables. Thus, this chapter is designed to cover the propagation techniques for the most commonly grown vegetables in home gardens.

VEGETABLE PROPAGATION TECHNIQUES

Many home gardeners enjoy the aspect of producing new plants from existing ones in the home garden, or from seed collected from plants previously grown in the garden. The propagation of plants by home gardeners is less expensive than buying plants from garden centers or other retailers. With fairly little effort and time, a home gardener can quickly fill their garden space with lots of vegetables at a minimal cost, if they are willing to grow specific varieties of vegetables and save seed for the following year. Most vegetables are seed propagated and the seed of many vegetable varieties can be collected, stored, and then planted the following growing season.

Plants are propagated by sexual (seed) and/or asexual (vegetative) methods. Most plants are generally propagated by one of the two methods, but some plants can be propagated successfully by both methods. Plant propagation is used to preserve and perpetuate desirable plant genotypes.

Seed Propagation. Sexual propagation involves the use of seeds that result from the fusion of male (pollen) and female (eggs) reproductive cells. Most vegetables are propagated by seed with the genetic quality of the seed regulated by private seed companies, who maintain careful control over their seed stock. Most vegetables and flowers sold by companies today are termed "hybrids." Hybrids are the result of two homozygous inbred lines (lines that "breed true" when self-pollinated) that are crossed with each other with the resultant offspring or progeny (termed F_1's) always identical to each other, as they have the same genetic makeup. Most vegetable seeds or plants sold today are F_1 hybrids. Vegetables that are F_1 hybrids have improved quality and yield characteristics compared to many of the older open-pollinated varieties (inbreds that breed true when self-pollinated). However, there are still many varieties of vegetables that can be found that are open-pollinated and today, many of these are sold as heirloom varieties (see Chapter 15).

Vegetative Propagation. The use of vegetative propagation techniques is made primarily on woody perennial plants that are highly heterozygous (or plants that do not breed true if self-pollinated). If heterozygous plants are self-pollinated, important phenotypic traits are not retained in the resulting offspring plants that develop from those seeds. To maintain genetic similarity of plants, the use of sexual propagation (seed) must be avoided to produce new plants. Thus, in many woody plants, vegetative tissues (e.g., leaves, roots, or stems) are often used to produce new plants by inducing the formation of adventitious roots, shoots, or both. Adventitious shoots can arise anywhere on the plant other than shoot meristems, whereas adventitious roots form any place on the plant other than root meristems. The primary advantage of using asexual propagation techniques is to maintain uniformity of the propagated plants, as this propagation technique will produce plants that are exactly the same as the mother plant. However, the use of adventitious tissue formation is never or rarely used to produce new vegetable plants. Although it is rarely used commercially, home gardeners can easily use this method to produce additional plants for the home garden. For example, stems containing leaves that have been pruned from a favorite tomato plant can be placed in a glass of water. Adventitious roots will form on these stems and can be used to produce a new plant that is identical to the mother plant from which it was taken. Stems cuttings can also be used to produce new plants for many other vegetables including sweet potato and pepper.

Although there are various propagation techniques that home gardeners can use, the use of a specific technique really depends on the vegetable plant being propagated. Most vegetables are seed propagated, but home gardeners should know that several vegetables are propagated by primarily by asexual methods and these will be discussed next.

Asparagus – Although asparagus can be propagated by seed, new plantings in the garden are generally started from crowns purchased at a local garden center or from a mail-order company selling horticultural products. Asparagus crowns are really just the base and roots of one-year-old or older plants. Most gardeners find it easier to grow from crowns, which are readily found in the spring at local garden centers. However, the roots should still look firm and fresh, and not be withered or mushy. Unlike most plants, the roots of asparagus crowns can withstand some air exposure.

Garlic – Garlic cloves (or bulblets) when considered collectively together form what we know as the bulb. The propagation of garlic generally involves planting individual cloves in the soil during early autumn, although there is another less utilized method for propagation. Bulbils offer an alternative asexual propagation approach to growing garlic. Bulbils will form if a scape is allowed to mature, and the scape is the stalk that will grow out of a bulb. Similar to cloves, bulbils can be used to propagate garlic vegetatively, and the bulbs that grow from them are clones of the parent plant. However, a major problem is that it takes 2 to 3 years (depending on the type of garlic) to grow full-sized bulbs from bulbils compared to cloves, which will form bulbs in about 9 months after they are planted in the autumn. Thus, garlic can be vegetatively propagated by two distinct methods (Fig. 4.1).

Horseradish – Horseradish is propagated exclusively from root cuttings. Small, pencil-sized-diameter roots about 10 to 12 inches in length are used to produce new horseradish plants. Roots can be either from an established horseradish plant in the garden or roots purchased from a mail-order company. When planting, the head of the root or the top of the root (thickest in diameter) should be raised slightly in the soil and then covered with about 6 to 8 inches of soil.

***Figure 4.1** Garlic bulb with clove dislodged (top) and garlic bulbils forming at the tops of scapes (bottom).*

Sweet potato – The sweet potato is another vegetable that is propagated exclusively through asexual means. The two most popular methods used to propagate sweet potatoes are sprouts from roots (often called slips) and vine stem cuttings. New slips are produced commercially by sprouting roots from the tuber in moist sand or sawdust, and then pulling the sprouts off the roots. Vine stem cuttings can be rooted and then planted in the ground. A method to produce a few slips for planting using both of these techniques is to submerge about one-third of a sweet potato root into a jar of water, place the jar on a heating mat, break off slips once they reach about 4 to 6 inches, put the slips in water to grow roots, and then plant into the soil outside.

Onion – The onion can be propagated by several different techniques, including seeds, transplants, sets, and bulbils. **Seeds** are the most difficult way to start bulbing onions, since they will require extra time in the garden to produce edible bulbs compared to the other methods described. **Onion transplants** are young onion plants grown from seeds that can often be found at local garden centers in the spring. They are ready to put right into the spring garden soil and will mature in early summer. **Onion sets** are small onion bulbs that are ready for planting. They are easy to find during the early spring at farm stores or nurseries and these immature onion bulbs produce large onions in a short period of time. Onion sets are used exclusively in home gardens for production of bulbing onions in both spring and fall. Unlike most onions, Egyptian walking onions (*Allium cepa* var. *proliferum*) reproduce by **bulbils** that grow at the top of their mature flower stalks. These tiny onions are often used for food, but they are also capable of being planted and producing more onions.

Potato – The potato is actually a tuber and not a root. Potato tubers are thickened modified stems that have buds (or "eyes"). The potato is asexually propagated by replanting small tubers or pieces of tubers known as "seed potatoes." These seed potatoes can be purchased in the spring from farm stores, local garden centers, nurseries, or seed catalogs and should be used to obtain the best results in a garden. Gardeners need to understand that each potato "seed piece" must have an eye when it is planted so that it will produce another potato plant (Fig. 4.2).

***Figure 4.2** These potato sprouts each originate from an "eye" on the tuber and have the potential to develop into a new potato plant when planted into the soil with all or a portion of the tuber.*

Chapter Five

VEGETABLE RELATIONSHIPS

It is important for home gardeners to understand the relationship between various vegetables, and this is best achieved by knowing which vegetables belong to a particular vegetable family. The knowledge of vegetable relationships is important so that gardeners can manage home vegetable gardens in a correct manner. Gardeners can apply this knowledge about vegetable families to either keep related vegetables together in a specific location in a garden bed, or to prevent them from following one another when practicing crop rotation.

Crop rotation will be discussed a little further in the pest management section (Chapter 9), but is basically rotating unrelated annual vegetable crops from one area of the garden to another, so that related crops are not grown in the same place in consecutive years. This is an easy way to control many insect pests and diseases in the garden, as members of the same family tend to attract similar pests, and tend to have similar disease problems. Oftentimes, insect pests or disease-causing organisms overwinter in or live in the soil. Thus, placing plants elsewhere the following season will help keep your garden healthy. A few important families are discussed next in further detail.

Alliaceae (or **Alliums**) – This is the onion family that contains vegetable plants such as garlic, leeks, chives, shallots, and onions. This is a cool-season family of vegetables that are generally grown primarily for their edible bulb, although in some cases, the foliage is plant part that is eaten (e.g., chives). These plants tend to have few, if any, problems with pests and disease in most home gardens. In fact, these are great crops to plant among some of your other vegetables to reduce insect pests, and do well following squash and lettuce.

Apiaceae (or ***Umbelliferae***) – This is a cool-season family of vegetable plants and is commonly known as carrot or parsley family. Most Apiaceae vegetables are classified as biennial herbs, and consist of mostly aromatic plants with hollow stems. The defining characteristic of this family is the inflorescence, which is either a simple or compound umbel. A carrot flower is shown in Fig. 5.1, which is a typical umbel type flower. Many members of this family are cultivated for their taproot (carrot and parsnip), although many are also grown for their foliage (parsley), or seed and foliage [dill, coriander (or cilantro)]. Thus, this family contains some of the most flavorful aromatic herbs. Almost every widely cultivated plant of this group can be used as plants for companion plantings. This is due to the tiny flowers that form the umbels, which are well suited for ladybugs, parasitic wasps, and predatory flies. These beneficial insects drink nectar from umbel flowers when not reproducing, and then prey upon insect pests on nearby plants.

Asparagaceae – Most plant members in this family are evergreen long-lived perennials. The best-known species in this family is the edible *Asparagus officinalis*, commonly referred to as *asparagus*. Asparagus is a perennial, cool-season vegetable that emerges from the soil in early to mid-spring. Asparagus was formerly classified as part of the *Liliaceae*, or the lily family, but has recently been reclassified into the *Asparagaceae* family.

Figure 5.1 A carrot flower. This shows the umbel type flower that is characteristic of the Apiaceae.

Asteraceae (or *Compositae*) – This family is commonly referred to as the aster, daisy, or sunflower family. The term Asteraceae comes from *Aster*, which is the most prominent genus in the family. *Aster* has its derivations from the Greek language, meaning "star," which is connected with the star form of the flowers. The globe artichoke produces a flower that is characteristic of the Asteraceae (Fig. 5.2). The *Asteraceae* was formerly known as *Compositae*, and many plants in this family are still commonly referred to as composites. The term "composites" refers to the compound nature of the flowers, since the disks of the flower heads are actually composed of many small flowers packed tightly together. Commercially important vegetable plants in the *Asteraceae* include lettuce, endive, chicory, globe artichoke, and Jerusalem artichoke. Sunflower and safflower are other important crops found in this family. Although plants like sunflowers and artichokes require warmer weather for growth, lettuce likes the cooler temperatures and is classified as a cool-season crop. Many members of this family are grown as ornamental plants primarily for the beauty of their flowers, including *Chrysanthemum*, *Gerbera*, *Calendula*, *Dahlia*, *Tagetes*, and *Zinnia*.

Figure 5.2 The globe artichoke flower has characteristics similar to other members of the Asteraceae.

Brassicaceae* or *Cruciferae – This cool-season vegetable family is also often referred to as the *Brassicas*, crucifers, cole crops, or mustard family. The name Brassicaceae is derived from the included genus *Brassica*, as many of the most common cruciferous vegetables consumed are in a single species (*B. oleracea*), and are not distinguished from each other taxonomically, except only by the horticultural category of groups. Cruciferae is an older name that means "cross-bearing" and describes the four petals of the flowers produced by plants in this family, which resemble a cross (Fig. 5.3). Cruciferous vegetables are one of the dominant food crops worldwide, and include vegetable crops such as cabbage, kale, radish, Brussels sprouts, mustard, turnip, broccoli, cauliflower, bok choy, arugula, and horseradish. This family of vegetables is considered to be a healthy food source, as they contain high amounts of vitamins, fiber, and multiple nutrients and phytochemicals.

***Figure 5.3** The cross-shaped, four-petal flowers typical of the* **Brassicaceae** *family.*

***Chenopodiaceae*, Goosefoot, or Beetroot** – This cool-season vegetable family includes plants such as beets, spinach, and Swiss chard. Most plants in this family produce edible leaves, and in some cases, edible roots. They tend to require high amounts of fertilizer and prefer moist soils. Many members in this family will also tolerate a certain amount of shade, so if a portion of the garden is partially shaded, many of these plants can be used in that location.

Cucurbitaceae or Cucurbits – This family of warm-season vegetables is also known as the vine crop or gourd family. The cucurbits are a family that produces many different economically important species, particularly those producing edible fruits. They are annual or perennial herbaceous plants native to temperate and tropical areas, and most species are vines that spread out over the soil surface or climb by spirally coiled tendrils. No member of this family will tolerate frost or cold soils. All species of cucurbits are extremely sensitive to freezing temperatures, which limits their native geographic distribution and area of cultivation. This family includes many economically important vegetables including cantaloupe, cucumber, gourds, pumpkin, summer squash, watermelon, and winter squash. Cucurbits generally have a low nutrient content, although one exception is the winter squash (some cultivars of *Cucurbita maxima*, *C. moschata*, and *C. pepo*), which have an orange flesh and high beta-carotene content.

Figure 5.4 Sweet potato flowers. This shows the typical funnel-shaped flowers of the Convolvulaceae family.

Convolvulaceae – The *Convolvulaceae* is commonly known as the morning glory family, and is dominated by twining or climbing woody or herbaceous plants with funnel-shaped flowers (Fig. 5.4). Other common characteristics of plant members in this family include heart-shaped leaves and that they exude a milky sap (latex) from stems when injured. The stems of these plants are usually winding, hence their Latin name (from "convolvere," which means to wind or twist). The sweet potato (*Ipomoea batatas*) is the most important vegetable crop found in this family.

Fabaceae (or *Leguminosae*) – This family is commonly referred to as legumes, pulses, or the bean or pea family. The Fabaceae is one of the largest and most important plant families. This family contains some important crop plants such as beans, peas, peanuts, lupines, alfalfa, and clover. These crops convert atmospheric nitrogen into a form that plants can use through a symbiotic relationship with nitrogen-fixing bacteria. Due to their nitrogen-fixing capabilities, legumes can be grown without supplemental fertilizer which is needed by most other kinds of agricultural crops. This family has vegetable crops that can be classified as either cool-season (common garden pea) or warm-season (beans and cowpeas).

Malvaceae – The *Malvaceae* is also commonly known as the hibiscus or mallow family. The most well-known members of this family include okra, cotton, and cacao. Although most numerous in the tropics, plant members of this family are widely distributed and are represented in all except the coldest parts of the world. Okra is typically cultivated for its edible green seed pods in tropical, subtropical, and warm-temperate regions around the world.

Poaceae – This family is also known as *Gramineae* or grass family. The grass family contains some of the most widely distributed groups of plants on Earth. It is important to note that agricultural grasses grown for their edible seeds are called cereals or grains, and three cereals—rice, wheat, and corn—provide more than half of all calories eaten by humans. In America, sweet corn is one of the most popular vegetables grown in home gardens. Sweet corn is often considered to be a vegetable, rather than a grain, since it is eaten fresh like a vegetable.

Polygonaceae – This family has one important member that is often coveted by home gardeners, and that is rhubarb (*Rheum rhabarbarum*). *Rheum* species are herbaceous perennials growing from fleshy roots. *Rheum rhabarbarum* is used to make pies, jellies, jams, and wine. All parts of the rhubarb plant contain the poison oxalic acid, but its concentration in the plant petioles (which are used for food) is very low, and their tart flavor is instead caused by non-toxic malic acid.

Solanaceae – The Solanaceae derives its name from the genus *Solanum*, and is often referred to as the solanaceous or nightshade family. Although this family contains many of the most important vegetable crops grown in home gardens, including tomatoes, peppers, eggplant, and potatoes, it has often been associated with plants that produce seed or fruit that have poisonous or narcotic properties. Solanaceous plants are known for possessing a diverse range of alkaloids, and humans were initially suspicious of foods like potatoes and tomatoes because of concerns about these toxins contained within some plant parts. **Solanine** is an alkaloid toxin found in most members of the nightshade family. Although the solanine levels in vegetables such as eggplant, potatoes, and tomatoes are too low to cause health problems, there are some circumstances in which solanine can be elevated. For example, unripe tomato fruit tend to have higher levels of the toxin, as do potatoes that have been damaged or exposed to the sun. Tomatine is another poisonous alkaloid found in the tomato plant, and it is found in the green parts of the plant, such as the leaves, stems, and unripe fruit. It is important to note that ripe red tomato fruit has only minimal levels of tomatine and is generally safe for human consumption.

Chapter Six

COOL-SEASON VEGETABLES FOR THE HOME GARDEN

Although family structure is important in understanding vegetable relationships, vegetables are often categorized as either **cool-** or **warm-season**, depending on the temperatures required to obtain the best growth. This allows gardeners to determine the time of the year that a particular vegetable can be grown. Cool-season vegetables are generally adapted to average daily temperatures below 70°F, while warm-season crops grow best when the average is above 70°F. Cool-season vegetable crops will withstand slight freezing temperatures, although some, such as potato, will be severely damaged or killed by frost. However, all warm-season crops are sensitive to cold weather and will be killed by frost conditions. Cool-season vegetables generally include those in which the edible portion is a root, stem, leaf, or immature flower part, while the edible portions of warm-season vegetables are fruits. However, there are exceptions such as the garden pea (a cool-season vegetable that produces fruit) and the sweet potato (a warm-season vegetable that produces an edible root).

COOL-SEASON VEGETABLES

There are many vegetables that grow best at lower temperatures that occur in the early spring and fall, and these temperatures generally provide ideal growing conditions for cool-season vegetables. In warm regions, cool-season vegetables grow best and develop the highest quality when planted from late summer to early autumn for a late autumn, winter, or early spring harvest. In the coldest regions where summers are generally cool (e.g., high elevations or Alaska), cool-season vegetables can be planted in May or June for a summer harvest. Many are able to withstand some light frost, and develop better flavors at cooler temperatures.

Cool-season crops such as lettuce, spinach, and peas can be planted in early spring, and again in late summer/early autumn for late-season cropping. This practice of doing more than one planting and harvesting at different times of the year is called succession planting. It is also important to note that cool-season vegetable crops need to be harvested before warm weather arrives in summer, as they tend to bolt and go to seed. Several cool-season vegetables will be discussed in further detail regarding production practices, including asparagus, carrot, celery, celeriac, cole crops, lettuce, onion, garlic, peas, potato, rhubarb, spinach, beet, and Swiss chard.

ASPARAGUS: *ASPARAGUS OFFICINALIS*

Asparagus is a perennial, cool-season vegetable that produces fern-like foliage, and is grown for the tender young shoots produced in the early spring (Fig. 6.1). Asparagus is becoming one of the most popular vegetables for home gardens since it is relatively easy to grow and maintain, and has a high cost when purchased at the grocery store. Once planted, this perennial vegetable can be harvested during the early spring for the next 25 years or longer if properly managed. Asparagus is the earliest vegetable that can be grown and harvested from the garden, and thrives in most temperate areas of the United States.

***Figure 6.1* White and green asparagus are available in most markets. Many consumers like white asparagus since it does have the strong flavor like green asparagus.**

Asparagus spears and new growth develop each year from crowns that are initially planted several inches into the ground many years earlier. One-year-old crowns are normally used to start asparagus production in gardens, but older crowns (two or three years old) can be purchased if the gardener is eager to start harvesting spears from the garden sooner. Typically, one-year-old crowns will need at least three years of growth before any spears should be harvested. Although asparagus seed can be purchased, gardeners should use crowns rather than seed. The planting of crowns allows spears to be harvested from a garden quicker compared to seed, and they also provide more male plants, which live longer and yield more than do female asparagus plants. Thus, crowns provide two distinct advantages compared to seed: 1) quicker start for harvesting spears out of the garden and 2) greater percentage of male plants. Once the asparagus crowns are in hand, the gardener should avoid allowing crowns and roots to dry out before planting, and crowns should be planted during the early to mid-spring.

Soil preparation is the really the key to successful asparagus production, since the foundation is being prepared for the next 25 years or more. Soil should be prepared to a depth of about 18 inches with lots of compost or organic matter and about 2 to 3 lbs of complete fertilizer that contains about 10% N per 100 ft^2 mixed into the soil. A trench should then be dug about 8 inches deep and 12 inches wide, with trenches placed 4 to 5 ft apart. Compost or manure should be placed in the bottom of the trench and covered with a few inches of garden soil. Crowns should then be placed in the bottom of the trench about 1 to 2 ft apart in the row and covered with 2 inches of soil. As new shoots emerge from the crowns, gradually fill the trench with soil until the trench is completely filled (Fig. 6.2).

Asparagus in the garden requires two fertilizer applications annually. At each application, apply about 2 lbs of complete fertilizer containing about 10% N per 100 ft^2. The application timings should be made before spears emerge in spring, and as soon as harvest is completed in late spring to encourage high amounts of fern growth. The ferns that grow and develop after harvest completion in the late spring should not be cut or mowed until they die naturally in fall. The fernlike foliage growth throughout the summer and autumn is required for the build-up of food reserves in the crowns (Fig. 6.3). This is essential for maximizing spear production during the following year. Compost or aged manure can be applied in fall and/

Figure 6.2 A visual progression showing the placement of an asparagus crown into a trench (A), and the gradual addition of soil to the trench as the asparagus plant grows until it is completely filled (B).

or early spring to increase organic matter content in the soil. An organic mulch (e.g., straw) can also be applied around asparagus to reduce emerging weeds, which can compete with the spears and reduce production. Hand-pulling can be used if any weeds appear after the asparagus is mulched.

White asparagus (shown in Fig. 6.1) is much more common in Europe than in the United States. White asparagus is produced by excluding light from the developing spears; this is done by mounding soil over the row and covering with a light-excluding plastic. Covers are removed and replaced daily during the production season, and spears are cut while still under the surface just before the spear breaks through the soil. A long knife is used to cut the spear and pull from the soil. Spears must also be handled in the absence of light to keep them white after harvest.

Figure 6.3 Commercial asparagus planting showing fern-like foliage in summer.

There are a couple of different ways to obtain white asparagus in a home garden planting. First, white asparagus can be produced under black plastic tunnels supported by wire hoops. One side of the tunnel is buried by soil, the other held down by some type of weight. Tunnels are then opened periodically for harvest and then closed again once harvest is completed. Another method is to pile wheat straw (or some other organic material such as leaves) about 2 feet deep over the planting. During the harvest, the straw is removed and then replaced once the daily harvest is complete. So, there is lots of extra work to get white asparagus.

It is important for gardeners to realize that spears should not be harvested anytime during the spring after the first season of growth if one-year-old crowns are planted, and only a few spears should be removed the second year. This allows the plant to build up large amounts of photosynthates in roots to provide for spear growth the following year. The third year after planting, the gardener should be able to harvest spears appearing in early spring for about a month. If two- or three-year-old crowns are planted, the time until they can be harvested for a full month can be reduced to only two years or one year after planting, respectively.

Typical schedule for establishing and harvesting asparagus from planted one-year-old crowns:

Spring – plant crowns
Spring 1st year – do not harvest
Spring 2nd year – harvest lightly (only a few spears)
Spring 3rd year and all springs after – harvest 4 to 6 weeks

Early in the season, spears can be harvested about 2 to 4 times per week, depending on the ambient air temperatures. However, as temperatures rise later in the season, it may be necessary to cut once or twice a day. Spears should be harvested when 6 to 8 inches in height by snapping or cutting at ground level. Fig. 6.4 shows asparagus spears emerging out of soil.

Healthy asparagus foliage, crowns, and roots must be maintained for optimum spear production. There are several insects and diseases that can reduce an asparagus plant's productivity. The most important insect pest is the **Asparagus beetle (Fig. 6.5)**. It has a metallic blue head with a red thorax having two blue spots and is about ¼ inch in length. Each bluish-black wing cover has three yellowish square spots. The larvae are light gray to brown with black heads and legs. Asparagus beetles and their larvae can completely destroy the foliage of an asparagus plant. Asparagus beetle eggs can commonly be seen on asparagus plants in the spring.

Figure 6.4 Asparagus spears emerging out of soil.

There are three primary diseases of asparagus that must be controlled to achieve optimal yields. These are asparagus rust, *Fusarium* crown and root rot, and *Phytophthora* spear rot. **Asparagus rust** produces reddish-brown masses of spores on all above-ground plant parts except the berries, and is most damaging

Figure 6.5 Asparagus beetle (left) and asparagus beetle eggs on flowers of asparagus (right).

during prolonged dry periods. It will weaken the plant and reduce marketable yield. Most new varieties of asparagus have resistance to this disease. ***Fusarium* crown and root rot** produces reddish-brown internal lesions at or below the surface of the soil on stems and within the crown, eventually causing tissue breakdown and death. Planting on well-drained soils is the best way to reduce problems caused by this disease. ***Phytophthora* spear rot** is characterized by soft, water-soaked lesions on shoots at, slightly above, or below the soil level. *Phytophthora* rot of spears typically occurs during periods of high rainfall, particularly at low-lying sites with poor drainage. Spears shrivel from the tips and lesions can occur anywhere along the length of the emerging spear and have a water-soaked appearance, ranging from green to light brown in color. These diseases can be somewhat controlled by selecting resistant varieties and most of the newer varieties are resistant to one or more of these diseases.

When selecting an asparagus variety, gardeners should consider only those that have a high number of male plants, as well as having multiple disease resistances. Suggested varieties:

'Mary Washington' – Probably the leading home garden variety. It is an older open-pollinated variety that it mostly female, but has some resistance to rust.

'Jersey King' – A versatile asparagus variety that is easy to grow at home with mostly male plants producing large tender green spears with purple bracts. It has high tolerance to rust and *Fusarium* crown and root rot diseases.

'Purple Passion' – A novelty variety having beautiful purple spears, although spears turn green when cooked. Good resistance to rust and tolerance to *Fusarium* crown and root rot. It is predominately male, but will have a smaller yield than other all male varieties.

'Jersey Knight' – Many gardeners say that it is the best asparagus variety for home gardens and has excellent spear quality and flavor. It is a predominantly male variety that is widely adapted to both warm and cold climates. It is highly productive in most soils, including heavy clay soil types, although it prefers alkaline soils. It is resistant to rust and highly tolerant to *Fusarium* crown and root rot, and slightly later than 'Jersey Giant'.

'Jersey Supreme' – Predominantly male hybrid with excellent yield potential. It is a versatile asparagus variety that grows in and adapts to most climates, and has greater yields than 'Jersey Giant'. This variety also has good resistance to rust and highly tolerant to *Fusarium* crown and root rot.

'Jersey Giant' – Predominantly male hybrid that produces uniform, extra-large green spears that are tender and flavorful. The spears produced by this variety are thicker than both 'Jersey Knight' and 'Jersey King'. Good resistance to rust and tolerant to *Fusarium* crown and root rot. It is also widely adapted to most climates and soil types, and popular with both commercial growers and home gardeners due to its high productivity and earliness. Unfortunately, this versatile variety has a relatively low yield.

'Millennium' – This high-yielding male hybrid is best adapted to heavier soils. It is a top producer noted for maintaining its high yield year after year, and its ability to withstand cold winter temperatures. 'Millennium' is a hardy, high quality asparagus ideal for the home garden.

CARROTS: DAUCUS CAROTA

Carrots belong to the Apiaceae family and are really an easy vegetable to grow (Fig. 6.6). They are adaptable and tolerant of poor gardening techniques, so they should be grown in most every vegetable garden. Carrots are a cool-season crop and prefer temperatures for growing in the 45°–75°F range. When planted in the early autumn and then harvested during the

Figure 6.6 Carrots are a nutritious, versatile vegetable, and provide an excellent source of vitamin A. They are fairly easy to grow in home gardens.

winter and/or early spring the following year, carrots will have an outstanding sweet flavor that you cannot find at most supermarkets. Some varieties work best for this purpose, and the placement of an organic mulch over the soil surface will also help to protect roots from cold winter temperatures. Any home vegetable gardener in temperate climate areas can have carrots year-round out of the garden if they just know how to manage these plants during cold weather.

The soil type or condition of the soil is probably the most important factor when considering growing carrots in the garden. Carrots grow poorly in heavy soils. Long-rooted varieties may be stunted, forked, or twisted. Growing conditions can often be improved by incorporating organic matter, such as compost, into heavy soils. Although any good garden soil will grow carrots, a deep, loose, and fertile sandy loam or peat soil with good moisture holding capacity and high organic matter will grow the straightest and smoothest roots. The soil should be tilled deeply and raised beds tend to work best. Compacted soil conditions will cause malformed or misshapen roots.

Carrots are light nutrient feeders, so limited sidedress fertilizer applications are typically needed. However, nitrogen applications should not be made close to harvest or once they are about three-fourths their mature size. Late applications of nitrogen will also cause roots to split. So, before planting, it is recommended to apply about 1 to 2 lbs of complete fertilizer containing about 10% N per 100 ft^2 of row and that's really all the fertilizer you need to apply for the entire season.

Crusting of the soil surface and dry soil conditions may result in poor germination. This is particularly true with late-summer plantings. Crusting may be prevented by lightly covering the seeded row with moist compost. During dry weather conditions, provide a source of water daily to the seeded carrots to promote germination.

For spring planting, seeds should be planted 2 to 4 weeks before last expected frost and for fall planting, seeds need to be planted approximately 2 months before first expected frost. Sow carrot seeds at a depth of ¼ to ½ inch. For a continuous harvest, succession planting should be used for carrots and additional plantings should be made every 3 to 4 weeks. Space rows 18 to 24 inches apart. Thin seedlings within a few weeks of germination, and after thinning, seedlings should be spaced 2 to 3 inches apart. The soil should be kept moist to prevent soil crusting before seedlings emerge. Carrots usually mature in 50 to 70 days after seeding. Carrots can, however, be harvested at any point that the roots are large enough to eat. Carrot crowns should be covered with soil to prevent greening of crowns, which cause roots to develop a bitter taste. Carrots are usually dug at any time after they obtain a good orange color and are greater than 1 inch in diameter. Generally, the best harvest period lasts about 3 weeks (or 5 to 7 weeks during cooler, fall weather).

Carrots need approximately 1 inch of water every week from rain or supplemental irrigation. Adequate water is essential for root enlargement. Watering should be reduced when carrots are three-fourths their mature size to reduce the chance of root splitting.

Generally, carrots do not have many insect or disease problems. However, problematic insects include the carrot rust fly (maggots chew roots and will causing plant stunting), the carrot weevil (will defoliate plants, and larvae tunnel into carrots), the parsley worm (will totally defoliate plants), and the wireworm (larvae chew and bore into roots). Disease problems include root-knot nematodes (cause plant stunting, yellowing of foliage, and galls on roots), damping-off of young seedlings (usually caused by various species of *Pythium* or *Rhizoctona solani*), and leaf blights (species of *Alternaria* and *Cercospora* can be important pathogens).

Weed control can be a major issue when growing carrots. It is best to control weeds when they are still in the seedling stage and this can be achieved with frequent, shallow cultivating (hoeing) or pulling weeds by hand in the immediate vicinity of the carrots. When the carrots are well established, the application of a mulch between the rows will help control weeds and conserve soil moisture.

There are numerous carrot varieties available to home gardeners. The main differences between varieties are the shape and size of the root. However, a few varieties have unusually colored roots. Carrot roots can be long and tapered, cylindrical, or even roundish. Most carrots are orange, although yellow, red, and purple varieties are also available.

Suggested carrot varieties for home gardeners include:

'Danvers Half-Long' – (75 days; uniform, 7- to 8-inch-long roots tapered to very blunt end)
'Danvers 126' – (75 days; smooth roots; foliage can withstand hot temperatures).
'Imperator 58' – (68 days; smooth, fine-grained, long, tapered roots; standard long, thin type of carrot)
'Little Finger' – (65 days; baby type with tiny tender roots; ½-inch-thick and 5-inch-long roots; golden orange; sweet and crisp)
'Purple Haze' – (70 days; novelty type; 2006 AAS winner and first Imperator-shaped purple carrot; 10- to 12-inch-long purple-skinned roots with vivid orange centers)
'Red-Cored Chantenay' – (70 days; heavy yielder; short, thick roots, broad at the shoulder, tapered to blunt tip)
'Royal Chantenay' – (70 days; broad-shouldered, tapered roots; bright orange)
'Scarlet Nantes' – (70 days; 6-inch-long bright-orange roots that are slightly tapered; standard for high-quality carrots)
'Short 'n Sweet' – (68 days; baby type with sweet flavor; 4-inch roots, broad at shoulder, tapered to a point; good for heavy or poor soil)
'Tendersweet' – (75 days; long, tapered roots; nice orange color; coreless)
'Thumbelina' – (60 days; 1992 AAS winner; round roots; good for planting in containers and in heavy, shallow, or rocky soil)

Carrots are a nutritious, versatile vegetable, and provide an excellent source of vitamin A. They are rich in carotene (the source of vitamin A) and high in fiber and sugar content. Carrots can be shredded, chopped, juiced, or cooked whole. They are delicious raw, roasted, boiled, steamed, stir-fried, and grilled, and they team up beautifully with almost any vegetable. Carrots boost the nutritional value of soups, stews, salads, and many other foods. Mature carrots can be stored for several months at a temperature of 32°F and relative humidity of greater than 95%.

CELERY: *APIUM GRAVEOLENS* VAR. *DULCE*

Celery is generally considered one of the more difficult vegetables to grow. Celery demands more time and attention than most vegetables and that is why it isn't seen very often in most vegetable gardens. Celery is a cool-season vegetable crop found in the Apiaceae family. It is a long-season crop that likes cool temperatures, constant moisture, and will not tolerate warm temperatures. Although it is typically transplanted, it can oftentimes be difficult to establish in a garden. Summer plantings in the north and winter planting in more southern locations make celery a plant that can be grown in many temperate climates. Many gardeners find that all the work is worth it when the crunchy, green stalks are harvested.

There are two main varieties of celery grown today. The most common is the pale green Pascal celery. Golden celery is another type that is grown under a layer of soil to prevent it from developing and turning green.

Seed should be sown ¼ inch deep (seeds are tiny) about two months before the expected transplant date and require about 15 to 21 days to germinate. Seeds germinate best at 55° to 70°F and for best germination, soak seeds in warm water overnight before planting. Plants should be kept moist by covering the flats or pots with shade cloth. Harden off seedlings before transplanting by reducing water slightly and keeping them outdoors for a couple hours each day. Transplant celery into the garden and provide lots of shade and moisture to the new plants. This is critical since plant survival and establishment in the garden is oftentimes difficult. If you don't start from transplants, plant seeds 2 to 3 weeks before the last expected frost in the spring or for fall planting, about 2 to 3 months before first expected frost. This really limits the regions in which they are adaptable, since they will not tolerate hot weather. For summer gardeners, a late-summer direct sowing is possible, but make sure that temperatures stay between 55° and 70°F throughout growing period. It is important to mulch and water directly after planting. In cool climates, plant celery every month for a succession of celery all season long, but even in these areas, celery prefers some shading in the heat of the summer. Celery grows best at temperatures of 65° to 75°F. Celery will tolerate a light frost, but not consecutive frosts.

Celery is a heavy nutrient feeder. Before seeding or transplanting, broadcast about 3 lbs of a complete fertilizer containing about 10% nitrogen per 100 ft^2 of garden space, and sidedress at monthly intervals with another 1 lb of the same fertilizer per 100 ft^2. Space celery plants 9 to 12 inches apart in rows with 12 to 15 inches between rows. Blanching of stems can be done to produce white petioles; this can be achieved by setting plants in trenches 6 inches deep and then gradually filling trenches as plants grow. However, there are some self-blanching varieties that can be found. Tie growing celery stalks together to keep them from sprawling.

Celery is also a heavy user of water and if celery does not get enough water, the stalks will become dry and small. Make sure to provide plenty of water during the entire growing season, especially during hot, dry weather.

Celery is harvested by cutting the taproot below the soil surface. Typically, the number of days to maturity is about 125 after direct seeding. The plant can be allowed to bolt, which will produce seed and is used as a seasoning in several dishes. The parts of celery that are harvested are mainly the stalks, which will be above ground. Harvest stalks from the outside in, and this can begin when stalks are about 8 inches in height. The darker stalks contain more nutrients, but texture changes with color and dark green stalks will be tougher.

Celery can be attacked by many insects and diseases. Insects that are problems in celery production include aphids (will turn foliage yellow and form honeydew on plants, which attracts ants), carrot rust fly (flies lay eggs in crown of plant and maggots will chew on roots), carrot weevil (larvae chew celery hearts and tunnel into tops), parsley worm (will defoliate plants), and celery leaftier (leaves will fold over and be closed by webs, with holes appearing in leaves and stems). Several diseases can disrupt celery growth including black heart (which is a physiological disorder due to lack of calcium), damping-off of seedlings (usually caused by various species of *Pythium* or *Rhizoctona solani*), early blight, *Fusarium* wilt, late blight, pink rot (pink growth at base of stalk), and celery mosaic.

Celery is a vegetable that is difficult to grow in many environments, since muck soils or well-drained, medium-textured mineral soils with irrigation are best suited for celery. The crop will withstand light freezes, but it is damaged by moderate freezes. Seedstalk development, rather than normal growth, will occur if young plants are exposed to temperatures below 55°F for 10 or more consecutive days.

There are few varieties of celery that can be grown successfully, and in most temperate climates, it is best to use slow-bolting varieties such as Conquistador (80 days, early variety, widely adapted), 'Utah 52-70R Improved' (100 days, has good disease resistance, widely adapted), and 'Ventura' (80 days, widely adapted).

CELERIAC (*APIUM GRAVEOLENS* VAR. *RAPACEUM*)

Celeriac is a form of celery grown that is grown for its swollen root, but the plant is smaller and the foliage is very dark green. Botanically, celeriac is a fleshy, globe-shaped enlargement of the hypocotyls and root tissues (Fig. 6.7). Celeriac is also called **turnip-rooted celery**, **knob celery**, or **celery root**. Celeriac should be grown in the same manner as celery: use transplants, needs continuous supply of water, heavy nutrient feeder, space plants 9 to 12 inches apart in row and 12 to 15 inches apart between rows, and is affected by the same insects and diseases as celery. Celeriac needs full sun conditions and consistent moisture during the growing season, otherwise the bulb will be small and tough.

***Figure 6.7** Celeriac growing in the garden. Notice the swollen enlargement of the hypocotyl and root tissues that form the bulbous root.*

The swollen root can be harvested once it reaches 2" or more in diameter, and this is done by undercutting the swollen root.

Only a limited number of varieties are available, including 'Brilliant' (105 days), 'Monarch' (100 days), 'Prinz' (95 days), and 'Snow White' (100 days).

FENNEL: FOENICULUM VULGARE

Fennel is a cool-season crop in the Apiaceae family and requires cool weather for optimum growth (Fig. 6.8). Fennel will grow about 2 feet high with broad leafstalks that overlap each other at the base, forming a bulbous enlargement that is firm, sweet, and white inside. Fennel is used much like celery, but has a licorice flavor. This crop is not as popular compared to some of the other vegetables, but gardeners should be aware of fennel.

Fennel can be planted as either a spring or fall crop. Seeds can be sown about ¼ inch deep in soil, and thinned to about 18 inches apart in rows with 1½ ft between rows. Fennel is not usually transplanted since root disturbance will cause bolting. When fennel plants are about half-grown (approximately 30 to 40 days after planting), garden soil can be mounded around the base of plants to blanch the bulb. Other cultural methods and insect/diseases are similar to celery.

A few varieties of fennel are available, including 'Rudy' (80 days, bolt resistant) and 'Zefa Fino' (80 days, bolt resistant).

Figure 6.8 Fennel growing in a garden.

COLE CROPS

The **Brassicaceae** is one of the more important vegetable families and is probably the most important family of cool-season vegetables. The family of vegetables is often referred to as the cole crops, and includes some of the most widely grown vegetables in the temperate zone and are excellent home garden vegetables. Vegetables in the *Brassicaceae* family include cabbage, cauliflower, broccoli, and Brussels sprouts. The members of this family that produce edible leafy greens and/or roots will be discussed by that grouping. Many cole crops are started from seeds indoors and are transplanted into the garden, although many, like the leafy cruciferous greens group and radishes are direct-seeded into the garden soil.

Cole crops grow best in cool weather and can survive several light frosts in the fall. Of this group, cauliflower is the least sensitive to excessive hot or cold conditions, while Brussels sprouts are the most cold tolerant. Most cole crops will grow at temperatures around 40°F, with optimums around 65°F, although little growth occurs over 80°F. All cole crops are frost hardy with most tolerating temperatures into the low 20s°F.

A soil pH in the range of 6.0 to 6.8 is optimum for cole crops, but a pH maintained closer to 6.8 will help control the soil-borne disease called clubroot. Cole crops are heavy nutrient feeders and do best in a deep, fertile soil enriched with plenty of organic matter. Before planting, gardeners should apply about 3 lbs of a complete fertilizer containing about 10% N per 100 ft^2, followed by two sidedress applications of 1 lb per 100 ft^2 of the same material at 3 and 6 weeks after transplanting. Since cole crops have a high sulfur requirement, it is preferable if one of these sidedress applications includes either ammonium sulfate or potassium sulfate.

Longtime gardeners know that the micronutrient boron is critical for production of cole crops. Cole crops have a moderate to high boron requirement for growth and proper development. The symptoms of boron deficiency vary, although most boron-deficient cole crops develop cracked and corky stems, petioles, and midribs. The stems of broccoli, cabbage, and cauliflower can be hollow and are sometimes discolored. Cauliflower curds become brown and leaves may roll and curl, while cabbage heads may be small and yellow. Cauliflower is by far the most sensitive cole crop to boron deficiencies. However, only a small amount of boron is normally required to correct a soil deficiency and soils can quickly go from being deficient to having toxic levels very quickly. Boron may also be applied as a foliar treatment to cole crops if soil applications were not made, and these foliar boron applications should be made prior to heading of cole crops.

Various insect and disease pests can be problematic in the production of cole crops. Crop rotation plays an important role in managing insect and disease pests in cole crops. Gardeners should not plant any cole crop in an area of the garden occupied the previous year by another cole crop family member, and two- or three-year rotations are even better.

Several insects are annual problems on cole crops and must be controlled to produce a usable plant product. The most important insects that will be found every year on these vegetables are foliar-feeding caterpillars, such as the larvae of the cabbage looper, diamondback moth, and imported cabbageworm. If caterpillars are not controlled, cole crops will become readily infested with these worms. The larvae will chew holes in leaves, bore into heads, and leave excrement on plant parts. Young plants can be killed or maturity delayed if damage is extensive. Severe feeding injury after heading on cabbage may cause head stunting or abortion. Signs of insect feeding and the presence of caterpillars (along with large amounts of excrement) on edible plant parts make them less desirable for eating. Insect barrier rowcovers left on throughout the growing season, or regular applications of Bt (*Bacillus thuringiensis*), a bacterial poison that kills only these caterpillars, will control cabbage worms of all species.

Other insects that are more sporadic problems are aphids, onion thrips, flea beetles, and cabbage maggots. Flea beetles and root maggots can cause serious damage to cole crops, feeding on the roots of young seedlings, which can result in wilting and eventual death. Covering beds or rows with floating rowcovers made of spun-bonded polyester for the first month or more of growth will prevent attacks by many types of insect pests.

Diseases are not typically as much of a problem as insects, but are worse in some years and will become a major problem if vegetables are not rotated around in the garden. Some of the more important diseases are black rot, club root, downy mildew, and fusarium yellows. Many of the varieties of the different cole crops have resistance to various diseases and if possible, these varieties should be chosen over others. With the selection of resistant varieties coupled with proper sanitation and vegetable rotation, a gardener will usually have little problem with diseases on plants in this family. Clubroot is a soil-borne disease that causes stunted plants to wilt even in moist soil; and, if this disease persists in the garden, raise the soil pH to 7.0 to 7.2.

Cole crops planted in a garden provide the gardener with food for a more healthy diet. Most cole crops are good source of potassium, calcium, iron, phosphorus, and vitamins A and C. Research has shown that other health benefits result from cole crops, such as reducing the risk of certain cancers and heart disease. Since cole crops are such an important part of home gardens, several commonly grown vegetables in this group will be discussed.

CABBAGE: *BRASSICA OLERACEA* (CAPITATA GROUP)

Cabbage is a cool-season vegetable that is widely grown in vegetable gardens. Cabbage is a leafy green biennial plant, grown as an annual vegetable crop for its dense-leaved heads. It is closely related to other cole crops, such as broccoli, cauliflower, and Brussels sprouts, and the cabbage that we know was developed gradually from leafy, non-heading types growing in the wild in various parts of Europe. Over a period of years, tighter-headed types were selected (Fig. 6.9).

Cabbage can be planted for either the spring or fall growing seasons. However, gardeners need to plan their plantings so that only a few heads become ready to harvest at the same time; otherwise, fifty cabbage transplants planted at the same time will produce fifty cabbage heads ready to harvest all at the same time. This can be overcome by succession planting, where a few transplants are planted at weekly intervals for about four to five weeks after the initial transplanting date. This method should allow for several cabbage harvests in both the spring and autumn, and prevent a gardener from having an excessive amount of cabbage that must be harvested from the garden at one time.

The plant spacing used for cabbage in the garden will determine the head size, and proper spacings must be used by the gardener to obtain the desired head size. To produce large heads, plants should be spaced about 2 ft apart in the row with 3 to 4 ft between rows. Smaller or

Figure 6.9 Note the overlapping and tight nature of the leaves inside a cabbage head.

more normal-sized heads can be developed at 12-inch plant spacings in the row. Organic mulches, such as straw, can be placed around the base of cabbage plants to increase productivity in many ways. First, cabbage plants like the cooler, more moist soil conditions provided by an organic mulch during the latter part of their growing season in spring when weather conditions can be quite warm; and lastly, an organic mulch will usually aid in moisture retention in the soil and provide some weed control.

Cabbage harvest can be started when heads are firm and a little larger than a softball (Fig. 6.10). Harvested heads should to be hard and solid. If a gardener starts to harvest a little earlier while heads are smaller than desired, a longer harvest season can be realized for cabbage planted at the same time. When harvesting, cut with a knife just beneath the head, and leave some basal leaves to support the new growth of small lateral heads. If weather conditions do not get too hot after the head is removed, then these smaller heads will develop and can be harvested several weeks later.

Splitting is a common problem that is often observed on cabbage grown in a garden. This is a physiological issue directly related to rapid new growth resulting from a sudden influx of water into the plant after the plant had been subjected to drought conditions or periods of little moisture. Any way to maintain consistent amounts of soil moisture, such as mulching, will help alleviate this problem.

Cabbage varieties differ by color, disease resistance, shape and size of mature head, and leaf type desired. Many savoy types are becoming popular with home gardeners (Fig. 6.11). There are many cabbage varieties from which a gardener can choose, but it is probably best to grow some of the newer hybrids since they will have a significant amount of resistance to the different diseases. Some suggested cabbage varieties are:

'Bravo' – This hybrid cabbage is widely adapted. It produces round, small- to medium-sized heads that are uniform and a beautiful blue-green color. In-row plant spacings at 10 to 12 inches are suggested for this variety, and will mature in about 80 days. It is also resistant to *Fusarium* yellows.

'Dynamo' – This variety is single-serving mini-cabbage with round, tight bluish-green heads and will mature in about 60 to 70 days after transplanting. The heads are only 4½ inches in diameter, with small cores and weigh about 5 to 7 lb. It is an AAS winner for 1997 and has resistance to *Fusarium* yellows.

Figure 6.10 Mature cabbage head in a garden.

'Royale' – This is an early red hybrid cabbage variety that produces high yields of 3 to 4 lb. round, uniform, bright-red heads. This variety is typically ready to harvest 64 to 72 days after transplanting. 'Royale' also has resistance to *Fusarium* yellows.

'Red Rookie' – This is an early maturing hybrid cabbage that will form medium-sized heads that reach about 3 to 4 lbs in about 70 days. 'Red Rookie' produces dense, uniform, short-cored heads that have a deep red internal color. It is also is resistant to splitting.

'Early Jersey Wakefield' – This is an older open-pollinated variety that produces blue-green, pointed (conical-shape) heads that average about 7 inches in length and weigh about 3 lbs. It is an excellent early variety, and takes anywhere from 55 to 70 days to mature.

'Mammoth Red Rock' – This heirloom (open-pollinated) cabbage from 1889 produces large, solid, round, 8-inch-diameter dark red heads that average 7 lbs. It is a late-season variety, as it matures in about 90 to 100 days.

'Late Flat Dutch' – This older open-pollinated heirloom variety was introduced by European settlers in the 1860s. It produces large, round, bluish-green flattened heads in the range of 12 to 15 lbs. This is a late-season variety that will mature in about 110 to 105 days.

'Copenhagen Market' – This is an excellent older open-pollinated heirloom variety that has been a favorite of gardeners since it was introduced into the U.S in 1909. It has been termed as an early, very solid, and immense yielder. 'Copenhagen Market' has blue-green 6- to 8-inch-diameter heads, weighing about 4 lbs each, and matures in 65 to 70 days.

Chinese Cabbage is an oriental specialty vegetable that combines the thin crispy texture of lettuce with the fresh peppery tang of juicy cabbage. Heading types (***Brassica pekinensis***) include the short, round varieties generally known as wong bok or napa, and the tall, slender varieties that form cylindrical heads. The non-heading types (***B. chinensis***) have celery-like stalks and are called pak choi or bok choy. Chinese cabbage, unlike lettuce or cabbage, will tolerate hot weather. There are many varieties available that are bolt resistant.

Chinese cabbage is similar to cabbage in that it can be grown in the spring or the fall. Plant seeds or transplants 12–18" apart in rows and 18–30" apart between rows. Chinese cabbage will grow well after the first light frost. Harvest the heading types once they reach usable size, and for non-heading types, harvest a few outer leaves at a time or cut the entire plant for a single harvest.

There are numerous varieties are of Chinese cabbage are available for home gardens. For the non-heading types, suggested varieties include 'Joi Choi', 'Mei Qing Choi', while for heading types, 'China Crunch,' 'Yuki', 'Cha Cha' are some of the recommended varieties.

Figure 6.11 Savoy type cabbage.

BROCCOLI: *BRASSICA OLERACEA (ITALICA GROUP)*

Broccoli means "little sprouts" in Italian, and with broccoli, the immature flower buds are eaten. Broccoli is one of the most popular cole crops with home gardeners and for good reason, as it can be used to make so many different tasty dishes. Broccoli is highly productive, and usually about 10 to 15 plants in a garden will supply enough for most households. The culture of broccoli is similar to cabbage. Broccoli will need approximately 60 to 80 days to mature from transplants (depending on climate) and another 2 to 3 weeks (75 to 100 days) if direct-seeded into the garden soil.

Figure 6.12 Broccoli head ready for harvest.

In most temperate climates, broccoli can be planted late winter/early spring or late summer for two possible crops. Transplanting is the best planting method to use for broccoli in the spring, so that gardeners can try to harvest heads before the summer heat causes the heads to bolt. Broccoli must be harvested when the center green bud cluster and the buds (often called beads) are still tight (Fig. 6.12). Most broccoli heads will reach a diameter of 4 to 8 inches at maturity. However, when temperatures in the late spring/early summer surpass more than 80°F, the head will bolt, causing yellow flowers to develop in the head (Fig. 6.13), and broccoli should be harvested before this stage. Although broccoli at this flowering stage is still edible, an excellent way to prevent broccoli flowers from developing is to harvest early and frequently before the weather becomes hot in the late spring. As long as the harvested shoots are not cut back to the main stem, smaller side shoots (2 inches or so in diameter) will develop after the main head has been harvested and a continuous harvest will occur for about a month; this works primarily for fall plantings, as the summer heat comes too quickly for the spring planting to always produce edible side shoots. During hot weather, buds will pass from green to yellow very quickly. Thus, without a doubt, fall-produced broccoli tends to have a higher quality.

Figure 6.13 Broccoli plant in flower.

For broccoli, space plants about 12 inches apart in the row with rows spaced 3 to 4 feet apart. Broccoli can be spaced closer than cabbage, since it requires less space to grow. However, it is important to note that the more space broccoli is given in a garden, the larger the head will be at maturity.

There are other types of broccoli and this often confuses home gardeners, so a brief description is provided next.

Sprouting Broccoli is grown for its long, tender shoots, which are prepared in the same way as asparagus, or used in stir-fries. This broccoli type needs frequent cutting to encourage steady production of new shoots. Sprouting broccoli is best grown from fall to spring in areas with mild winters and from midsummer to fall in most climates.

Broccoli Raab is a fast-growing type of broccoli, also known as turnip broccoli, which forms multiple small heads. It is a branching, as opposed to heading, type of broccoli. It forms multiple small heads, instead of one large central head. For best texture and flavor, the tender shoots should be harvested before the flowers open. This hardy broccoli loves cool weather and produces consistently for several weeks. Its leaves and stems are also edible.

Gai-lon (or Chinese broccoli) is a type of Asian green that's smaller in size, darker green, grows more quickly, and has a stronger flavor than broccoli. This type doesn't form traditional heads, but instead, the whole plant is eaten.

There are many broccoli varieties available, but two hybrid varieties that are seen most often in stores that sell products to home gardeners are 'Packman' (matures 50 days from transplanting, widely adapted) and 'Green Comet' (matures 55 days from transplanting, widely adapted, produces 7-inch-diameter heads with numerous small heads on side branches, heat tolerant). From my experiences, these two varieties do not necessarily produce the highest quality and yield compared to many other varieties, but they are the two that can be most commonly found in the springtime at local garden centers. Other suggested varieties include 'Cruiser', 'Green Goliath', and 'Belstar'. For sprouting broccoli types, 'Apollo', 'Calabrese', and 'Purple Sprouting' can easily be found and are widely adapted varieties.

CAULIFLOWER: *BRASSICA OLERACEA* (BOTRYTIS GROUP)

For many gardeners, cauliflower remains one of the most challenging crops to grow in a home garden, as cauliflower is restricted by climate more than either cabbage or broccoli. Cauliflower is less cold-tolerant and will not head properly in hot weather. Being a cool-season plant, cauliflower struggles in hot climates and needs to grow at a slow, steady rate to develop a well-formed mature head (Fig. 6.14). Stresses provided to the plant, such as temperature fluctuations or soil nutritional imbalances, are really the main reason that gardeners often produce small, stunted plants that do not produce useable heads. However, with a little patience and attention to specific growing conditions, many gardeners can effectively grow this vegetable in their home garden.

Figure 6.14 Mature cauliflower in garden.

Cauliflower performs best when started indoors about 6 to 8 weeks before the last frost date in the spring or mid to late summer for an autumn planting. These plants are then transplanted into the garden soil. Also, varieties that mature in 50 to 60 days are ideal for the home garden—remember, the approximately 2 months needed to mature must be cool. So, plant early spring or late summer in most temperate climates, and in most cases the spring crop will not produce good heads due to the excessive heat. Planting transplants in the spring also provides the extra time needed for plants to mature faster before the weather gets too hot in the summer. If old or stunted, or root-bound transplants are used, then these plants will typically produce "buttons" (or development of small heads). Fall crops tend to be of higher quality, similar to most other cole crops.

Transplants, or if seeds are directly placed into the garden soil, should be spaced about 1½ to 2 ft apart in the row with rows 3 to 4 ft apart. If closer spacings are used, smaller head formation will result. It is important to mulch the soil around cauliflower to help it remain cooler than the surrounding bare soil during the warmer months and to keep the soil moist. A consistent supply of moisture is essential for good head development, so water cauliflower on a regular basis; watering is especially critical during head formation. Try to ensure that the soil receives adequate moisture at this time. Dry conditions will result, disrupting cauliflower growth and preventing optimum development of the head. Cauliflower has a shallow root system and the plant can quickly go from having sufficient moisture to being deficient. Cauliflower grows best in full sun conditions at 70°F, and temperatures above 80°F may cause a rough texture to develop or discolored head.

Although cauliflower has high nutrient requirements like other cole crops, it has other nutritional needs that can adversely affect its growth. Magnesium, calcium, boron, and molybdenum are other elements likely to be limiting for cauliflower production. To ensure proper growth and development of this crop, some of these nutrients will most likely need to be applied sometime during production. Cauliflower requires high magnesium levels and

Figure 6.15 'Violet Queen' cauliflower in garden ready for harvesting.

shows deficiency symptoms readily when soils are too acid or the element is in short supply. Magnesium is routinely applied as dolomite limestone or Epsom salts. Calcium is usually not limiting, especially if soils have been properly limed. Boron is usually deficient in most soils and will require soil or foliar applications to correct this deficiency. Otherwise, home gardeners can mix 2 tablespoons of borax with 5 qt of a complete fertilizer and apply this to 100 ft of row. Molybdenum deficiency is common in cauliflower and may be prevented by applying as ammonium or sodium molybdate, and can be found in certain pre-plant fertilizer mixes. Cauliflower is one of the most sensitive of crops to molybdenum deficiency. Molybdenum is the only plant essential micronutrient that becomes less available as the soil acidifies—the opposite of copper, zinc, and iron. Typical symptoms include rolling of leaves, or slight amounts of leaf tissue along the midribs of leaves. Often, the entire plant becomes stunted, the growing tip is destroyed, and no head is produced.

When cauliflower heads reach 2 to 3 inches in diameter, they need protection from sunlight to ensure that the head does not become discolored. So, when the curd (or head) starts to develop, it is best if the gardener lightly ties the inner leaves together over the head with a rubber band or string. This should be done to block out any sunlight, but leaving just enough room for the cauliflower head to develop to maturity. This is called "blanching," because when light is excluded, the curd will remain white and tender. If the cauliflower curds are exposed to sunlight during development, the curd color will turn yellowish rather than pure white. Failure to blanch cauliflower not only discolors the head, but will also alter the taste and

Figure 6.16 Romanesco cauliflower sold at a farmers' market in Champaign-Urbana, Illinois.

texture. Self-blanching varieties are available and develop curled leaves that protect the head naturally from direct sunlight. Although there are self-blanching varieties, in which leaves curl naturally over the head, oftentimes they do not, so leaves should be tied over the developing head anyway to guarantee that the head does not yellow. Plants are ready to harvest when they reach 5 to 8 inches in diameter and before the segments begin to separate. Unwrap leaves every week to check for hiding insects, but then tie leaves back together. If the weather turns hot during development, mist or sprinkle the cauliflower plants with water to keep them cool.

Several varieties of cauliflower are available to home gardeners including 'Snowball Self-Blanching' (70 days; self-blanching), 'Snow Crown' (60 days; resistant to Fusarium yellows; tolerant of heat and cold), 'Snow King' (55 days; very early; huge white heads; heat tolerant), and 'Violet Queen' (54 days; purple heads; no blanching required; see Fig. 6.15). There are also several varieties of "Broccoflower" available (cauliflower–broccoli hybrid) including 'Green Goddess' (65 days, no tying required; lime green; good taste; easy to grow) and 'Veronica' (85 days; produces large, 6- to 7-inch heads with swirling conical light green florets; amazing variety; known as Romanesco cauliflower; see Fig. 6.16).

BRUSSELS SPROUTS: *BRASSICA OLERACEA* (GEMMIFERA GROUP)

Brussels sprouts (named after Brussels, Belgium, where the first sighting of this mutation occurred and is also where the vegetable first became popular) are a hardy, slow-growing, long-season vegetable (matures 90 to 110 days after transplanting, depending on climate) that belongs to the cole crop vegetable group. Brussels sprouts are the most cold-tolerant member of the cole crop family, and are best grown as an autumn/early winter crop in most temperature climates. If planted in the spring, transplants must be used and anyway, there are generally not enough growing days for it to mature before the temperatures become too warm for its production.

Brussels sprouts are grown much like the related cole crops cabbage and broccoli. Space transplants about 1½ to 2 ft apart in the row and space about 2½ to 3 ft apart between rows. If direct seeding, place seeds about 4 inches apart and thin to 1½ to 2 ft between plants at the 2- to 3-leaf stage. For fall production, seeds or transplants should both work equally well, if sufficient moisture is maintained in the soil to allow germination. Similar to other cole crops, Brussels sprouts are a high nutrient feeder. After the initial broadcast application of a complete fertilizer, gardeners need to apply one sidedress application of a nitrogen-based fertilizer when the plants are 12 inches tall and then every three to four weeks thereafter. It is also important to provide sufficient water to plants to maintain vigorous plant growth, especially during the heat of late summer when plants are just being established.

The sprouts form in the axils of the leaves (the space between the base of the leaf and the stem above it) (Fig. 6.17). The sprouts mature in sequence from the bottom of the plant upward. Brussels sprouts can be harvested from after the first frost until early winter. A light frost or two improves their flavor. Harvest the sprouts when they are firm and before they open up. Pick sprouts that are firm and have a decent size and shape. Brussels sprouts typically form heads that are 1 to 2 inches in diameter. Break off the leaf beneath the sprout as it is harvested, since this will prevent disease build-up and encourage heavier production. The upper sprouts will continue to form and enlarge as the lower ones are harvested. In late fall/early winter, leaves can be removed after the plant has been pulled from the ground, and the plant can be hung in cool basement to provide Brussels sprouts throughout the winter months. Harvest the sprouts before the leaves turn yellow.

For a once-over harvest of sprouts, pinch off the growing tip when the plants are 1½ to 2 ft in height, as this will promote uniform development and maturity of the sprouts. A full stem of uniform-sized sprouts will develop 4 to 5 weeks later, and then the entire plant can be cut and harvested (see Fig. 6.17). This technique is particularly useful in areas where low freezing winter temperatures come in late fall or early winter. However, pinching off the growing tip will reduce yields since fewer sprouts will be produced over the season.

Figure 6.17 Brussels sprouts on stalks, taken at a farmers' market in Massachusetts.

There are only a few varieties of Brussels sprouts available to home gardeners. A few recommended varieties are 'Jade Cross' (100 days; 60–100 sprouts per plant), 'Long Island Improved' (90 days; closely set sprouts that are tight, dark green, and about 1½ inch in diameter), and 'Oliver' (90 days, early, well adapted to many climates; excellent producer of solid, bright green, 2-inch diameter sprouts on compact plants).

LEAFY CRUCIFEROUS GREENS (MUSTARD, COLLARDS, KALE, TURNIP, AND RUTABAGA)

There are many different leafy cruciferous greens grown in home gardens. Typically, they are sown in late summer and grown during the cool autumn weather. Vegetables such as mustard, collards, kale, turnips, and rutabagas are included in the leafy greens group, although two—turnips and rutabagas—are often grown for their swollen edible root. All leafy cruciferous greens prefer to be grown under the cool conditions of autumn and develop the best flavor at this time of the year compared to being grown in the spring to early summer.

All are cool-season crops and should be grown in early spring or fall to optimize productivity. It is important to note that kale is the most cold-tolerant of the leafy greens group and can withstand temperatures as low as 18° to 20°F, while the rest of the leafy green group can withstand only a few light frosts before dying.

MUSTARD: (*BRASSICA JUNCEA*)

Mustard greens, while oftentimes not as popular as kale, collards, and other cruciferous leafy greens, make an excellent addition to the home garden. Their strong, spicy hot flavor is probably the reason that many gardeners neglect them, but there are many good reasons to provide a little space for them in a garden. Mustard is one of the easiest leafy vegetable crops to grow, as this vegetable requires little care once established. Cooked mustard greens are a very popular dish in the Southern part of the U.S.

Mustard is a cool-weather, short-day crop and bolts to seed very early in the spring. It can be harvested in as little as 30 days after seeding. If planting in the spring, sow seed as early as possible (3 weeks before the frost-free date and then again 3 weeks later). Seeds can also be planted from midsummer on for autumn harvesting. Plantings that mature in the autumn are usually of higher quality because they mature under cooler conditions in most locations. Mustard often bolts during the hottest part of summer; and, to prevent bolting, keep plants cooler by frequent misting or using rowcovers to protect the plants from the more intense rays of sunlight.

Sow mustard seed about ⅓ to ½ inch deep and plant in rows 12 to 18 inches apart with seedlings thinned to 4 to 5 inches apart in the row. Seeds are also often broadcast over an area instead of being planted in rows. Seeds will sprout in about four days, if the soil is moist and warm. Mustard plants grow quickly if plants are watered on a regular basis.

It is best to harvest leaves before they are fully grown, since they will be more tender at a younger age. If grown under hot conditions, the leaf texture will become tough and develop a strong flavor. Leaves can be individually harvested or the entire plant can be cut and harvested. Again, leaves can be harvested approximately 30 to 40 days after seeding.

Mustard typically has few disease problems, although the most important insect pest is aphids. These insects will often build up significant numbers on the undersides of leaves and become a nuisance since leaves are the product being consumed.

There are several varieties of mustard, including smooth- and curled-leaf types, and mustard can be found in green, purple, and red leaf varieties. The more colorful varieties are often grown as ornamentals to add color to a landscape. 'Southern Giant Curled' is a widely grown mustard variety, although other varieties such as 'Florida Broadleaf' and 'Green Wave' work well in most gardens. Ornamental mustard varieties include 'Osaka Purple' and 'Red Giant'.

COLLARDS: *BRASSICA OLERACEA* (ACEPHALA GROUP)

Collards have been a favorite of southern gardeners for many years, and are one of the vegetables that are often associated with southern cuisine (Fig. 6.18). They are nearly as cold resistant as kale, more cold tolerant than cabbage, but unlike kale, collards can withstand considerable heat. So, collards are somewhat heat and cold tolerant, and this is something that doesn't occur too frequently in plants.

Cultural requirements are similar to cabbage, so they are heavy nutrient feeders and follow similar fertilizer recommendations for cabbage. Sow collard seed in early spring (4 weeks before the last expected frost) for a summer harvest, or in mid-summer (8 to 10 weeks before the first expected frost) for a late-autumn harvest. The seeds need to be planted ¼ to ½ inch deep with seedlings thinned when they are about 2 inches tall to about 1 ft apart in rows placed 3

Figure 6.18 Bundle of collards ready to sell.

ft apart. Water frequently to keep the seed bed slightly moist until germination, and seeds will germinate in about 1 week. Drip irrigation often works best for home gardens as seeds must be kept continually moist prior to germination and plants are much more productive if high soil moisture is maintained.

Collards can be harvested continuously by cutting the outer leaves when they are about 12 inches in height, leaving the inner layers of leaves to continue growing. The entire plant can also be harvested at the end of the growing season. In frost-free climates or climates that have only light frosts, collards may produce throughout the winter months.

Widely adapted varieties that are recommended to grow in many areas are 'Top Bunch Hybrid' (50 days, early, tall plant produces medium green, slightly savoyed leaves) and 'Champion' (60 days, has long, broad, wavy, tender leaves).

KALE: *BRASSICA OLERACEA* (ACEPHALA GROUP)

The leaves of kale are powerhouses of nutrients and are gaining popularity for those looking for vegetables to include in healthy diets. Kale is very rich in iron and other nutrients like vitamins A and C. Kale is grown only for its leaves, and is the most cold-hardy of the leafy greens. These plants grow best during the cooler autumn weather, and frost improves the flavor of kale. Kale develops bitter flavors under hot weather conditions. Along with cooler temperatures, moist soil helps keep kale leaves sweet and crisp, rather than tough and bitter. Thus, an organic mulch placed around kale plants in the garden will moderate soil moisture, and during times of warm weather, will provide cooler soil conditions.

Kale plants prefer to grow in a soil that has high organic matter and is slightly acidic pH (5.5–6.5 pH). These plants require high amounts of soil fertility, so fertilize similar to cabbage.

Kale can be direct-seeded in the garden or started indoors and set out as transplants. Direct seeding is the best planting method to use in fall, and transplanting is best for spring plantings. At planting, seeds should be covered with about ½ inch of soil and constant moisture should be provided to reduce the chance that seeds will dry out before germinating. Kale seeds germinate quickly in warm soil and should be up within 5 to 8 days. Gardeners can direct-seed in the spring as soon as the soil can be worked and the soil temperature is at least 45°F, but the optimal soil temperature for planting is 60° to 65°F. Start plants indoors about 6 weeks before your last expected frost date. In warm climates, kale can also be direct-seeded in late summer/early fall and grown as an autumn/winter crop.

Kale is recommended to be grown about 4 to 8 inches apart in the row with 1½ to 2 ft of spacing between rows, although seeds are often broadcast like turnips and mustard instead of growing in rows. Kale matures quickly, in about 2 months or less. Most gardeners who grow kale will generally have multiple crops by planting every few weeks. A winter crop of kale in warmer climates can be much sweeter than a summer crop.

Harvest kale any time leaves are large enough for intended use. Tender young leaves are best for eating fresh and older leaves for cooking. When harvesting, twist off the outer, older leaves when they reach a length of 8 to 10 inches and are medium green in color. Large, dark green leaves are usually overly mature and are more likely to be tough and bitter. New leaves will form and grow, providing a continuous harvest for several months. There are many varieties available depending upon a specific gardener's preference. Many gardeners prefer colored varieties to add color to their landscapes during the usually drab winter months as well as providing a source of fresh greens or a colored garnish. Recommended varieties include 'Winterbor' (has curled, blue-green leaves), 'Redbor' (50 days, magenta kale, finely curled leaves, vigorous and cold-hardy edible landscape plant), 'Red Russian' (50 days, bright purple stems with slate-green, deeply lobed foliage; see Fig. 6.19), 'Red Peacock' (center of plant is deep pink to red, and is more intense in cooler weather), and 'White Peacock'

***Figure 6.19** 'Red Russian' Kale. Notice the purplish venation in the leaves.*

(white centers, and white color is more intense during cool weather). All varieties listed will produce edible leaves in about a month after planting, but will take about approximately 2 months to form a mature plant.

TURNIPS: *BRASSICA RAPA* (RAPIFERA GROUP) RUTABAGAS: *BRASSICA NAPUS* (NAPOBRASSICA GROUP)

Turnips and rutabagas are among the most commonly grown and widely adapted root crops for vegetable gardens. The scientific name of turnip is *Brassica rapa*, while rutabaga is *Brassica napus*. The two are similar in plant size and other general characteristics. Turnips and rutabagas are often considered together since both foliage and roots are eaten. Generally, turnips have white-fleshed roots while most rutabagas are yellow-fleshed (Fig. 6.20). The leaves of turnips are rough, light-green, hairy, and fast growing, while rutabaga leaves are smooth, bluish-green, thick, waxy, and develop slowly. Furthermore, the roots of turnips generally have little or no neck and a distinct taproot, while rutabaga roots are often more elongated and have a thick, leafy neck and roots originating from the underside of the edible root as well as from the taproot.

Turnips and rutabagas are both cool-season crops and have best root growth during temperatures that range from about 40° to 60°F. Either one of these plants can be grown as a spring or autumn crop; however, rutabagas require a longer growing season (about 90 days) and should be planted as early in the season as possible. Early maturing varieties of turnips can be ready to harvest in 40 days, while late-maturing varieties take up to 75 days. It is best to grow these crops during the autumn, and sow seeds about six weeks before the first frost. When planted in the spring, hot weather conditions often make leaves tough and the roots woody and bitter to the taste.

These crops do best in well-drained soil on a site with full sun. A moderately deep, highly fertile soil with 6.0 to 6.5 pH is best for growing turnips and rutabagas. Although culture is similar to that used for other crucifers, less nitrogen is needed for these crops and recommendations are to apply about half the amount used for other crops in this family.

Seed can either be broadcast over an area or can be grown in rows. Sow seeds about 6 weeks before the first expected frost. Plant the seed of turnips and rutabagas about ½ to ¼ inch deep; and, if planted in rows, the rows should be 1½ to 2 ft apart. Plants should be thinned to 12 to 18 inches apart in the row for root development; and, if they are being grown only for greens, space plants 3 to 4 inches apart in the row.

Harvest leaves or roots anytime they are large enough to use. Roots can be harvested and stored in a cool basement to be eaten during the winter months. Harvest roots before temperatures drop below 20°F. Rutabagas and potatoes can be mashed together to make a hearty dish during the autumn or winter months.

Turnips and rutabagas are usually not attacked to a great extent by insects or diseases. However, leaves are often infested by aphids and require removal before cooking. Varieties differ mainly in color and shape of root. Again, most turnip varieties are white-fleshed and most rutabaga varieties are yellow-fleshed. Widely adapted recommended varieties are 'Purple Top White Globe' turnip (58 days from seeding; bright purple crown, white below the crown, 5 to 6 inches in diameter, globe-shaped; see Fig. 6.20) and 'American Purple Top' rutabaga (90 days,

Figure 6.20 Turnip (on left) and rutabaga (on right).

deep purple crown; yellow below the crown; globe-shaped root; 5 to 6 inches in diameter with yellow flesh color; medium size; see Fig. 6.20).

KOHLRABI: *BRASSICA OLERACEA* (GONGYLODES GROUP)

Kohlrabi is an unusual and little-known vegetable, although many gardeners think that it deserves to be grown and appreciated more in gardens. Kohlrabi has not achieved great popularity in the United States like it has in many parts of Europe. The word "kohlrabi" is a German word and was adapted into the English language without change, where "kohl" means cabbage and "rabi" means turnip. It somewhat resembles a turnip and it is often accepted as a substitute for turnip. However, the edible structure is not a root, but is a compressed swollen stem that somewhat resembles a root (Fig. 6.21). The flavor of kohlrabi is similar to both turnip and cabbage, but milder and sweeter than either.

This is a vegetable that is very easy to grow. Kohlrabi is a cool-season vegetable that prefers to be grown in full sun. Kohlrabi grows best when temperatures do not exceed 75°F. It prefers fertile, well-drained soils having high organic matter content for best growth. Most garden soils are suitable for kohlrabi production. Cultural and fertility practices are similar to turnip. The kohlrabi plant has a relatively short growing season and will mature in about 40 to 50 days after seeding. However, it is important to remember that kohlrabi is the least hardy of the crucifer group of vegetables.

Kohlrabi is direct-seeded into the garden soil. Seeds should be planted about ¼ to ¾ inch deep and thinned to the final stand when plants have 3 or 4 true leaves. Plants should be thinned to 6 inches apart in rows and 1 ft apart between rows. It can be seeded in the garden in early spring, so that most of the growth is complete before the summer heat, or seeded in late summer to harvest in fall. High summer temperatures reduce growth, decrease quality, and cause the enlarging stems to become tough and woody. In hot areas, it is best to grow kohlrabi as a spring or autumn crop. For fall-maturing kohlrabi, select early maturing cultivars and

Figure 6.21 Kohlrabi growing in garden. Notice the straw mulch being used for weed control and to help maintain sufficient soil moisture.

plant 50 days before the anticipated maturity date, and the maturity date can be timed for 2 to 3 weeks after the first frost in the autumn.

It is important to irrigate regularly, since roots are shallow, to avoid water or fertilizer stress during growth. About 1 to 2 inches of water are required per week and some type of drip irrigation works best to conserve water. Many gardeners will also apply a mulch around kohlrabi plants to help conserve soil moisture and reduce weed growth (Fig. 6.21). Moisture fluctuations will cause the stems to become tough and woody.

The thickened stems of kohlrabi should be harvested when the stem enlarges to 2 to 3 inches in diameter. Larger stems tend to become tough and woody. The young leaves can also be eaten like cabbage or kale. Kohlrabi can be stored for 2 to 3 weeks at 32°F and 95% relative humidity. The outer skin of the kohlrabi is peeled off when prepared and the inner flesh is eaten raw or cooked. Many will use kohlrabi like turnips: peel off skin, boil, add seasonings, and enjoy.

There are two primary types of kohlrabi: white (which is really light green) and purple. A few varieties are available for home gardeners, including 'Grand Duke' (55 days, smooth, round bulbs, tolerant to black rot), 'Purple Vienna' (60 days, reddish-purple colored bulb, with white tender flesh), 'Early White Vienna' (55 days, slightly flattened medium-green colored globe), 'White Danube' (51 days, pale-green globe), and 'Purple Danube' (54 days, purple-colored globe).

SPRING RADISH: (*RAPHANUS SATIVUS*) DAIKON OR ORIENTAL RADISH: *R. SATIVUS* (LONGIPINNATUS GROUP)

Radishes are one of the easiest vegetables to grow, and probably should be grown in every garden. The descriptive Greek name of the genus *Raphanus* means "quickly appearing" and refers to the rapid germination of these plants. Radishes are a cool-season vegetable in the Brassicaceae family.

There are two radish types: **spring radishes** (25 to 30 days from seeding to maturity) and **winter radishes** (often referred to as Asian or Daikon radishes, which take 50 to 70 days to mature). Spring radishes are ready to harvest in less than a month, while the winter radishes take around 2 months or longer. Spring and winter radish types are shown in Figs. 6.22 and 6.23, respectively.

Spring radishes can be grown throughout the year except the coldest months in cooler areas, and in warmer areas, all except the coldest and hottest months. Spring radishes mature rapidly, with many varieties germinating in 3 to 5 days, and reaching maturity in only 3 to 4 weeks. Spring-type radishes are a common garden crop and the quick time from seeding to harvest makes them a popular choice for children's gardens. These radishes come in many different sizes, from round or oblong to elongated, as well as in many different colors from red, pink, purple, white, and bi-colored.

Winter radishes are slower growing, larger, and will keep for a longer time compared to spring radishes. The Asian radish varieties are often referred to as "winter radishes," because they mature late in the season and can be stored over winter. "Daikon" is a Japanese word meaning "great root" and some of these radishes can indeed grow to a great size, although not all Asian radishes are long. The black winter-type radishes tend to be the most pungent. Winter radishes prefer the cool temperatures and short days of late summer and fall, and are known for their ability to be stored for several months.

Spring types can be started very early in the spring, as soon as the ground is relatively dry, and again toward the end of summer, for an autumn harvest, and even in late fall, with some protection. In temperate climates, it is normal to plant spring radishes every two weeks from early spring until a few weeks before the first frost, except during periods of hot weather. In warm-weather climates, spring radishes are normally planted in the fall. Winter radishes are normally seeded in the late summer. Both radish types can be planted in succession for an extended harvest.

Radishes are directly seeded into the garden soil. Sow seeds about ½ inch deep for spring types and about 1 to 1½ inches deep for winter types, as the depth at which seeds are planted affects the size of the root. Radish seeds prefer to be planted in cool to warm soils (60° to 80°F) in rows 6 to 12 inches apart. Spring radish plants can be thinned soon after emergence to 2 to 3 inches apart, depending on the final root size that is desired at harvest. Roots will begin to expand when only 2 weeks old and need space to develop, so it is important to thin as quickly as possible. For a

Figure 6.22 White and red spring radishes displayed at a farmers' market.

continuous supply of spring radishes, repeat plantings every 2–3 weeks. Winter radishes require more space since they develop larger roots. Plant winter radish seeds about 2 inches apart, and thin to 4 to 6 inches apart in the row, with rows spaced about 1½ ft apart. Harvest in 50–70 days, and winter radishes harvested during the fall can be stored for several weeks.

Although radishes tend to have few pest problems, flea beetles are often the most destructive pest in home gardens. Their initial injury is observed as small holes in leaves, although they do not injure the root. Monitor a radish planting in a garden to catch these insects before they completely defoliate the plants. This insect can be almost entirely avoided in a home garden by growing radishes under insect-barrier rowcovers.

Spring Radish Harvest – The fast maturing, round radishes are ready for harvest in 3–4 weeks. When the radish appears to be almost an inch across, it is ready to be pulled from the soil. Radish roots can be harvested by gently pulling or by loosening the surrounding soil. It is important to note that when mature radishes are not pulled, they will get woody or spongy and develop a more pungent flavor.

Winter Radishes Harvest – The long, Asian radish varieties take longer to mature, sometimes up to 3 months. They should be harvested before the ground freezes and can be stored for several months.

After harvest, the green tops should be removed. If the foliage is left attached, they will draw moisture and nutrients from the radish roots. The globe-shaped spring radishes can be

Figure 6.23 Winter radish roots.

stored in the refrigerator, in plastic for about a week, while winter radishes will last about 2 weeks in the fridge or several months in cold storage.

Recommended **spring radish varieties** include 'Cherry Belle' (22 days, cherry-sized radishes), 'Early Scarlet Globe' (22 days, red with round roots), 'Champion' (27 days, large), 'White Icicle' (28 days, white, long and tapered, more tolerant to hot conditions), and 'Easter Egg' (25 days, oval-shaped, various colors-white, pink, red). **Varieties of winter radishes** include 'China Rose' (52 days, red skin with white flesh), 'Round Black Spanish' (55 days, black skin and white flesh), 'Daikon' (60 days, long, white and mild), 'Minowase' (50 days, 8- to 10-inch-long uniform roots, white inside and out, mild in flavor), and 'Long Black Spanish' (60 days, long 9-inch black roots have pure white flesh that is crisp and pungent).

HORSERADISH: *ARMORACIA RUSTICANA*

Horseradish is a hardy, perennial plant that belongs to the *Brassicaceae* family and is thus related to cabbage, mustard, and other cruciferous vegetables. Horseradish has been cultivated for more than 2,000 years for its white, thickened, and pungent roots. The pungency of horseradish roots results from the sulfur-containing compounds in root tissues. Horseradish is a flavorful pungent vegetable that has been used for centuries to enhance the flavor of food, aid in digestion, and to improve health. Today, horseradish is best known for adding zesty flavor to all kinds of condiments that are most often used with different types of meats (Fig. 6.24).

Typically, horseradish is grown in gardens for its pungent roots that are used to make various types of horseradish sauces. The plant grows 3 to 5 feet tall and has large, glossy, oval-shaped leaves that are serrated along the edges. The leaves can grow 20 inches or longer. Horseradish is hardy in USDA plant zones 2 to 9. Although horseradish grows in any garden soil, the best roots develop in soils that are moist and have high fertility. The fertility recommendations are similar to cabbage and require high amounts of nitrogen. Also, make sure that the soil has adequate amounts of boron.

Horseradish is planted in the spring and requires full-sun conditions. Horseradish is propagated in an asexual manner by planting small pieces of secondary plant roots in the spring that were collected from the previous year's crop. These small roots are typically smaller side roots taken off the main root and range in size from 0.5 to 0.8 inch diameter and 10 to 16 inches in length. Once you have secured a piece of root, whether from a gardening friend, mail-order catalog, or some other source, select a spot in a garden where the soil is at least 2 ft deep. Most gardeners will grow only a few plants in a garden and plants should be spaced so that they each have about 6 to 10 ft^2 of growing space. Many gardeners will form a ridge of some sort and plant the roots into the raised ridges to a depth of about 6 inches. Since it is a root crop, make sure the soil is deeply tilled before planting to ensure the formation of large, straight roots. From the planted roots, foliage will emerge and a large marketable primary root will increase in diameter over the growing season. During a single growing season, horseradish plants in a cold-temperate climate will develop a large marketable root (Fig. 6.25). Horseradish is usually planted in the early spring with roots harvested beginning in late autumn to early winter, depending on soil conditions. Harvesting can also be done once the soil thaws in late winter to early spring, as quality is not reduced by this late harvest.

It is important that a gardener plant horseradish roots in an area of the garden where it can grow by itself, as it needs a little space to spread out—many term horseradish an invasive plant species. Horseradish will often out-compete other plants growing nearby and aggressively take up more than their share of the garden. Root pieces left in the soil will develop new plants the following spring. So, once planted, remember that the horseradish plant will be there for a long time; in fact, if gardeners are not careful when digging and leave the smaller roots in the soil, the horseradish plant will expand into the garden in every direction.

For large straight roots, push back the soil from around the crown of the plant when the leaves are about 12 inches in height. The smaller secondary roots coming out the side of the main root should be cut and removed from the plant. Leaf removal is also important so that the number of sprouts at the crown is limited to 2 or 3, and this should be done a few times during the growing season. However, many gardeners do not "lift and strip" horseradish, but this will provide a gardener with a large, high-quality root.

The flea beetle is probably the most important pest of horseradish in home gardens. However, research has indicated

Figure 6.24 Horseradish is commonly sold as freshly ground product in distilled vinegar. This condiment is often eaten with meat, such as prime rib.

Figure 6.25 A freshly dug horseradish root, showing the large primary root and the smaller secondary roots that will be saved as planting stock for spring planting.

that even at high populations in gardens, they do not cause economic damage or significantly reduce the size of harvested roots.

Horseradish is vegetable crop that can add much to a garden and it also makes a nice ornamental plant for the landscape as well. Variegated types can be found that make it even more of an attractive plant for the garden. There are several types that can be found:

Common or Maliner Kren

The 'Common' or 'Maliner Kren' horseradish has crinkled and wide leaves, and produces a large nice primary root. While the root quality is high, commercial growers generally prefer other varieties because 'Common' is vulnerable to diseases like white rust and turnip mosaic virus. However, confusingly, some smooth-leaved horseradish plants are sometimes called 'Maliner Kren' as well.

Bohemian

The Bohemian-type horseradishes include such cultivars as 'Bohemian Giant', 'New Bohemian', 'Sass', and 'Swiss'. Bohemian-type horseradish plants also are characterized by smooth leaves that taper at the base. This type arrived in the United States, especially the Midwest, with German and other Central and Eastern European immigrants—hence the name "Bohemian," which means "Czech." Some Bohemian types are sold as Czech. While they vary widely in their resistance to diseases, they all have fleshy roots, although smaller than those of 'Common'.

Big Top Western

'Big Top Western' is a type developed for disease resistance, particularly to turnip mosaic virus. 'Big Top Western' has large, somewhat rough roots and wide, smooth leaves with a tapering base. Agricultural scientists have recently bred many new strains of this type, most of which are numbered lines rather than being named, and thus are probably of much more interest to researchers and farmers than to home gardeners.

Variegata

The highly appealing foliage of 'Variegata' emerges from the center of the plant. 'Variegata' has large, wavy leaves with splashes of cream and dark green intermixed. However, it may take two to three years after the initial root stock planting to achieve such a display for the variegated foliage. This variety is grown primarily as an ornamental for the home landscape.

GLOBE ARTICHOKE: (CYNARA SCOLYMUS)

Globe artichokes are thistle-like herbaceous perennials in the Asteraceae family that are grown primarily for their immature flower bud (Fig. 6.26). Although in most cold temperate climates it is grown as an annual, globe artichoke plants can get fairly large, so plant this vegetable in an area of the garden where it will not get in the way of the culture of other vegetables. The "globe" is the flower bud, which includes the edible parts: the fleshy bases of the outer bracts, the inner bracts, the receptacle, and portions of the floral stem. This vegetable can be cooked in a wide variety of dishes, and artichoke hearts are considered a delicacy by many cultures.

Figure 6.26 Globe artichoke growing in a home garden with edible flower buds ready to harvest.

Globe artichokes were once propagated asexually primarily by root divisions, since seed propagation would result in variable plant types and yields. However, now plant breeders have developed varieties that come true from seed. When propagating from seed for cold-temperate climates, remember that artichoke seedlings require a vernalization period to produce flowers the first year. "Vernalization" is a term used to describe the chilling process required of certain seeds, plants, or bulbs that must be done to allow the normal flowering process. Artichoke plants must be vernalized at temperatures <50°F for a certain length of time for flowering to occur. The time needed below this temperature really depends on the particular cultivar, with 'Imperial Star' needing about 200 hours (8 to 10 days), 'Emerald' almost none, and 'Green Globe' about 1,300 hours (about 55 days). So if a gardener works back and takes the length of time required for vernalization, they can determine when to sow seed and this can be possibly as early as January.

Sow artichoke seeds about ¼ inch deep and ¼ inch apart in a lightly moistened soil-less growing mix. It is best to use a good-quality, sterilized soilless potting mix. Germinate artichoke seed at a temperature of about 75°F; heating pads can be used to obtain the desired temperature to maximize seed germination. Soon after emergence, transplant seedlings into pots that are about 4 inches in size. It is best to try to keep the transplants growing at temperatures as close to 65°F during the day and 55°F at night. The day and night temperatures can be plus or minus 5°F, but try to keep the day/night difference at around 10°F. When the outdoor daily highs reach about 50°F, seedlings can then be hardened-off outside in a protected structure of some sort. Transplant them after a week or several weeks, but this time is directly related to the length of time required at about 50°F for their vernalization period to be satisfied. Also, remember that they must be transplanted after the risk of frost is over. It is generally recommended to plant seedlings about 3 ft apart in rows, although some recommendations go up to as much as 6 ft. While the overall number of artichokes is increased with narrow spacing, the average size of the chokes is most times reduced.

Globe artichokes are adapted to a wide range of temperatures with 60° to 65°F (day) and 50° to 54°F (night) optimum for the highest-quality flower bud development. Plants can tolerate temperatures above 85°F, but high temperatures will cause faster development of the flower, which decreases flower bud quality. Place artichokes in a garden location where they will receive at least 6 hours of sun each day.

Figure 6.27 'Emerald' artichoke showing the edible flower buds.

Artichokes require a really good soil to thrive. Before planting seedlings, form a well-tilled bed to obtain a well-drained soil, and mix compost or manure into the soil, as this will also improve drainage. If the soil is heavy (clay), work more organic matter and even some sand into the soil. A slightly acid soil (pH 6.5) works best for artichokes. This vegetable is somewhat tolerant of drought conditions, but needs approximately ½ to ¾ inch of water per week. Fertilize twice a year: apply ½ lb of a compete fertilizer containing about 10% nitrogen per plant in the spring at planting, and then apply another ½ lb of the same fertilizer per plant in mid- to late summer.

Mature buds can be cut from artichoke plants from July to October. To avoid harvesting overdeveloped buds during high temperatures, it is best to harvest every 4 or 5 days. But during cooler periods, harvests may only need to be done every 2 to 3 weeks.

'Imperial Star' is the variety that is grown by most home gardeners, although other varieties can be found. This cultivar was developed to be grown as an annual, and will mature in approximately 85 days after planting. Other varieties that are can be grown include 'Emerald' (Fig. 6.27) and 'Green Globe'.

LETTUCE: LACTUCA SATIVA

There are only a few vegetables in the **Asteraceae** family grown to any extent in home vegetable gardens and the most important by far is lettuce. Lettuce is widely grown in most vegetable gardens and can easily be grown as a container vegetable where space is limited. Other lesser grown vegetables in this family include endive (*Cichorium endivia*), Chicory (*Cichorium intybus*), Globe artichoke (*Cynara scolymus*), and Jerusalem artichoke (*Helianthus tuberosus*). Globe artichoke has previously been discussed in this section, while Jerusalem artichoke is a warm-season plant and will be covered in that section. Endive and chicory will be covered following the discussion of lettuce.

Figure 6.28 Lettuce that has bolted in a garden. Notice the elongated stalks that should normally have leaves around a compressed stem.

Lettuce is a cool-season vegetable with optimum temperatures required for growth of between 60° and 70°F. At temperatures greater than 80°F, lettuce will bolt (flower and produce seed) and develop bitterness in the leaves (Fig. 6.28). However, lettuce will tolerate a few days from 80° to 85°F, provided that the nights are cool (<70°F). Lettuce seed will germinate at 35°F, but optimum germination occurs between 70° and 75°F. If plants are sufficiently hardened, they will withstand freezing temperatures. Repeated exposure to below-freezing temperatures will severely injure or kill lettuce.

There are numerous types of lettuces, although there are four generally recognized morphological forms of lettuce: **Butterhead**, **Crisphead**, **Cos (Romaine)**, and **Looseleaf**. All are commonly grown in home gardens. There are however, two other lesser grown

Figure 6.29 The crisphead (or iceberg) types growing in garden. Notice the round head that has formed by leaves overlapping each other.

types: **Batavian** and **Chinese** lettuces. These are rarely grown **in home gardens.**

The **crisphead** (or **iceberg**) types are easily distinguished by their firm heads (Fig. 6.29). The leaves form a head by overlapping one another in a regular manner. There are prominent main leaf veins in this lettuce. Heads are usually 6 or more inches in diameter. This is the most important commercial type of lettuce grown in the U.S., and is the lettuce in most salads.

The **butterhead** is another type of lettuce and is distinguished by its soft pliable leaves and delicate buttery flavor. The leaves form a head by overlapping one another in a regular manner (Fig. 6.30). The veins are finer and less prominent than those on crisphead types, and the head is less firm and somewhat smaller than that of the crisphead types too. Some classify **buttercrunch lettuces** as a different type but here they are placed in the same grouping as butterhead. **Buttercrunch lettuces** are crosses between butterhead and romaine varieties. They have a more upright structure, so fewer leaves are in contact with the soil. Buttercrunch lettuces are less subject to the leaf rots that can damage butterhead lettuces in late fall. **'Buttercrunch'** is the main variety grown and is open-pollinated with a maturity of about 48 days after planting. It forms a tight, compact green head of fan-shaped leaves, and has excellent bolt resistance. It is a good choice for growing lettuce into the summer months.

The **Romaine** (or **cos**) lettuce types are easily recognized by the upright character of the plant, the long loaf-shaped head, and the long narrow leaves, with a prominent midrib veins in the leaf (Fig. 6.31). Romaine is another common shipping type of lettuce, but has a higher nutrient content compared to the crisphead type.

Leaf (or **looseleaf**) lettuce does not form heads. The leaves are clustered or pressed together, but only the youngest leaves near the center of the plant overlap to any extent (Fig. 6.32). The older leaves are arranged loosely around the stem. Leaf lettuce varieties are popular with many gardeners, since there are many looseleaf varieties that can't be found in the supermarket but are easily grown in the home garden.

The other two, **Batavian** and **Chinese** lettuces, are less popular and tend to not be as common in home gardens. Batavian lettuces, also known as ***Summer Crisp Lettuces,*** have thick, crunchy leaves that hold better in the heat than do other varieties. Chinese lettuces are stiff, strong-flavored, and grown for

Figure 6.30 Butterhead lettuce growing in a garden.

their stalks as well as their leaves. Their slight bitterness is covered up when used in stir-fries and soups. 'Celtuce' is a common celery-flavored Chinese lettuce variety.

Figure 6.31 A mature head of romaine lettuce growing in a commercial field near Watsonville, CA.

Looseleaf, butterhead, buttercrunch, and romaine lettuce types can withstand greater fluctuations in the environment and have a shorter growing season than crisphead types and tend to do better in temperate climates. Lettuce can be grown as a spring crop or as a fall crop in most temperate climates, but not as a summer crop due to the excessive heat. Seeds are normally directly seeded into the garden soil. However, many gardeners will start seed of heading types indoors and transplant into the garden, allowing the plant to mature before the onset of hot weather. Successive crops can be grown planting at 2- to 3-week intervals beginning mid- to late summer for an extended autumn harvest.

The seed of heading lettuce types should be sown ¼ to ½ inch deep in rows that are 18 to 24 inches apart. It is best to thin plantings to 12–14" between plants. Lettuce plant thinnings can be eaten or transplanted for a later harvest. For leaf lettuce types, sow seeds at the same depth and at the same in-row spacing as heading types, but thin to 6 to 8 inches between plants. The leaf types of lettuce do not require as much space to grow and develop as the

Figure 6.32 Various colored looseleaf lettuce varieties growing in a garden soil.

heading types, so they can be grown at closer spacings. Again, thinnings can be eaten or transplanted for a later harvest.

Lettuce grows best on well-drained, high organic matter soils. Adequate amounts of nutrients and a continuous moisture supply are essential to vigorous growth and development of the lettuce plant. If the growth of a young lettuce seedling is stopped or reduced by a lack of nutrients or moisture, it will never fully recover. Fertilize soil before planting with 2 to 3 lbs of a complete fertilizer containing at least 10% nitrogen per 100 ft^2. It is best to sidedress with 2 lbs per 100 ft^2 of the same fertilizer after plants start to grow rapidly in approximately 3 to 5 weeks. It is important to also apply water to maintain a continuous moisture supply for plants.

Harvesting differs between the leaf and heading types of lettuce. With the leaf types, harvest the older, outer leaves as soon as they are 4 to 6 inches in length. If the leaves develop to be much longer than 6 inches, then they often become tough and develop a bitter flavor, especially under hot conditions. Harvest the heading lettuce types when the heads are firm and before seed stalks form.

Several insects can be found on lettuce in the home garden, with aphids, cabbage loopers, leafminers, cutworms, and wireworms usually causing the majority of problems. The insect pressure on fall-grown lettuce is typically much greater than on spring crops. The placement of floating rowcovers made of spun-bonded polyester over lettuce plants will prevent most insect feeding. Several diseases can also be a problem on home garden lettuce including damping-off of young seedlings (usually caused by various species of *Pythium* or *Rhizoctona solani*), downy mildew (will usually be seen under wet, humid conditions), mosaic diseases on foliage (such as lettuce mosaic virus), and Fusarium yellows.

There are many different lettuce varieties from which a home gardener can choose. Several factors need to be considered before choosing a variety, including the type of lettuce, heat tolerance, disease resistance, and leaf color. A home gardener should be aware that there are considerable differences among lettuce varieties for heat tolerance. If the gardener lives in a region that becomes hot quickly after a short spring, then it is recommended to grow a variety that has heat tolerance.

Crisphead lettuce varieties that are widely adapted include 'Ithaca' (72 days, heat tolerant) and 'Summertime' (70 days, heat tolerant). There are several Butterhead varieties available to gardeners including 'Buttercrunch' (65 days, widely used by home gardeners, slow bolting) and 'Sangria' (49 days, red butterhead, slow to bolt). Romaine (or Cos) varieties available include 'Paris White Cox' (76 days, heat tolerant), 'Cimarron' (60 days, red-tinged, resists bolting), and 'Little Gem Hybrid' (50 days, space-saver, 5 inch diameter heads). Leaf lettuce varieties include 'Green Ice' (45 days, slow to bolt), 'Red Sails' (45 days, red crinkled leaves), 'Black Seeded Simpson' (46 days, widely adapted), and 'Salad Bowl' (45 days, heat resistant, slow to bolt).

There are also many different salad lettuce mixes that can be purchased and it just depends on the preference of a particular gardener. Some are spicier than others, while others have more coloration or are blends of different types of lettuces. Some mixes are more specific to growing in containers, while others are better adapted to being grown in a garden with more space.

ENDIVE (*CICHORIUM ENDIVIA*) AND CHICORY (*CICHORIUM INTYBUS*)

Endive and chicory are cool-season crops and are members of the Asteraceae family. Both species are grown primarily for their foliage, which is used mainly in salads. The foliage can be quite bitter in some instances. Blanching (or the removal of light) is used to reduce the bitterness and to increase the tenderness of the leaves. These vegetables can be grown as spring crops, but will do better if planted for a fall harvest. Frost appears to improve the flavor of these vegetables and make them less bitter.

The foliage tends to resemble looseleaf lettuce types but they are more prostrate in their growth. **Endive produces narrow, deeply cut, curled leaves, whereas the broad-leaved type is referred to as chicory (Fig. 6.34)**. Both types can be blanched by pulling the outer leaves together and tying with a string or placing some structure over the center of the plant, such as a terra cotta pot or saucer (Fig. 6.33). In about 2 to 3 weeks, the blanching process should be complete.

Endive and chicory grow best at temperatures of 60° to 65°F, with endive slightly more tolerant of hot conditions compared to chicory. Cultural production practices are similar to lettuce. Endive and chicory prefer to be grown in well-drained garden soils with high organic matter. For spring plantings, start indoors 4 weeks before the last average frost date; transplant to garden soils after the last average frost date, and plant seedlings about 6 inches apart. A few weeks later, thin seedlings to about 12 inches apart. It is important to note that these

Figure 6.33 Endive that has been blanched by preventing light from reaching leaves through use of terra cotta pots or saucers.

Figure 6.34 Chicory growing in a garden.

vegetables do not grow well in the heat of the summer, and should be harvested before late June in most temperate climates. If they are allowed to grow into the hot summer months, they will most likely need to be shaded and/or misted in some manner to keep them cooler; and, even then, they are likely to bolt and become bitter in flavor. Similar to many other cool-season vegetables, endive and chicory are best grown as fall crops. For fall plantings, seed directly into the garden soil about 1½ months before the first frost date.

Make sure that plants receive adequate water as they are growing in the garden and do not let plants become water stressed, since leaves will become tough and bitter with inadequate water. Endive and chicory require approximately 1 inch of water per week during the period that they are most actively growing.

These plants can usually be harvested about 2 to 3 months after planting. Similar to loose-leaf lettuce, individual leaves or entire plants can be harvested. When harvesting entire plants, cut stems with a knife as close as possible to the soil surface. Suggested varieties of endive are 'Fine Green Curled', 'Salad King Improved', and 'Tres Fine Maraichere', and recommended varieties of chicory are 'Broadleaf Batavian' and 'Florida Deep Heart'.

ONION, GARLIC, AND OTHER *ALLIUMS*

These cool-season vegetables are some of the most widely grown vegetables in home gardens for their edible bulbs, although the foliage for most can also be eaten. Onions are by far the most important *Allium* grown in home gardens, followed by garlic. A brief discussion will also focus on a few of those *Alliums* that are not as popular with home gardeners.

Onion: *Allium cepa*

The onion is a hardy, cool-season vegetable grown primarily for its edible bulbs. It is a member of the *Amaryllidaceae* family, which also contains many other related edible bulb species in the *Allium* genus (see Table 6.1). Most people are accustomed to finding edible members of the Amaryllidaceae family in their kitchens every day in many of the meals served. Two members of this family (onions and garlic) are the most important, as they have been an integral part of the cuisine in numerous cultures for millennia and will be discussed in detail.

Onion is certainly the most popular *Allium* cultivated in the world today. Although the foliage can be eaten, it is mostly grown for its edible bulb. This is a popular vegetable grown in home gardens, as it is generally an easy-to-grow vegetable. Onions are resistant to cold weather conditions, and can help keep a garden productive during much of the year, except during the coldest and hottest months.

Although onions prefer to grow foliage during cool weather conditions and form bulbs under warm conditions, the timing of bulb formation is controlled by both temperature and day length. Onions are day-length sensitive and are divided into "short-day," "day-neutral," and "long-day" varieties. It is important to use varieties designated for your area. Short-day onions are planted in the southern U.S. in the autumn, as a winter crop. They develop bulbs as days lengthen to about 12 hours in early summer and are harvested usually in June. Long-day onions are more adapted to be grown in northern areas and these types require 14 to 16 hours of daylight to form bulbs. When long-day types are planted in the early spring, they will form bulbs under the long days of summer. It is important to note that long-day onions are typically more pungent and noticeably firmer compared to short-day onions.

Table 6.1 Vegetables commonly grown that are in the Allium genus.

Latin name		Common name
Genus	*Specific epithet*	
Allium	*ampeloprasum*	leek, elephant garlic
Allium	*cepa*	onion, shallot
Allium	*fistulosum*	scallions
Allium	*sativum*	garlic
Allium	*schoenoprasum*	chives
Allium	*tuberosum*	garlic chives

Figure 6.35 'Candy' day-neutral type onions grown in garden soil. The wheat straw works great to reduce weed competition and to provide consistent soil moisture to this shallow-rooted crop.

Day-neutral onions set bulbs based only on temperature and day length has no effect on bulb formation.

Onions need continuous soil moisture due to their shallow root systems and prefer to be grown without alternating temperature extremes of hot to cold or cold to hot conditions. Onions grow best at temperatures of 55° to 75°F. A steady moisture supply is essential, particularly during bulb formation (Fig. 6.35). An irrigation regime of two ½-inch applications of water per week will ensure larger bulbs and higher yields. Under hot conditions, onions will bolt (flower and form seed). Onions that bolt will produce small bulbs with large necks. Onions with large thickened necks are hard to cure and are generally not usable.

Onions can be started from seeds, transplants, or sets (which are immature onion bulbs, <¾ inch in diameter) depending on a gardener's needs. Sets are generally available in late winter to early spring from gardening catalogs and in gardening supply stores. Seeds sown in the garden are generally the least popular since it takes several weeks longer to get a usable bulb compared to transplants or sets. Transplants can either be purchased from garden stores and mail-order companies, or grown from seed. For transplants from seed, sow seed about ½ inch deep indoors in soil approximately 6 weeks before the outdoor planting date. Transplants are popular because they produce a large bulb over a short time (<60 days).

Onions are heavy nutrient feeders, and some fertilizer should to be worked into the soil before planting. This plant prefers a fairly neutral soil pH (~6.5 to 7.0). Soil that is too acidic or too alkaline can cause slow growth and late maturity. The addition of 2 to 3 lbs of a complete fertilizer containing about 10% nitrogen per 100 ft^2 prior to planting should provide sufficient nutrients for early-season plant growth. Onions will also need to be sidedressed with

½ lb of a similar complete fertilizer 2 or 3 times during growing season at monthly intervals. However, do not apply within 3 weeks of harvest, since heavy nitrogen applications late in season may result in increasing neck size.

Raised beds are ideal for onions and should be used if at all possible. Onion spacing is important since plants spaced to close to each other will result in small bulbs. Rows should be made about 2 ft apart. Place sets or transplants, or thin seedlings to 3 to 4 inches apart (4 to 6 inches apart for larger bulbs and 2 to 3 inches for smaller bulbs). For a large bulb harvest, start working soil away from bulbs about a month before harvest, and stop pulling soil away from bulb when about ⅓ of the bulb is exposed. This will help with bulb and neck drying and allow the gardener to keep bulbs over a longer period of time.

Harvest onions when about half the tops have died and fallen over naturally. After the tops have fully wilted (extra water removed from bulb), remove tops about 1½ inches above bulb. Prepare bulbs for storage by curing in a mesh bag for 2 to 3 weeks in a well-ventilated area out of the sun. Clean bulbs by removing soil and outer loose dry skins, and store where air is dry and between 35° and 50°F.

Several insects and diseases can be potential problems on onions in a garden. The onion maggot (white maggots that burrow into onion), seed maggot (damage similar to that of onion maggot, but usually occurs earlier in the season), and thrips (cause silvery patches to appear on the leaves and plants may wither and collapse) can often be found feeding on onions in a home garden. Several diseases can pose problems to onions in a garden including seed rot or damping-off of young seedlings (usually caused various species of *Pythium* or *Rhizoctona solani*), downy mildew (occurs under wet or humid conditions), and pink root (roots are light pink, later deeper pink to red, and finally purple).

There are many different varieties from which a home gardener can choose. Due to the wide variety of onions, it's helpful to know what characteristics to consider in selecting them for your garden. It is of foremost importance to first select the appropriate variety for your area from either a short-day, long-day, or neutral-day type. The short-day types of onions are adapted for use in areas south of 30°N latitude in the U.S. Long-day types require long days with 14 to 16 hours of sunlight to form bulbs and do better when grown in more northern areas of the U.S. With day-neutral types, day length has no effect on bulb formation, and these types are adapted to grow in areas from 30°N latitude to about 38°N latitude in the U.S. Many of the newer long-day and day-neutral onion varieties are bolt-resistant, and the gardener should check to make sure the variety has bolt-resistance before buying. Several bulbing onion varieties are available, including 'White Sweet Spanish' (long-day type, 115 days), 'Yellow Sweet Spanish' (long-day type, 100 days, 1 lb size onions), 'Copra' (long day, 104 days, stores well up to a year), 'Candy' (day-neutral, 85 days, widely adapted, most popular day-neutral; see Fig. 6.35), 'Red Candy Apple' (day-neutral, 95 days, 3 inch diameter deep red, flattened globe shaped, sweet bulbs), and 'Walla Walla Sweet' (short-day, 270 days, overwintering type).

To get the small scallion or green onions, grow a standard onion variety for your area and harvest when the bulb is small. In addition to the bulbing type of green onion, there are perennial bunching types (*Allium fistulosum*) that do not produce bulbs, but continue to divide at the base to form new shoots throughout the growing season. These types are winter hardy and will produce throughout the growing season. Shoots are crisp and mild early in the season, becoming more pungent later. Varieties of bunching onions that are available include

'He-shi-ko' (70 days, can be left in soil all winter) and 'Evergreen White Bunching' (120 days, 4–9 plants per cluster).

Garlic: *Allium sativum* or *A. ampeloprasum*

Almost everyone is familiar with the scent and taste of garlic, and in many countries, it is the most widely used culinary herb. It is now available in processed forms and value-added products including dietary supplements and garden pest deterrents. Garlic is readily available fresh, braided, bagged, pickled, minced, canned, dehydrated, granulated, and powdered. Garlic has gained attention in last few years as a medicinal herb primarily for being able to reduce cholesterol levels and thus reduce the chance of heart disease. It is also used as a potent antibiotic, and will kill many different kinds of pathogenic and nonpathogenic bacteria. Garlic is widely used in cooking and should be grown in every garden to maintain a supply of fresh garlic. Store-bought garlic just can't compare to that grown in the garden!

Garlic is a hardy perennial cool-season bulb in the *Amaryllidaceae* family that prefers full sun for growth. Garlic has flat gray-green leaves that are about 1 inch wide and 12 inches in length. Garlic is propagated by bulblets often referred to as "cloves." The cloves are tightly clustered bulblets, with each wrapped in a papery skin to hold them together. The garlic bulb consists of numerous cloves packed tightly together. Individual cloves can be planted in the early spring or autumn. So, there are two different times during the year in which garlic can be planted. However, those planted in the autumn will produce larger bulbs than those planted in the spring. Garlic is vernalized by the cold winter weather and is thus triggered to produce a bulb the following year, whereas spring-planted garlic does not typically form a bulb. Freshly harvested fall-planted garlic that is being allowed to dry is shown in Fig. 6.36.

***Figure 6.36** Freshly harvested garlic from the garden.*

Cloves should be planted about 1 inch deep, and 6 to 8 inches apart in the row, with 1 to 1½ ft between rows. Garlic requires a fertile, well-drained high organic matter soil with a pH of 6.5 to 7.0. It is best to avoid planting garlic in areas that have poor water drainage and thus, raised beds work great for garlic production. Cover planted garlic with an organic mulch; this is especially important for fall-planted garlic. **Garlic** bulbs planted in the fall start root growth as soon as they're planted, making these plants ready for top growth the following spring when temperatures rise. Prior to planting garlic in the autumn, apply about 2 to 3 lbs of a complete fertilizer containing about 10% nitrogen per 100 ft^2; in the spring as plants start to grow and again about 4 weeks later, apply 2 lbs of a complete fertilizer containing about 10% nitrogen per 100 ft^2. Garlic is a shallow-rooted crop and requires a constant moisture supply during the growing season.

Two to three weeks before harvesting garlic, allow the soil to dry around plants since this will prevent rotting due to the high soil temperatures at this time and it will also prepare them for storage. When most of the leaves have turned brown, gently dig or pull up bulbs, and place plants and bulbs on top of a table or in a mesh bag for 2 to 3 weeks in a shady spot that has good air circulation. When plants are dry, rub off dirt and excess scales, and place in mesh bag in an airy, cool location such as basement. Some gardeners will keep out the largest bulbs to plant again in the autumn or spring.

The flower stalks (or scapes) reach 2 ft or more in height and produce clusters of mauvish-white flowers that are really asexual propagation structures called bulbils. Scapes should be removed when they first appear in late spring to divert most of the energy into bulb formation (Fig. 6.37).

Figure 6.37 Flower stalks (or scapes) emerging from garlic and require removal to maximum bulb size. Notice the typical curling pattern of the scapes.

There are three types of garlic that are available: 1) **Softneck type** (*Allium sativum* var. *sativum*), 2) **Stiffneck type** (*A. sativum* var. *ophioscorodon*), and 3) **Elephant** (*A. ampeloprasum var. ampeloprasum*).

Softneck types are those most commonly found in supermarkets because they have a longer shelf life than the hardneck varieties (Fig. 6.38). Softnecks store well and generally do not produce flower stalks. The softneck varieties can be recognized by the papery white skin on the surface of the bulb and the abundance of cloves, often forming several layers around the core. Normally there are anywhere from 10 to 20 small to medium-sized cloves produced on a bulb. The softneck types have necks that are soft at maturity and can easily be braided. Softnecks are well adapted to the more temperate climate of the southern U.S., and because they do not bolt easily, can flourish through the erratic temperatures of southern winters. There are three main types of softnecks: **Artichokes** are very vigorous and produce a large bulb with many cloves. Plants are shorter than hard-neck varieties with leaves that spread rather than grow upright. The leaves are broader and a darker green than any other type. The bulb wrappers on the artichoke varieties are coarser than the silverskin and may have some purple blotches. 'Inchelium Red' and 'Italian Softneck' are popular artichoke-type varieties, and other varieties include 'Red Toch' and 'Susanville'. **Silverskins** are the type generally found at the supermarket due to their very long storage life. They are the highest-yielding softneck type, typically having 12–20 cloves per bulb in about 3 to 5 different layers. Outer cloves are usually flat and wide, while inner cloves are tall, narrow, and concave. Silverskins have long been the most popular garlics for braiding because of the smooth, shiny skin and symmetrical shape (Fig. 6.38).

Figure 6.38 The silverskin softneck type of garlic commonly found in most supermarkets.

Silverskins do well in a wide range of climates. 'Idaho Silverskin' is a popular variety. Other varieties that can be found include 'Silver White', 'Nootka Rose', and 'S&H Silver'. **Asiatics** and **Creoles** are generally classified as weakly bolting softnecks that are well suited to mild winter areas and are generally the first garlic to be harvested each year. These bulbs mature rapidly in their final weeks, so make sure to harvest these before the heads become over mature and start splitting. There are many different varieties of each and really have not been adequately trialed to determine which would work best in home gardens. In the Asiatic group, 'Russian Redstreak' and 'Pyong Yang' are suggested varieties, while in the Creole group, 'Ajo Rojo', 'Burgundy', and 'Creole Red' are suggested for trial in home gardens.

Stiffneck types are the most cold-hardy garlic type and plants have a hard, stiffneck. The bulbs have large outer cloves and no inner cloves. This type will produce scapes (hard-stemmed flower stalks that bear aerial bulbils), but will and have a variety of complex flavors. They generally prefer the cold winters and long, cool springs of more northern temperate climates. Stiffnecks have fewer and larger cloves than softnecks have, which makes them easier to peel in the kitchen, but they do not store as well as softnecks.

There are three main types of hardnecks: Rocamboles, Porcelain, and Purple Stripe. **Rocamboles** tend to grow better in cold winter climates and produce large tan or brown colored cloves that are very easy to peel. Their loose skins, however, give rise to their major disadvantage: a shorter storage life than most other varieties. By the end of January, most Rocamboles show signs of dehydration or begin to sprout. Longer storage is possible if bulbs are well grown and well cured before storage. However, they have a deep full-bodied flavor, considered by some to be the finest-tasting garlic of all. Suggested rocamboles varieties are 'Carpathian', 'German Red', 'Killarney Red', and 'Spanish Roja'.

Porcelains produce a beautiful plant and an impressive bulb. Porcelains are becoming increasingly popular as gardeners learn about them. They have a satiny white bulb wrapper with four to six large cloves around a sturdy scape, and are frequently mistaken for elephant garlic. Because of their smooth, tight bulb wrappers, they store longer than Rocamboles. Porcelains are easy to peel and great for cooking. Their flavor is outstanding, rivaling that of rocamboles. As a group, the porcelains have the highest yields of allicin, the sulfur compound most associated with garlic's therapeutic benefits. The porcelain types we grow are 'German Stiffneck', 'Georgian Crystal', and 'Music'.

Purple Stripes are named due to the vivid purple striping on the bulb wrappers and clove skins. These are really one of the most attractive garlics. Their taste is strong, complex, and rich, without being overly sulfurous. Many gardeners regard purple stripes the best type for roasting. Most strains have 8 to 12 cloves per bulb (more than Rocambole), so clove size is slightly smaller. Cloves are noticeably somewhat elongated and crescent shaped. They store slightly longer than Rocamboles and peel almost as easily. Purple Stripe plants are distinctive from Rocamboles because their leaves grow at wider angles to the stem. A few suggested varieties to try in a home garden are 'Chesnok Red', 'Metechi', and 'Persian Star'.

Elephant garlic often produces bulbs of ½ lb or more, with very large cloves and have a tender, mild, slightly sweet flavor. It has a very mild taste compared to other garlic types, but is less hardy compared to softnecks and hardnecks. Some like it because it can be eaten raw and used in cooking for a hint of garlic flavor without being overwhelming.

Elephant garlic should be mulched heavily for fall plantings, and it is recommended that 6 to 8 inches of some type of an organic mulch be applied to soil in most cold-temperate climates before the onset of winter. Plant elephant garlic 4 to 6 weeks before the first expected frost, and the roots will have plenty of time to establish over the winter.

Elephant garlic is not a true garlic, but actually a variant of the species to which leek belongs. It has a tall, solid, flowering stalk and broad, flat leaves much like those of the leek, but forms a bulb consisting of very large, garlic-like cloves. The flavor, while not exactly like garlic, is much more similar to garlic than to leeks, is milder than other garlics, and much more palatable to some people than garlic when used raw.

Unlike most garlics, elephant garlic does not have to be harvested or divided each year, but can be ignored and left in the ground without much risk of rotting. The plant, if left alone, will spread into a clump with many flowering heads. These are often left in gardens as an ornamental and to discourage insect pests. Of course, once they get overcrowded, the plant may not do as well, and growth will be stunted, and there may be some rotting. The most common variety for home gardens is probably 'Elephant', although varieties such as 'Giant French Mild' can also be found.

Leek: *A. ampeloprasum* var. *porrum*

Leeks are a large, upright, non-bunching type of onion (Fig. 6.39). However, leeks do not form bulbs like onions, are hardier than onions, and have a much wider, flatter leaf compared to onions. The leek is a cool-season plant and a member of the *Amaryllidaceae* family. Leeks are a delicate type of onion, and add a rich but subtle flavor to everything from creamy sauces to soups and stews. Leeks are a must for any gardener interested in the culinary arts. They are popular in Europe, and are known in France as "the asparagus of the poor." Leeks are sometimes steamed like asparagus, but most often they are used in soups or stews.

Figure 6.39 Leeks bundled and ready for purchase. Notice the blanched lower stems and bulb region.

Leeks are propagated by seed, and seeds can be directly planted into the garden soil about a month before the last frost date, or started indoors and transplanted outdoors at the time of the average last frost. For developing transplants, sow seeds in flats anytime during February, and then transplant seedlings to larger containers such as cell packs. Leeks have a long growth period, so many suggest that they should only be transplanted and not be directly seeded into a garden. Leeks will typically take around 90 days to mature from transplants and 120 to 150 days from direct seeding. When plants reach 12 to 18 inches in height, transplant into the garden. Plant these seedlings about 6 inches apart in the row and 2 ft between rows. If direct seeding, plant early in the spring and thin to the spacings previously described. Leeks prefer to be grown in full sun.

To get longer white stems, blanching can be achieved by first planting in trenches 4 to 6 inches deep and later hilling or mounding soil up against the stems as the plants grow. For specifics on cultural practices, refer to the section on onions.

This frost-hardy plant can be harvested long after many other garden vegetables have disappeared from the garden. Begin harvesting leeks when the stem is about 1 inch in diameter. In mild-winter regions, leeks can be overwintered in the garden and harvested into the spring months. In cold areas, mulch with 4 to 6 inches of straw to protect overwintering plants. There are a few varieties available to home gardeners, and suggested varieties are 'King Richard' (100 days from seeding), 'Arkansas' (120 days from seeding), and 'American Flag' (130 days from seeding).

Shallot: *Allium ascalonium*

The shallot is also a member of the onion and garlic family. It tends to grow as a cluster instead of one large bulb like the onion. Shallots are used like onions in most dishes and can readily be found in most supermarkets (Fig. 6.40). In many recipes, 3 to 4 shallot bulbs can

Figure 6.40 Shallot bulbs for sale in supermarket.

be substituted for one medium onion. Many gardeners also grow them for their green tops to place in different dishes and are often used like chives in food preparation.

Shallots are grown primarily for the young green foliage produced, and a popular use is as a green onion, but they can also be utilized for their bulbs. Shallots are a multiplier type of onion, since they will divide into a clump of smaller bulbs. The tops and bulbs produced by shallots are somewhat similar to but smaller than onions. The bulbs produced by shallots are pear-shaped, and have reddish, yellowish, or brownish skins (Fig. 6.40). Shallots are hardy and will overwinter as perennials; for best results, remove clusters of bulblets (as many as 10 to 15 bulblets may be produced in a cluster) at the end of each growing season and replant the smaller ones in the fall. Bulblets should be planted about 1" deep, 2–4" apart in the row and 12–18" between rows. It is important that disease-free bulblets be used for propagation, and these can be easily obtained from many different mail-order vegetable seed companies or local garden centers in the spring months. Bulbs are typically harvested when tops die down in late summer. For specifics on cultural practices, refer to onion.

There are different types of shallots that gardeners should be aware of. **French shallots** have pinkish-brown or brownish-colored skin and pinkish-purple flesh. **Dutch shallots** have an orangey-yellow skin. They are larger than the French types, round in shape, and have a stronger flavor than other shallots. Dutch shallots are the type most commonly found in supermarkets. **Banana shallots** are the largest type of shallot, and are often known as the echalion shallot. They are named because of its size and shape, and the large shallot's tan skin hides a delicate flavor. **Potato shallots** are also known as the potato onion, and the shape of the potato shallot definitely resembles a potato. It is much different than a traditional French shallot, as the potato shallot is round and its leaves grow in one clump rather than in shoots. The potato shallot has a bronze-red skin.

Varieties suggested include 'Dutch Red Shallot' (coppery-red bulb wrappers and purple-pink bulb flesh, 1 to 2 inches in diameter), 'Yellow Moon Dutch' (high-quality golden skinned, round bulbs produce heavily and mature early), 'Holland Red' (coppery-red outer skin with bulbs that are round, flat, short and fat; peels easily), and 'Picasso' (red-brown skins and round bulbs with pink flesh, resists bolting, easy to grow).

Chives: *Allium schoenoprasum*
Garlic Chives: *Allium tuberosum*

Chives and garlic chives are also members of the *Amaryllidaceae* family and both can be found in home gardens. These plants are the favorite herbs of many home gardeners, although they often struggle to keep the plants green and healthy during the winter months.

Chives are a versatile plant and can be utilized throughout the garden, such as in clumps or as edging plants. A clump of chives makes an excellent edging for a flower bed or plant border. It will make compact clumps of grass-like foliage that grow to about 12 inches in height (Fig. 6.41), and will become covered with pink or purple flowers in early summer. Chives are a hardy perennial and will do best in full sun or partial shade. This herb will do well in containers or in window boxes, and is often grown indoors, by or in a window. Many people just grow chives in a small pot in a kitchen window to have some readily available to go into certain dishes. Chives can be propagated by dividing plants or by seed. If chives

Figure 6.41 A clump of chives flowering in the early spring near a water feature on the Southern Illinois University campus in Carbondale.

are grown to be used primarily in food preparation, remove flowers that develop to allow more foliage growth to occur. Harvest young leaves by pinching them off at the base. For chives grown outdoors, the harvesting period can be extended to last about nine months if plants are protected in the early spring and late fall. Chives are an easy plant to grow and are well worth the effort. Chives are often an underestimated herb since this plant has many other uses besides putting on a baked potato. They will add an excellent flavor to soups or stews. Chives can be best described as having a mild onion-type flavor and they make a great addition to cream sauces and egg dishes when only a slight onion flavor is desired.

Garlic chives are also known as Asian or Oriental chives and grow much larger than common chives. The leaves of garlic chives are longer and flat, whereas the leaves on the others are thinner and rounded. Garlic chives have a much stronger flavor than the common chives. Garlic chives can quickly become invasive, so gardeners should be made aware to watch for the seed heads and harvest them before they mature. The plant also becomes covered by the masses of white flowers produced in the early summer.

PEAS: *PISUM SATIVUM*

Peas belong to the **Fabaceae** (or legumes) family, which is quite an interesting group, since they are the only vegetables that have the ability to fix their own nitrogen. Several vegetables in

Figure 6.42 Snap beans (left) and sugar snap peas on trellis (right) growing at an organic farm near Carbondale, Illinois.

this family are widely grown, including snap beans, lima beans, cowpeas, and peas (Fig. 6.42). However, peas are the only cool-season vegetable member in this family.

The pea is a frost-hardy, cool-season vegetable that can be grown throughout most of the United States, and is best adapted to wherever there is at least two months of a cool growing season. Peas are relatively easy to grow and should be grown in most all vegetable gardens, whenever possible. Peas are one of the first vegetables that can be planted and harvested in the early spring, although it can also be grown as fall crop. There are different types of peas, and a decision must be made by the gardener as to the type that will be grown. The different types of peas have the same cultural requirements.

For gardening purposes, peas are classified as garden peas (English peas), snap peas, and snow peas (sugar peas). The **Garden/English Pea** is a shelling-type pea and pods are not meant to be eaten. There are both wrinkled-seeded and smooth-seeded types. Wrinkled peas are generally sweeter and preferred for home gardens, whereas smooth types have more starch reserves and are generally better for an early spring planting due to the high starch content. The smooth-seeded types are used most often to produce ripe seeds that are used like dry beans and to make split-pea soup. **Snow Peas/Sugar Peas** produces edible flat, pods that are often used in Asian stir-fry dishes. Snow peas are meant to be harvested as flat, tender pods before the peas inside develop. Strings may need to be removed from larger pods before they are eaten. The **Snap Pea** is a type of pea that produces a round, edible pod. Snap peas have been developed from garden peas to have low-fiber pods that can be snapped and eaten along with the immature peas inside. Pods are plump and used like snap beans. Again, strings may need to be removed from larger pods.

Peas should be directly seeded into the garden soils and not be transplanted since they do not transplant well. Pea plants can survive light frosts, but will not tolerate temperatures over 75°F. Two crops can be typically grown each year in cold-temperature climates; the first should be planted 4 to 6 weeks before the last expected frost and up to about 2 to 3 weeks after the first frost date, and the last crop should be planted 12 weeks before the first expected frost in the autumn. Seeds should be planted 1 to 2 inches deep in the soil, and 60° to 70°F is the optimal soil temperature for germination, although seed germination will occur at lower temperatures around 40°F or so. There are two types of plant-growth habits: low-growing types (bush or dwarfs) and climbing or vining types. The low-growing or dwarf varieties do not require staking or trellising, and are the easiest to grow. They can be planted in rows spaced about 2 ft apart. Climbing types should be trellised with rows spaced 3 ft apart; generally, one row is planted on each side of a trellis. Plants should be thinned to 4 to 5 inches apart in the row.

Peas grow best in well-drained, high organic matter loamy soils. Peas are nitrogen-fixers, so they are light nitrogen feeders. It is recommended to apply about 2 lbs of complete fertilizer containing about 10% N per 100 ft^2 before planting and sidedress once with ½–1 lb of the same fertilizer material at 30 days after planting.

Garden peas should be harvested when pods are well filled with peas, but still bright green in color. The lower pods should be harvested first. The harvest period for garden peas lasts several weeks and consistent picking will increase production. To get the sweetest flavor and crunchiest texture with snap peas or snow peas/sugar peas, harvest before peas in pods touch each other. The harvest period will also last just a few weeks for these pea types.

There are several insects and diseases that often need to be controlled in peas. Insect pests on pea include wireworms, cucumber beetles, aphids, thrips, and pea weevils. Diseases that can become problematic if not controlled properly include Fusarium wilt, powdery mildew, root rot, and mosaics. It is recommended to plant disease-resistant varieties if available.

There are many different pea varieties from which to choose. Suggested garden peas varieties include 'Maestro' (57 days, slender 4-inch pods, can be grown without trellis), 'Knight' (56 days, 2 ft vines, no trellis needed, medium sized peas in 3½-inch-long pods), 'Bounty' (61 days, a freezer pea, 2 ft vines, medium sized peas), and 'Little Marvel' (62 days, dwarf 18-inch-tall vines, 3-inch-long pods with 7 to 8 peas/pod). For flat-podded sugar peas/snow peas, recommended varieties include 'Mammoth Melting Sugar' (68 days, 4½ ft tall vines requires support, 4½-inch-long pods), 'Oregon Giant' (60 days, 2½ ft tall vines, no support needed, 4½-inch-long pods), 'Oregon Sugar Pod II' (60 days, 2½ ft tall vines, no support needed, 4-inch-long pods). For the round-podded snap peas, two varieties are suggested: 'Sugar Snap' (65 days, 5 ft tall vines require support, 3-inch-long pods) and 'Sugar Ann' (56 days, 2 ft tall vines, no support needed, 2½-inch-long pods).

POTATO: *SOLANUM TUBEROSUM*

The common "white" or "Irish" potato is a cool-season plant that is a member of the important **Solanaceae** family. The **Solanaceous** family is often referred to as the nightshade family, and this family contains some of the mostly widely grown vegetables in home gardens, including tomato, white potato, pepper, and eggplant. In almost every home garden, at least one or more

Figure 6.43 There is significant variation in shapes and skin coloration among different potato varieties.

of these vegetables can commonly be found, and the potato is especially important since they are used in the kitchen in so many different ways.

The potato originated in the Andes Mountains of Peru, and even today many wild types (that look like knotty tubers) are grown in the mountains for consumption by the native populations. Potatoes were introduced into Europe in the 1500s after the Spanish conquest of Peru. The cultivation and the importance of the white potato as a staple crop to the Northern European diet grew quickly. But the importance of the potato to the diet was not really known until the Irish potato famine. It is important to understand that potatoes were the diet of the Irish sharecroppers and they ate potatoes for every meal. This dependence on the potato resulted in starvation and mass exodus from Ireland in 1845 to 1847 when late blight, caused by the fungus *Phytophthora infestans*, destroyed the Irish potato crop for two years in a row. Ideal conditions for the development and spread of the disease occurred during these years. Besides this, the winters during these years were excessively cold, as many in the populace froze or died from starvation. As a result of this large crop failure, more than one million Irish died and almost 2 million immigrated to the U.S. Furthermore, it is thought that during colonial times (1600 to 1700s) that potatoes were introduced into the U.S. by Irish immigrants, and this is why we often refer to white potatoes as Irish potatoes.

Potatoes are herbaceous annual plants grown for their edible tubers. Thus, the edible potato is not a root but a swollen underground stem called a tuber. Tubers are the enlarged, fleshy tips of underground modified stems. They are storage organs and are high in carbohydrates.

The eyes found on a potato are buds from which lateral branches will grow. Each tuber has numerous buds that can produce normal vegetative branches.

Potatoes are a cool-season crop, slightly tolerant of frost, but are damaged or killed by freezing weather. Potatoes require a frost-free growing season of approximately 90 to 120 days, depending on the climate and the variety grown. Potatoes planted for harvest in the autumn should be planted approximately 120 days before the first killing frost. The ideal climate for growing potatoes has a relatively cool summer. Spring potatoes should be planted a couple of weeks before the last killing frost in the spring. Potato sprouts will develop at low temperatures, but will most likely be injured if exposed to frost when above ground. It is usually safe to plant potatoes when the soil is warm enough to be worked in the spring, which is about 45°F and before the danger of the last spring frost has passed. Maximum yields of high-quality tubers are produced when mean temperatures are between 60° and 65°F during the growing season. Tubers form best at 65°F, and tuber formation is reduced above 70°F and inhibited at temperatures above 85°F. When planting fall potatoes, 'Red LaSoda' is the cultivar to plant because of its heat-resistant qualities. Potatoes germinate and emerge best when soil temperatures are above 50°F.

A few proven techniques used by the home gardener can make a large difference in growing potatoes. When growing potatoes in a home garden, the gardener needs to be aware of several important factors that can influence productivity, including variety, seed certification, seed piece size, seed piece planting and spacing, hilling, soils, fertilization, irrigation practices, planting dates, pest control, mulching, and harvesting. All of these factors are discussed in further detail next.

Variety selection is an important part of growing potatoes successfully in a home garden. There are many different skin and flesh colors to choose from, as well as many different sizes and levels of disease resistances. Suggested varieties are 'Dark Red Norland' (red skin with white flesh, early productive variety, medium to large in size, resistant to scab and several viruses), 'Yukon Gold' (yellow skin with yellow flesh, medium to large in size, many consider this the best eating potato, early- to mid-season variety with some viral resistance), 'Red Pontiac' (thin red skin with white flesh, does well in heavy soils, widely adapted, mid- to late maturity), 'Kennebec' (thin, smooth skin with shallow eyes, late maturing potato, good for baking, stores well, resistant to early and late blight, as well as some mosaic viruses), 'Red LaSoda' (heat resistant, red skin with white to cream-colored flesh, round to oblong tuber), and 'All Blue' (novelty has purple skin and flesh). There are many other varieties available, and the gardener needs to check seed catalogs for type of potato desired. There are many different potato skin colors and shapes to choose from to plant in a home garden (Fig. 6.43).

Buy **certified disease-free "seed" potatoes** from garden centers or seed catalogs once a variety or varieties have been chosen to grow in the garden. Since potatoes are propagated vegetatively, they are more likely to accumulate and harbor plant diseases, especially viruses, compared to seeds. This is important because potatoes can harbor many diseases that reduce growth (like viruses) but are otherwise undetectable in most gardens. It is best to avoid bringing potato diseases into a garden. Potatoes bought from a grocery store should not be placed into a home garden, since they are not certified as being disease free. These potatoes will often carry diseases and may have been treated to prevent sprouting anyway. To avoid this problem, certified seed should be purchased for the garden. Certified seed potatoes have been inspected by a certification agency at least twice during the growing season to ensure that any disease present is within strict tolerances. Gardeners will need about 12 lb of seed potatoes for each 100 ft of row.

Seed Piece Size is an important consideration to maximize potato production in a garden. Seed pieces that have adequate size will increase the chances of obtaining the highest possible potato yield. Large seed potatoes should be cut into about 1½ inch square pieces and each piece should weigh at least 1½ oz. It is important to make sure that each seed piece has at least one good eye (Fig. 6.44), and to cut them a few days prior to planting to allow cut surfaces to heal over. Try to make cuts to expose the least amount of potato flesh surface area. Gardeners will often dip the healed cut seed pieces in a diluted bleach solution or a fungicide to prevent seed pieces from rotting in the soil. During cutting, discard any potatoes that show dark rings or discoloration inside. Whole small seed potatoes can be planted, which will avoid rot altogether.

Seed piece planting and spacing in a garden is also critical to maximize potato productivity. Place seed pieces, with the cut side down and eyes up, about 4 inches deep and 10 to 12 inches apart in rows with rows spaced 3 ft apart. To obtain smaller potatoes, plant seed pieces closer together in the row, and for larger potatoes, plant farther apart.

Hilling is an important cultural practice that should be used for potatoes to maximize their yield potential. Although potatoes can be planted on flat ground, forming a hill around the plant is preferable. Hilling is a simple method, and just pulling soil up around stems once they emerge will suffice. The stems will then put out new stolons that will form additional tubers in this new additional area of the soil. Thus, hills provide room for more tubers to develop and will also prevent sunlight-induced greening of tubers.

It is important to not allow tubers to receive sunlight. If tubers do receive sunlight, they will become green in color and develop a bitter taste. Although the bitterness results primarily

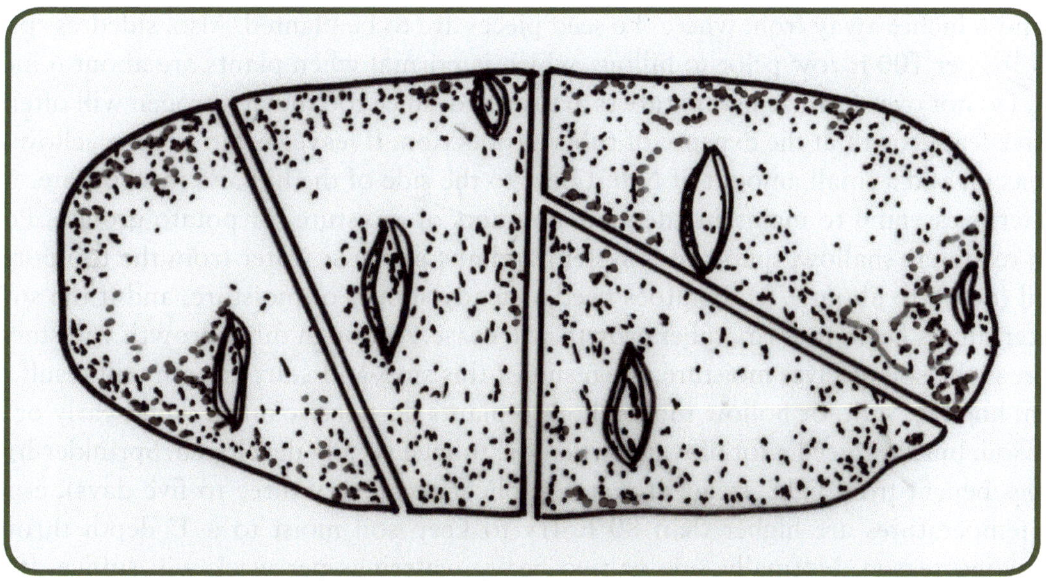

Figure 6.44 When cutting a potato tuber into seed pieces, try to make sure that each piece contains at least one eye.

from chlorophyll development in the tubers, the chemical **solanine** will also build up in tubers that receive sunlight and can be toxic if many green potatoes are eaten. Thus, when plants are about 6 inches tall, hill up garden soil around growing stems to increase tuber formation on the plant and to protect the tubers from sunlight.

Hilling potatoes once or twice during the season will greatly improve potato productivity. Potato hills can be formed at planting time by mounding soil around the seed piece, or they can be formed after the plants are growing. Most of the time, hills are formed a few weeks after the plants have emerged from the soil.

Soil is another factor that influences potato productivity in a garden. Although the potato is adapted to a wide variety of soils, potatoes grow best in a rich, loose, slightly acid soil that is well drained and has a high level of organic matter. Potatoes do not grow well in heavy or wet soils. Since potatoes are tubers and grow underground, they grow better and have an easier time expanding in loose, loamy soil compared to heavy, clay soils that tend to become compacted and prevent plant roots from obtaining adequate amounts of air, water, and space they require to grow. However, the application of organic matter in the form of compost or similar materials to heavy clay soils will over time improve the soil structure for better air- and water-holding capacity.

The recommended soil pH for potatoes is slightly acidic, about 4.8 to 5.4. If the soil is not acidic, **potato scab**, a devastating disease that causes brown corky tissue to develop on the potato skin surface, may develop. Gardeners should not normally apply lime to soils in which potatoes will be grown, and lime only if soil pH is below 4.8, which does not occur very often.

Fertilization is needed, as potatoes are heavy nutrient feeders, requiring large amounts of nitrogen fertilizer. Broadcast a complete fertilizer containing about 10% nitrogen at the rate of about 10 lbs per 100 ft of row before planting and incorporate into the top 6 inches of soil. Some gardeners prefer to place the fertilizer in bands below and slightly to the side of seed pieces at planting. If this method is used, place the fertilizer in an open trench 4 inches deep and 6 inches away from where the seed pieces are to be planted. Also, sidedress potatoes with 5 lbs per 100 ft row prior to hilling, which is normal when plants are about 6 inches in height. Do not over-fertilize before tubers are formed, since too much nitrogen will often cause excessive leaf growth at the expense of tuber production. If leaves begin to turn yellow before mid-season, add a small amount of N fertilizer to the side of the hill and incorporate.

Water is essential to maintain adequate amounts of moisture for potato growth. Potatoes have a relatively shallow, sparse root system and absorb most water from the top portion of the soil (down to about 1 ft). Potatoes need a steady supply of moisture, and if the soil dries out after tubers begin to form, tuber growth will cease. Although tuber growth will start again as soon as the soil receives moisture, the result of this stop-and-start growth will result in misshapen, knobby, split, or hollow tubers. Potato plants do not use much water early or late in the season, but they need a lot of water when the foliage is fully developed. Sprinkler-irrigated potatoes benefit from light, frequent water applications (every three to five days), especially when temperatures are higher than 80°F. Try to keep soil moist to a 1' depth throughout the growing season. Normally, one or two heavy waterings per week will suffice. It is also important to remember that extremely wet soils at planting will increase the possibility of seed-piece rot and decrease the numbers of potato plants that emerge from the soil in a garden.

Once the plants begin to yellow and the lower leaves start dying, reduce the irrigation rate prior to harvest. Too much water late in the season will oftentimes predispose the tubers to rot. The crop should be allowed to dry down to mature the tubers, to set the skins, and to reduce tuber damage (such as bruising) at harvest. The vines should be dead for one week prior to harvest. Withholding water late in the growing season will also help potatoes store better.

The **planting dates** utilized for growing potatoes is a key to their successful production. High soil temperatures (greater than 85°F) will prevent tuber formation. Thus, planting dates should be chosen to keep tubers from forming during hot weather. Thus, since the potato is a cool-season crop, it is planted in early spring or late summer. For spring production, it is recommended to plant about 4 weeks before the last killing frost, and for fall production, plant mid-July for an October harvest. Remember that fall production is really not possible in many warmer climates due to the problem of poor potato plant emergence at this time of year. The planted seed pieces will often rot in the soil when planted in mid- to late summer due to the high soil temperatures.

Pest problems can devastate a potato planting in a short period of time. Potatoes are susceptible to many insects and diseases. To reduce losses, use disease-free seed pieces, treat seed pieces with fungicides, rotate potatoes in a garden with other non-solanaceous vegetables, and use proper sanitation techniques in garden (such as removing infected or infested plant materials from the garden). A gardener should be prepared to control insects and diseases by chemical pesticide methods, whether organic or non-organic, if necessary.

Numerous weed, insect, and disease pests can afflict potatoes grown in a home garden. Potato plants should be examined every few days to detect and monitor pest outbreaks in a garden before the pest problem becomes excessive. Gardeners need to be aware that many pests can be controlled without using chemicals, especially if detected early in the garden. In most cases, disease control will not be necessary in the home garden.

Figure 6.45 Colorado potato beetle eggs (left), larva (middle), and adult (right) on potato.

Weeds can reduce yields by competing with potato plants for water, soil nutrients, and sunlight. The most critical period for weed competition for the potato plant is during early season growth and until plants reach about 12 inches in height. Although most weeds can be controlled by cultivation, hand-pulling, or hoeing, remember that once plants reach their full size, cultivation must be shallow to avoid pruning potato roots near the soil surface.

Insects can quickly defoliate a potato plant, transmit diseases, or cause irreversible damage to the edible tubers. Insects that are pests of potato plants in the home garden include the Colorado potato beetle, aphids, cutworms, and wireworms. But of all insect pests, the Colorado potato beetle is generally the most troublesome insect in the home garden since a few can defoliate a potato plant in a short period of time. Gardeners may first observe adult beetles feeding on foliage as soon as the potatoes emerge from the soil. The adult beetles overwinter in the soil and emerge in the spring to lay eggs on potato plants. Adult beetles have round, hard, cream to yellow-colored wings with 10 black stripes and lay bright orange-yellow eggs on the potato foliage (Fig. 6.45). The eggs hatch, producing larvae, which will do the most damage to potato plants, and larval feeding can defoliate potato plants. The larvae are brick red with two rows of black spots on each side (Fig. 6.45). To control the adult beetles, hand pick them from the potato plants before they lay their eggs on the underside of the leaves. Eggs can also be removed or destroyed prior to hatching. If the eggs hatch, pesticides such as *Bacillus thuringiensis* are available to kill the larvae (see Chapter 9 – Pest Management for Vegetables in the Home Garden or Chapter 10 – Organic Gardening).

Besides insect pests, there are numerous diseases that can be found on potatoes, but fortunately, most of them do not occur in the home garden. Many common seed-borne potato diseases can be avoided simply by purchasing and planting certified seed. However, diseases that home gardeners will frequently encounter include scab, early blight, late blight, *Fusarium* and *Verticillium* wilt diseases, and black scurf. Garden practices such as using certified potato seed, sanitation (removal of diseased plants from garden), and crop rotation are designed to prevent disease development from occurring.

Mulching is another practice that can be used in the home garden to provide a more conducive environment for potato growth and development (Fig. 6.46). Regardless of the mulch type, mulching potatoes will prevent the likely occurrence of solanine production—those green areas on potatoes—which is caused by potatoes being too close to the soil surface and getting sunlight exposure. Both organic and inorganic mulches can be used to enhance potato growth and development, but which to use depends on a particular gardener's needs and location.

Organic mulches can also be used to reduce soil temperatures during late spring/early summer, which will provide growing conditions more suitable for potato growth and tuber formation in those areas that have higher soil temperatures at this time of year. The most common way to use organic mulches for potatoes grown in the garden is to apply a thick layer of mulch over the soil surface once the last hilling process is completed. Mulches should also be placed among potato plants and on both sides of the hilled beds. Besides providing some weed control and conserving moisture, mulching potatoes in this manner for hot climates will also provide cooler conditions for potatoes. However, many gardeners choose to use organic mulches in a different way and apply a thick layer of organic mulch that is used

Figure 6.46 Potatoes mulched with wheat straw.

to grow tubers instead of growing them in the soil. The seed piece is planted either into the soil or laid into the bottom of the mulch layer. Instead of hilling, mulch is placed around the developing plants and new tubers form in the mulch layer. This method is just as effective as growing in hills and the potatoes are cleaner and there is no need to till the soil.

Plastic mulch can also be used in some gardens to grow potatoes more successfully. To obtain the best yields from potato plants, gardeners will often plant potatoes into plastic mulch. The use of black of mulch will warm the soil and allow potatoes to be planted earlier than normal in certain regions, increase productivity, and in some cases, maximize their productivity in less time. Plastic mulch benefits potato plants by retaining moisture and heat while preventing weeds from competing with potato plants.

Harvesting is the final step in the home garden potato production process. The best time to harvest tubers is after the foliage has died, but before the summer temperatures become really hot or before the first frost of the fall; this really depends on the planting dates used for potatoes. To harvest potatoes, gently dig under the plants with a shovel, spade, potato fork, or similar implement working carefully to avoid damaging the potatoes by cutting or puncturing the tubers in the process. Once under the plant with the digging implement, lift the entire plant and shake to remove the soil, and then pull potatoes from the vine. The smaller, younger tubers should be used immediately (called "new potatoes"), while the larger potatoes can be stored for later use. New potatoes can also be harvested from potato plants as soon as the potatoes flower. New potatoes are simply any potato harvested before full maturity, and are smaller, more tender than regular potatoes but will not store for long periods. Potatoes to be stored need to reach full maturity on the plant.

If potatoes are to be stored, it is best to not rinse them with water but to leave the few soil particles attached. To cure mature potatoes, store in the dark for a week at 70°F, as this will set the potato skin and heal minor tuber wounds. After curing, store potatoes between 35°F and 45°F, keeping the humidity fairly high. Mature potatoes can be kept for as long as six months, as long as you keep them in a cool, dark place that has enough humidity.

RHUBARB: *RHEUM RHABARBARUM*

Rhubarb is a cool-season, perennial vegetable in the **Polygonaceae** family that grows well in most areas of the United States and is often coveted by home gardeners. It is grown for its leafstalks (or petioles) that have a unique tangy taste that are used in making pies, tarts, and sauces. Besides the culinary uses of rhubarb, it makes a very beautiful garden plant, with huge, extravagant, lush green leaves with green, pink or red stalks (Fig. 6.47). Rhubarb should be planted in an area of the garden where it will not be disturbed since it can be productive for several years (five years or longer). Just a few plants will satisfy the culinary needs of this plant for an average-sized family. It is important to note that while the leaf petioles are edible, the leaves contain oxalic acid and should not be eaten.

Rhubarb will grow and produce in most soils, but it will grow best in fertile, well-drained soils that have good organic matter content. If the soil is carefully prepared, rhubarb will stay healthy and productive for many years. The planting area should be free of weeds, especially the more difficult-to-control perennial weeds. The ability to harvest rhubarb early in the season is more likely with a southern exposure. Since rhubarb is a perennial, it should be planted in an isolated part of the garden, so that it does not interfere with the planting and growing of other vegetables. The rhubarb plant is a great ornamental plant for the garden, due to its size, texture, and color. Some gardeners find it suitable to include in a perennial flower border. It is also important to note that rhubarb leaves can be added to compost piles. Although the leaves contain oxalic acid, leaves will rapidly decompose in the compost pile and have no negative effect on the quality of the compost.

Figure 6.47 Rhubarb plant growing in garden (left) and freshly harvested stalks with leaves (on right).

Cultural Requirements

Rhubarb is normally propagated by planting a division of a root from an older, healthy plant. It is best to plant or divide rhubarb roots in early spring while the plants are dormant. Rhubarb is generally purchased as crowns or divisions, rather than propagated from seed, and planting seeds is not recommended. Purchase rhubarb crowns from a local nursery, garden center, or from seed catalogs. Gardeners should try to plant the crowns as soon as possible before they dry out. Although rhubarb crowns do best if planted in early spring while roots are still dormant, it can also be planted in the fall once plants go dormant again.

At planting, gardeners should make sure that each root division has at least one strong bud. Old roots may be dug and divided to make new plantings, and in this case, cut roots into four to eight pieces. Many gardeners divide the older rhubarb plants and establish a new planting when the crown is about 6 to 10 years old, which will improve plant vigor and leaf size. Older plantings tend to become overcrowded and begin to lose vigor and productivity. It is best to dig roots of the most vigorous and healthy plants to establish a new bed as early in the spring as possible. To dig roots, insert a shovel about 6 inches into the soil next to the base of the plant and lift out the entire crown. Some roots will break off and should be left in the ground. Try to break the crown into fist-sized pieces using your hands, or some other tool such as a knife or hatchet, and replant the new divisions as soon as possible to prevent drying of roots. If planting the recently dug crowns is delayed for some reason, store them in the refrigerator, and rehydrate the divisions before planting by soaking in water for at least two hours or overnight, if possible.

In order for the rhubarb plant to become well established, leafstalks should not be harvested the first year after planting and only a few during the second year should be harvested. Newly planted rhubarb plants need all foliage possible to build a strong root system. Stalks may be harvested for 1 or 2 weeks during the second year, and for 8 to 10 weeks (a full harvest season) during the third and subsequent years. Thus, from the third and subsequent years, rhubarb can be harvested in late spring for 8 to 10 weeks. It is best to stop harvesting leafstalks when the plant begins to produce slender stalks, which is a sign that its food reserves are low.

Harvest during the fall only when rhubarb plants are to be removed the following season. To harvest, pull the leafstalks from the plant and trim off the leaf blades. The leaf blades contain large amounts of oxalic acid and should not be eaten. To keep the plants healthy, vigorous, and producing at a high level, remove only about one third of the leaves from a plant at any one time. In order to preserve enough foliage to sustain the crown, never harvest more than one third to one half of the plant's stalks. The stalks are most flavorful when fairly young, so harvest them soon after the leaf expands. Harvest by grasping each leafstalk near the base and pulling it slightly to one side; stalks can also be cut with a knife. After the last harvest in early July, the plants should be allowed to grow until killed by frost.

Good soil drainage is essential in producing rhubarb in the garden. Planting rhubarb in raised beds also ensures against crown rot, which can kill all rhubarb plants in a garden quickly. Thus, it is essential to place rhubarb in a well-drained area of the garden. Each plant will require approximately 10 to 15 ft^2 of space. Prior to planting, the soil should be loosened to a depth of about 1 ft. The addition of compost or composted manure into the bed before planting will also greatly increase production. So, for each plant, prepare the soil by adding 3 to 4 inches of compost or composted manure and a handful of a complete fertilizer that is relatively high in phosphorus and potassium, such as 5-10-10. Gardeners should plant roots with the crown bud about 2 inches below the soil surface and cover crowns with no more than an inch or two of soil. Planting rhubarb crowns too deeply will delay production. Press the soil firmly around the roots and water well. As soil and air temperatures begin to warm, new buds will push up through the soil. It is best to place roots about 3 ft apart in rows and at least 5 to 6 feet apart.

Once the plants begin to grow, the addition of a 3- to 4-inch layer of straw or similar mulching material should be added around plants to help control weeds and conserve soil moisture for plant growth and development. Organic mulches can be used to help moderate soil temperature and moisture, but do not use on poorly drained soils. Irrigation should be used especially during extended dry periods.

Flower stalks should be removed as they appear, since they will deplete reserves from the crown that supports vegetative growth. Rhubarb, like most vegetables, requires regular irrigation during dry weather conditions; try to maintain a moist but not soggy soil. Overwatering is very detrimental to a rhubarb planting in a garden. Annual applications of fertilizer should be applied to obtain optimal growth. To maintain a productive planting, a handful of a complete fertilizer (e.g., 10-10-10) should be applied to each rhubarb plant in the spring and also during midsummer.

Infertile soil, extreme heat or cold, drought, or long days that expose plants to too much light may cause bolting. Older plantings also tend to bolt more frequently than younger plantings. If seedstalks and flowers develop during the spring and summer, cut them from the base of the plant as soon as they appear. The petioles (leafstalks) are of the highest quality (maximum color, flavor, and tenderness) during the early spring, and they should be crisp and fairly thick at this time. Yield and quality are highest in petioles that have just reached full size and are harvested before any coarse fibers can develop. Medium-size stalks are generally more tender than larger ones, which are often stringy. For storage, trim and remove leaves from stalks. The freshly harvested stalks can be kept in the refrigerator, unwashed and wrapped tightly in plastic, for up to three weeks. Fresh rhubarb can be stored for two to four weeks at 32° to 36°F and 95% relative humidity.

The tops of rhubarb plants are usually killed in the first heavy freeze in the fall, but roots survive and produce new tops the following spring. After the tops are dead and the ground is frozen, plants can be covered with 2 to 3 inches of straw mulch. Gardeners should remove about one half of the mulch before growth starts in the spring. Rhubarb requires winter temperatures below 40°F to break dormancy and stimulate spring growth of leaf petioles. Summer temperatures averaging less than 75°F are needed to maintain vigorous vegetative growth.

Rhubarb injured by a hard frost or freeze in late spring can still be eaten, provided the stalks are firm and upright. However, if the stems are soft and spongy, they should not be eaten, since severe cold injury may cause the oxalic acid crystals in the leaves to move into the stalks, increasing the chance of poisoning. Cut and remove damaged stalks and place in compost, then allow new stalks to develop.

Pests are usually not a problem on rhubarb grown in the home garden. A gardener can cultivate shallowly around rhubarb plants as needed to remove weeds. For insect pests, the rhubarb curculio (a snout beetle about ¾ inch long) can cause serious damage by boring into the leafstalks, crowns, and roots of rhubarb plants. Curly dock is an alternate host for this insect, so plants of this weed near the garden should be removed. Under heavy infestations, treat the base of curculio-infested rhubarb plants with a recommended insecticide. Crown rot is a disease that can be a problem in commercial production, and can also be seen in the home garden. Symptoms of rhubarb crown rot include plants that fail to leaf out in the spring or plants that leaf out only to die abruptly. If diseased plants are dug, both roots and crowns are oftentimes rotted. The best control is prevention, such as planting in well-drained soils.

Rhubarb stalks grow in a spectrum of colors from red to green, while there are also speckled types as well. There are several varieties that are recommended for production and are generally separated into those that produce red or green petioles. Several varieties are discussed in further detail next.

'Canada Red' often produces shorter, more slender stalks than other varieties, but is tender and very sweet with good red color. It also tends to produce few seed stalks.

'Cherry Red' (also known as 'Cherry' or 'Early Cherry') has long, thick stalks that are a deep, rich red color inside and out. This vigorous producer is juicy, tender, and sweet.

'Crimson Red' (also called 'Crimson', 'Crimson Cherry', or 'Crimson Wine') produces tall, plump, brightly colored red stalks.

'German Wine' produces green petioles, similar to 'Victoria' but it is slightly more vigorous and typically has a darker pink speckling.

'MacDonald' is probably the most common variety available. It has large, brilliant red tender stalks and is highly productive, extremely vigorous, and wilt and root-rot resistant. It is excellent for use in canning, freezing, and making pies.

'Riverside Giant' is a cold-hardy, vigorous producer with large-diameter, long green stalks.

'Valentine' has broad, deep red stalks that retain a good rosy red color when cooked. This vigorous cultivar has much less acid than green stalked or other red varieties. It also produces few or no seed stalks.

'Victoria' is a speckled type that produces medium-sized stalks of excellent quality with long, round smooth ribs having good flavor. It develops pink speckling on a light green stalk with the pink color being more intense at the bottom of the stalk, fading to a solid green near the top.

SPINACH, BEET, AND SWISS CHARD

These vegetables all belong to the *Chenopodiaceae* (goosefoot) family. Two of these vegetables are leaf crops (spinach and Swiss chard), while the other (beet) is grown primarily for its root, although the foliage of this plant can also be eaten. Most vegetables in this family are fairly easy to grow if the guidelines outlined in this section are followed. All are popular spring and autumn plants for home gardens.

Spinach: *Spinacia oleracea*

Spinach is a cool-season crop and is a popular vegetable to use in fresh salads, and in recent years, it has been gaining in popularity with consumers primarily for this fresh use. Spinach is a plant that will reach edible maturity quickly (~30 to 45 days after seeding) and does best

Figure 6.48 Semi-savoy spinach growing in a garden in early spring.

during cool, moist weather conditions. During periods of warm weather and long days, it will produce flowers and seeds (bolt). Bolting in spinach is controlled by day length and influenced by temperatures. One of the biggest problems for home gardeners when growing spinach is the tendency of spinach to bolt, since bolting will stop the production of usable foliage. Bolt-resistant varieties should be used, especially during the spring growing season.

Spinach is a cold-hardy vegetable and can withstand hard frosts with accompanying temperatures as low as 20°F (Fig. 6.48). Thus, gardeners in many areas will often grow spinach in protected structures such as coldframes (see Chapter 11 – Gardening Aids) from the autumn until the spring. Harvesting fresh vegetables from the garden during this time of the year is a pleasure and not a chore, since there is little if anything being harvested from the garden at this time. From early spring until mid-May, successive plantings can be made every 2 weeks to supply a gardener's spinach needs for several months in most warm temperate climates. Several crops of spinach can usually be grown per year, with spinach grown in all except the coldest and hottest months. Gardeners should plant in the spring as early as possible to avoid the heat of the summer and make fall plantings about one month before the average date of first frost.

Spinach grows best on a well-drained, loamy soil high in organic matter, and often grows best in raised beds. Spinach is a heavy nutrient feeder; before planting, broadcast apply about 2 lbs of a complete fertilizer containing about 10% nitrogen per 100 ft^2 of row and work into the soil. Spinach will need to be sidedressed once a month thereafter with ¼ lb of a similar fertilizer material per 100 ft^2 of row. Spinach is sensitive to acidic soils, and gardeners who have acidic soils should apply 3 lbs lime per 100 ft^2 of row. However, if the pH is too high, leaves may develop a yellow color referred to as chlorosis, and a fertilizer that reduces the pH such as ammonium nitrate should be used. Spinach does not need as much water as most vegetables, and for best results, apply ½ to 1 inch of water per week, with less in the beginning when plants are small and more when the plants are actively growing.

Spinach plants are shallow-rooted and require adequate soil moisture. Plant spinach seeds 4 to 6 weeks before the last frost in spring and/or 6 to 8 weeks before the first frost in autumn. It is best to soak in water overnight prior to planting in the garden soil to soften seed coats and increase germination rates. Gardeners should direct-seed spinach at about a ½- to ¾-inch depth, with seedlings thinned to 4 to 6 inches between plants in the row and plants spaced in rows 1 to 1½ ft apart.

Under ideal growing conditions, spinach is ready to harvest in anywhere from 35 to 50 days from planting, depending on the variety grown. Spinach is ready to harvest as soon as leaves reach edible size, and must be harvested before leaves turn yellow or seed stalks develop. Harvest by cutting leaves off at the base when leaves are 4 to 6 inches in length. New leaves will regrow after the older leaves are harvested, which provides many additional harvests. The entire plant should be harvested from the garden at the first sign of bolting. To obtain the highest mineral content in the spinach leaves, harvest early in the morning. This time of day is when the highest mineral concentration occurs in leaves; minimum mineral content is found in leaves during in the late afternoon.

Insect and disease pests are typically not a problem on spinach. However, during certain years, pests are more problematic and can cause extensive injury to spinach. Important insect pests are aphids and leafminers, and common diseases observed on spinach are downy mildew, *Fusarium* wilt, and white rust. A gardener should plant disease-resistant spinach varieties if available.

There are different types of spinach based on the leaf morphology. Spinach varieties can be found that are savoy, semi-savoy, or flat-leaf types, and there are numerous cultivars of each type. **Savoy spinach** is productive and will tolerate cold-weather conditions better than most other types of spinach. It has deeply crinkled leaves and a low growth habit. Two suggested savoy varieties to grow in the home garden are 'Long Standing Bloomsdale' (42 days, large succulent leaves, heat tolerant, recommended for spring planting) and 'Samish' (45 days, dark green with a triangular shape, can be overwintered, but bolts with high temperatures, tolerant of white rust). **Semi-savoy spinach** varieties tend to be more upright with less-crinkled leaves than the savoy types (Fig. 6.48). They also tend to have better disease- and bolt-resistance, so this is usually the best type to grow at home. Two popular and widely grown varieties are 'Melody' (42 days, produces large, thick, dark green leaves on a large upright plant, highly disease resistant) and 'Tyee' (37 days, fast growing, high yielding, very slow bolting, heat and cold tolerant). **Flat-leaf**, also known as smooth-leaf spinach, can be distinguished by its smooth and broad leaves that are easier to rinse with water and clean compared to the crinkly savoy varieties. This is typically the type of spinach used in frozen and canned spinach, baby foods, soups, and other processed foods. Some of varieties of flat-leaf spinach that can be found for home gardens include 'Giant 157' (50 days, extra-large, smooth leaves, very productive, grows as spring or autumn crop) and 'Red Cardinal' (30 days, red veins in leaves and deep-red stems, bolts quicker than green-leaf types).

Figure 6.49 New Zealand spinach growing in a garden.

Spinach cannot be grown during the hot conditions of summer, but there are a few warm-season spinach alternatives that can be grown during hot weather. **New Zealand spinach** *(Tetragonia tetragonioides)* and **Malabar spinach** *(Basella alba)* are botanically unrelated to true spinach, but are good substitutes where summer heat limits the growth of true spinach.

New Zealand spinach is a hot-weather leafy green (warm-season vegetable) that will mature in about 70 days after planting (Fig. 6.49). This plant is not a true spinach, but the tender young shoot tips are used in similar ways as spinach would be used. New Zealand spinach produces small, tasty green leaves until frost. It grows rapidly during the summer, has many branches, and prefers a well-drained loamy soil that has high amounts of organic matter. Gardeners should plant New Zealand spinach once the soil has warmed and after the chance of frost has passed. It is best to soak seeds in water overnight before planting to improve germination. Other cultural practices are similar to spinach, although plants should be spaced 12 to 18 inches apart in the row.

Malabar spinach is a tropical, vining plant (warm-season vegetable) that grows best during the hot-weather conditions of summer. This plant is also easily grown from seed, and will make an attractive vine that should be trellised to prevent it from twining all over the soil surface. There are two leaf types, red and green, and individual leaves or the tender young shoot tips can be used as a hot-weather spinach substitute. Since it grows best with trellising, a technique often used by gardeners is to plant next to peas on a trellis and let them take over the trellis as the peas finish in late spring.

Beet: *Beta vulgaris* (Crassa group)

Homegrown beets are a tender and delicious treat best enjoyed fresh from the backyard garden. Beets can also be found at most local farmers markets, depending on the time of year (Fig. 6.50). They are another cool-season vegetable in the *Chenopodiaceae* family, and develop their best color and root quality when grown during cool conditions. Beets evolved from leafy plants without enlarged roots, and today, improved types of these plants are now known as Swiss chard. Although beets prefer cool weather conditions, they will tolerate a wide range of temperatures. Beets are easy to grow in almost any situation and every climate, except all but the heaviest soils. In most warm temperate climates, beets can be grown as a spring or fall crop. Beets grown during the fall can stand some mild freezing, but the roots must be removed from the ground in the fall before a hard freeze. Beets are not as cold tolerant as other cool season crops like broccoli, but they can tolerate a few light frosts. If grown during hot summer weather, plants may need misting to reduce plant temperatures.

The production of beets in a home garden works best if they are planted into loosened soil or in raised beds. Beets will grow best on well-tilled, well-drained, high organic matter soils. It is difficult to get a good stand in high-clay soils or those that tend to crust after a light rain. Beets will germinate best at soil temperatures of 50° to 85°F, but will germinate at temperatures between 40° and 95°F. Thus, beets have a wide range of temperatures in which germination will occur. Direct seeding is the best method for planting in the garden. Early beet plantings can be made 4 to 6 weeks before the last spring frost and fall plantings should be made about 8 to 10 weeks before the first expected frost. Plant seeds at ½-inch depths in loosened soil. Seed should germinate in about 7 to 10 days. Beet seedlings should be thinned to 4 to 6 inches in the row and 12 to 18 inches between rows. Beets can be planted in succession every 3 weeks, for a longer harvest. Successive crops can be planted until the weather begins to get hot in late spring.

The most difficult aspect of growing beets is oftentimes getting them established in the garden. They will often germinate poorly and once they do, seedlings often cluster together.

Figure 6.50 Beets are often sold in bunches, such as this bunch at a farmers' market.

A gardener should always use new, fresh beet seed to obtain the best results and soaking them in water overnight will improve germination.

Beets grow best at temperatures of 60° to 65°F. Cultural production practices for beets are similar to spinach and Swiss chard. Beets are relatively light nutrient feeders, with 2 lb of a complete fertilizer containing about 10% nitrogen broadcast applied at least 7 days before planting per 100 ft^2. About 4 to 5 weeks after seeding, one sidedressing of ¼ lb of a similar fertilizer material should be applied per 100 ft^2. It is also crucial to ensure that beets receive at least 1 inch of water per week, especially while they are actively growing. If watering is reduced and then later increased, splitting of roots will often occur, so be consistent in the amount of water applied.

Although beets are known as a root crop, all parts of the beet plant are edible. Tender beet greens can begin being harvested when thinning a row of beets. Leaves can also be eaten and are similar to the leaves produced by Swiss chard. It is best to harvest beets when they are fairly small, about 1 to 3 inches in diameter, as larger roots will be more woody and fibrous. Harvest dates differ between varieties, but normally beets are ready to harvest about 50 to 60 days after seeding. Harvest by gently pulling or digging plants. Beets are a great root cellar vegetable and can be stored for 3 to 4 months at near freezing temperatures with high humidity (>95%).

Most beets grown in home gardens are insect and disease free. The most common insects are leafminers and flea beetles, with carrot weevils, whiteflies, and wireworms causing sporadic damage. Some of the more common diseases include scab (corky spots on roots), damping-off of young seedlings, downy mildew, and leafspots (such as those caused by *Cercospora*). There are many beet varieties that a home vegetable gardener can grow. The most commonly known root beets are red, but golden and striped varieties are also available. Recommended varieties include 'Detroit Dark Red' (60 days, widely adapted, standard for many years in home gardens), 'Ruby Queen' (52 days, early, long harvest period), 'Golden' (55 days, yellow flesh, orange skin), 'Cylindra' (55 days, rich red cylindrical beet that is similar to carrot in shape), and 'Bull's Blood' (35 days: leaves, 60 days: roots; often grown for their reddish purplish leaves to add to salads).

Chard or Swiss Chard: *Beta vulgaris* (Cicla group)

Swiss chard is a close relative of beets and is grown for its large, edible leaves and stalks instead of its roots (Fig. 6.51). Swiss chard is a cool-season vegetable and grows best at 60° to 70°F. Although Swiss chard prefers cooler temperatures, a major advantage of growing this vegetable is its ability to grow at high summer temperatures while other cool-season leafy vegetables, such as spinach, mustard, and lettuce, bolt at these higher temperatures. A few plants of Swiss chard will provide a constant supply of greens throughout the spring, summer, and fall. It can also withstand a few light frosts.

Swiss chard grows best in well-drained, loose, high organic matter soils. The use of organic mulches and raised beds will increase production. Furthermore, mulching around plants will help to overwinter in temperate climate areas with mild winters. This vegetable will also tolerate partial-shade conditions. Swiss chard is a light nutrient feeder like beets and fertilizer guidelines are identical to beets. Swiss chard needs at least 1 inch of water per week from rain or irrigation, which is also similar to the water requirements for beets.

Swiss chard should be planted at the same time as beets, which is in the early spring and/or late summer. In the spring, plant 2 to 3 weeks before the last expected frost, and for fall production, plant seeds about 6 weeks prior to the first scheduled frost. Sow seeds about ½ to ¾ inch deep in rows that are 1½ ft apart, thin seedlings to about 12 inches apart. Seed can also be broadcast onto a tilled area and lightly covered with soil.

The large, crinkly leaves and fleshy stalks can be cut near the base of the plant, and new leaves will grow, which can provide a continuous harvest of leaves for several months. Even if all leaves are removed about 1 to 2 inches above the crowns on healthy plants, new leaves will appear. It is recommended to start a new planting for autumn production, as a spring planting of Swiss chard will look fairly ragged by the autumn and will not provide the productivity that many gardeners seek.

Swiss chard is relatively insect and disease free, but some pest outbreaks will occur. The pests are similar to those discussed for beets (refer to beets).

Swiss chard is a beautiful plant in its own right for a home garden and works well just as an ornamental plant. The dark red petioles of 'Ruby Red' or multi-colored petioles of 'Rainbow' or 'Bright Lights' can add a spectacular burst of color to an otherwise boring green vegetable garden. Thus, recommended varieties are 'Ruby Red' (59 days, red stem with green leaves), 'Bright Lights' (55 days, multicolored stems: white, pink, red, orange, yellow), and 'Rainbow'.

***Figure 6.51** Swiss chard growing in a home garden. Notice the brilliant red-colored stems that can add a magnificent splash of color to any garden!*

Chapter Seven

WARM-SEASON VEGETABLES FOR THE HOME GARDEN

The most important group of vegetables grown during the summer months is definitely the warm-season vegetables. Warm-season vegetables require lots of heat and at least 6 to 8 hours of full sun each day to grow sufficiently and to maintain their highest productivity. Warm-season vegetables require both warm soils and high temperatures to maximize their productivity. Thus, warm weather and frost-free nights bring the warm-season vegetable gardening season into full swing. Remember that warm-season vegetables are grown for their fruits rather than leaves or roots like cool-season vegetables.

Oftentimes in cooler climates, specific practices are used in the garden to provide a more favorable environment for warm-season vegetables to grow and develop, especially when temperatures are still cool in the springtime. Thus, black plastic mulch or rowcovers (see Chapter 12 – Season Extension Techniques for Vegetables Grown in the Home Garden) are often used to modify the temperatures near the warm-season vegetable plant to allow it to grow and develop at a faster rate. Many gardeners will protect warm-season vegetables seedlings from frost and cold air by using rowcovers, hot caps, or other insulation methods. This protection gives a gardener a head start on the gardening season. Soil temperatures of at least 70° to 85°F are required for warm-season vegetable seeds to germinate sufficiently. Oftentimes, the seeds of many warm-season vegetables are planted indoors and seedlings that develop from these seeds are planted into the garden soil to get an early start on the growing season. Hardened warm-season vegetable transplants can be planted through plastic mulch or directly in soil any time after the last average frost date.

There are many warm-season vegetables that are widely grown in home gardens during the summer, including cowpeas, cucumbers, eggplant, green and lima beans, melons, okra, peppers, pumpkin, squash, sweet potato, tomatoes, and watermelons. Many of these will grow under cool conditions, but will not generally produce fruit when night temperatures are in the 50°F range. They need warm temperatures to grow and mature, and once the fruit begins to form, it needs at least an additional month of frost-free weather. Warm-season vegetables are not suited for high-elevation gardens or for areas that have short growing seasons. Several warm-season vegetables will be discussed next in further detail regarding their production practices for the home garden.

CUCURBIT CROPS

Cucurbits refer to those plants in the *Cucurbitaceae* family and are also called "vine crops," and include many widely grown vegetables such as pumpkins, squash, cucumbers, cantaloupes, watermelons, and gourds. All vegetables in the cucurbit family are warm-season crops and will not tolerate any amount of frost. In home gardens, these are some of the most widely grown vegetable crops during the summer.

There are several practices that gardeners can do to prevent the occurrence of problems commonly associated with the growth and development of these vegetables in home gardens. These include variety selection, soils, crop rotation, fertilization, moisture, planting date, plant spacing, sunlight, pollination, pest control, and harvesting.

Variety Selection. The selection of a particular variety plays a major part in how well a cucurbit vegetable will grow and develop in a given area. For example, in certain regions that have cooler summers, gardeners can grow varieties that would perform poorly in warmer and more humid environments. It is best to try to locate varieties that are widely adapted or that have performed at a high level over numerous years at a particular location before they are selected to grow in a home garden. However, hybrid varieties are generally the best to grow since they typically produce higher yields and have a high level of resistance to certain diseases. Specific variety selection will be discussed under each particular cucurbit vegetable.

Soils. The cucurbits are not exacting in their soil requirements. Although most cucurbits prefer to be grown in well-drained soils, they will grow sufficiently on a wide variety of soils. Most cucurbits do, however, prefer a sandy loam soil with a high organic matter content. Sandy loams soils warm up quicker in the spring and are preferred to obtain an earlier maturing crop, but cucurbits can be grown on heavier soils if they are properly managed. Heavier soils (e.g., clays) have greater water-holding capacity, and can withstand drought conditions much better than lighter soils. If a soil is particularly clayey, organic matter such as compost or rotted manure can be added to improve soil structure. It is important to note that during fruit development, all cucurbits require low amounts of nitrogen and high levels of potassium. Gardeners need to be careful not to over-fertilize with nitrogen, as this will encourage a healthy vine and excess foliage growth at the expense of fruit development. It is always best to take a soil sample before fertilizing and to determine the amount of nutrients present in soil before fertilizing.

Soil pH is critical to the production of cucurbits, as they are very sensitive to acidic soils. They require a pH that is slightly acidic to neutral (6.0 to 6.8). To increase soil pH in highly acidic soils, gardeners should treat the soil with a liming material prior to planting.

Rotation. Rotation of garden spaces among different families is a major part of maintaining success in a garden. Diseases or insects that attack one cucurbit will more than likely injure most other cucurbits (e.g., anthracnose is caused by a fungal pathogen and is a disease of most cucurbits; cucumber beetle is an insect pest of most cucurbits). It is best not to grow cucurbits year after year in the same place in the garden, but to rotate to other areas of the garden on at least a 3- to 4-year cycle before returning to an area that was used to produce cucurbits.

Fertilizer. It is difficult to be specific about fertilizer recommendations because of great variation in soil types, soil fertility, and other soil conditions. However, it is generally recommended to initially broadcast approximately 3 lbs of a complete fertilizer containing about

10% nitrogen per 100 ft of row. Three weeks after planting and before vines tip over, a sidedress application is recommended with 1 lb of a similar fertilizer material used per 100 ft of row. If weather conditions are very hot at the sidedress application and a high-analysis ammonium fertilizer is used, it is important to make sure the sidedressed fertilizer is incorporated into the soil, otherwise fertilizer injury may occur to plants as ammonium is released.

Water. The watering of cucurbits should be slow and deep, since most cucurbits have shallow root systems. Deep watering encourages downward root growth that can offer some "protection" from drought. By maintaining an adequate moisture supply for cucurbit growth and development, yields will often increase by 50% or more. It is important to water at least twice a week during the summer months and to provide at least an inch of water a week to the growing plant. This is especially important since the fruit of many plants, such as cucumber, that do not receive adequate moisture during fruit development will develop a bitter flavor. The lack of water during pollination and fruit set can also lead to fruit abortion or misshapen cucurbit fruits. It is best to water early in the morning if possible to avoid disease problems, as diseases tend to be more of a problem if plants are watered in the late afternoon or evening. Gardeners should remember to keep soil moist for optimum cucurbit plant growth and yields.

The leaves of cucurbits will often wilt in midday, indicating lack of water, but this is not always an indication of a lack of soil moisture. Plants will often recover as the temperature becomes cooler.

Planting Date. Cucurbits are warm-season crops, and must be grown during the warmest part of the year. Some are relatively short-season crops, such as cucumbers and summer squash, which can be harvested in approximately 35 to 40 days after seeding. For these shorter-season crops, multiple crops can be grown during the summer months in those climates that have a relatively long summer. However, for later-season plantings, remember that disease and insect pressure will be higher during this time of year. It is also important that these crops are not planted too early in the growing season, since seeds will germinate slowly when soil temperatures are below 65°F. Also, at soil temperatures of 60°F or lower, seeds will often rot in the cold and wet soils. Plant seeds when the soil is warm enough to enhance germination, usually a week after the last average frost. Direct seeding is the best planting method, and plant seeds approximately 1 inch deep. Transplants can also be utilized for cucurbits, and this is used primarily to get an earlier crop. In seedless watermelon, transplants should be used since the seed has poor germination under field conditions. Home gardeners can direct-seed seedless watermelon, but should expect less than 20% germination under field conditions, even with newly purchased seed.

Seeding and Plant Spacing. Cucurbit seeds are relatively large, and should be covered to a depth of 1 to 3 inches. After planting, cover with soil and lightly tamp, but not so firmly as to create a crust. In lighter, sandy soils that tend to dry out rapidly, seeds should be planted at greater depths than in heavier soils. Remember that cucurbits are often referred to as vine crops and most need lots of space to grow, except for the determinate or bush types. Plant spacing is an important component in cucurbit production, and if plants are spaced too closely, then lower yields or smaller fruit will generally result. Those cucurbits harvested before maturity (e.g., cucumbers or summer squash) will have lower yields if spaced too closely and those harvested at maturity (e.g., pumpkins, watermelons, or muskmelons) will be smaller than normal if spaced too close. Thus, plant-to-plant competition in the row or between rows will result in lower yields and/or smaller fruit. Cucurbits vary in their spacing requirements and will be discussed under each cucurbit vegetable.

Sunlight. Sunlight is one of the most important factors to consider when growing vegetables and is probably the one that gardeners have the least control over. So, when growing vegetables, the gardener must consider the amount of light a given area in the garden will be receiving each day. All cucurbits require full-sun conditions, which means they need at least 8 hours of full sunlight each day to grow and develop properly. These plants do not tolerate growing under even partial-sun conditions. If cucurbits receive full morning sunlight, they can, however, tolerate receiving less than three hours of afternoon shade. Furthermore, if a garden is mostly shaded, then it is recommended to grow those crops that can tolerate those sunlight conditions, which tend to be more toward cool-season root or leaf crops.

Flowering and Pollination. Cucurbits typically have male (staminate) and female (pistillate) parts in separate flowers on each plant. Since male and female parts are in different flowers, pollen must be transferred from male to female flowers to achieve pollination and fruit set. This typically occurs with the help of honeybees and other bees, which harvest nectar or pollen from the flowers, and transfer pollen from one flower to another flower in the process. Thus, the cucurbits require cross-pollination for fruit set to occur in the garden.

The plant type based on flower morphology is important in the cucurbit family. Most plants in this family are **Monoecious**, meaning that they have male and female flowers on one plant (Fig. 7.1); however, some plants can be **Dioecious**, which means that only male or female flowers occur on each plant, and others that are more rare, can have **perfect** flowers, meaning that they have both male and female parts in each flower.

There are seven different plant types that can be found in the *Cucurbitaceae* family:

Gynoecious – only female flowers
Androecious – only male flowers
Hermaphroditic – only perfect flowers
Monoecious – male and female flowers
Gynomonoecious – female and perfect flowers
Andromonoecious – male and perfect flowers
Trimonoecious – male, female, and perfect flowers

Figure 7.1 Male and female flowers on a pumpkin plant. Notice the small pumpkin fruit below the female flower.

As previously mentioned, most cucurbits are **monoecious.** So, in most cucurbits, no fruit will set unless the pollen is transferred from male to female flowers. There are exceptions, such as some varieties of cucumbers and squash that are parthenocarpic (set fruit without pollination), but parthenocarpy is very rare.

Gardeners should observe the number of female and male blossoms found on their cucurbit plants in a garden. A female flower can be identified by checking the base of the flower; it is a swelling that resembles a miniature cucurbit fruit (e.g., squash or cucumber) with flowers attached (Fig. 7.2). Male flowers do not have this feature. Most cucurbit plants will typically produce only male flowers early in the season. So, no fruit development in the garden will occur until female flowers appear.

Figure 7.2 Female (left) and (male) flowers of a cucumber. Notice the small cucumber fruit below the female flower on the left.

Although a small fruit resembling a particular cucurbit will be attached to an open flower, it must be pollinated with pollen from a male flower to develop properly. If the female is not pollinated, it will shrivel and oftentimes fall to the ground; or if the female flower is pollinated poorly, it will be malformed (Fig. 7.4). Cucurbit flowers are viable for only about 6 to 8 hours during the morning to early afternoon on the day they open. The female is receptive for effective pollen deposition and pollination for only a short period of time. During cool or cloudy weather, the flower viability period can be extended a few hours, while on hot summer days, the time of flower viability is much shorter. However, in an active garden with lots of pollinators, most of the pollen has been deposited on female flowers within a couple of hours.

Figure 7.3 Bumblebee visitation to an open, receptive female pumpkin flower.

The effective pollination of cucurbits requires that the gardener keep an eye on these plants in the garden, and to observe pollinator activity and fruit development. A gardener should try to check plants early in the morning and observe whether or not any bees are visiting the flowers (Fig. 7.3). If there are lots of pollinators, a gardener can oftentimes just hear the buzzing of the bees as they walk through their garden. If no buzzing is heard, then there may be a lack of pollinators in the garden. Furthermore, if female blossoms drop off without developing into fruit, or developing fruit are malformed, then something may be need to be done to achieve effective pollination, such as hand-pollinating the plants.

A gardener should not worry when the first flowers fail to set fruit. Normally, male (staminate) flowers will be produced first. Some female flowers may open first before there are male flowers open to supply the pollen necessary for fruit set. These female flowers will dry up and abort. However, as pollen becomes available, female flowers will be pollinated and fruit will develop on the plant.

Pest control. Insect, disease, and weed pest-control issues will most likely have to be dealt with at some time in the garden when growing cucurbits. Similar to most other vegetables, the cucurbits are attacked by many different insect and disease pests. Scouting of plants in the garden for insects and diseases is a key to the successful control of these pests, although using practices to avoid insect or disease appearance in a garden is the best overall method to control these pests. However, once a pest is identified, then most gardeners will implement specific control measures, and this prevents the overuse of pesticides in the garden (or spraying by the calendar method), when their use is really not warranted.

Figure 7.4 Misshapen cucumber (on left) and cantaloupe (on right) fruits resulting from ineffective pollination.

Specific insects and diseases will be discussed under each cucurbit vegetable, but remember **scouting** and **identification** of the insects and/or diseases in the garden is the key to their effective control once they appear.

There are few registered **herbicides** for cucurbit vegetables, and really not many that a home gardener can use. So, if possible, select an area for planting in a garden where weed populations are low. It is important to start with a clean (or weedless) garden before planting, which can be achieved by tilling or spraying a non-selective herbicide, such as glyphosate, prior to planting. Otherwise, any emerged weed will have a significant advantage over the young cucurbit seedlings competing for water, nutrients, and sunlight.

Weeds can be removed by **cultivating** between rows with a rototiller or removing with a hoe. **Hand weeding** is really a good method of weed control for small areas such as the home garden, since many different vegetables are being grown and herbicide use is generally not warranted or feasible for a home garden. The use of **mulches** (either organic or inorganic) can be utilized to control weeds in the row, and they also provide other benefits as well.

Harvest. Cucurbit vegetables differ in whether they are harvested at **physiological immaturity** or **maturity**. Cucumbers and summer squash are harvested at physiological immaturity (seeds have not developed fully in harvested fruits), whereas, cucurbits such as watermelons, pumpkins, muskmelons, and winter squash are all harvested at physiological maturity (seeds have fully developed in harvested fruits). The timing of harvest differs with the specific cucurbit vegetable being grown, and more about harvesting is discussed under each cucurbit, as harvesting at the optimal time differs among the cucurbit vegetables.

CUCUMBERS (*CUCUMIS SATIVUS*)

Cucumbers are grown in most home vegetable gardens. This vegetable is relatively easy to grow, and will often produce fruit with little care. However, the fruit must be picked on a regular basis to ensure future production. The fruit have many different uses, and this contributes to its popularity.

The cucumber is a tender, warm-season vegetable that grows rapidly and generally requires substantial space in a garden. Cucumbers are vining plants that thrive on trellises, although compact varieties are more self-contained and do not require as much garden space. Thus, vertical training and new dwarf varieties can reduce the amount of space required to grow this crop in a garden.

Cucumbers are fast-growing, shallow-rooted plants and need regular watering, particularly after the fruit begins to develop. Because of this, cucumbers benefit from a nice layer of black plastic mulch to warm the soil and hold in moisture. Be sure the soil is well drained and rich in decomposed organic matter. While cucumbers can be harvested at any size, they become bitter as they turn yellow.

There are four primary types of cucumbers (pickling, slicing, beit alpha, and greenhouse - see Fig. 7.5), although there are an additional two types (Asian trellis and specialty) that are not as widely grown (Fig. 7.6). **Pickling cucumbers** have warty, light green skin interspersed with white specks, are blocky and short in shape, and have white or black spines. Spines are the small stickers that protrude from the warts of a cucumber skin. White-spine cucumber fruit turn cream-white when mature, whereas black-spine cucumbers turn yellowish-orange at maturity. White-spined types are preferred over black-spined types, since black spines can be easily seen and turn orange at maturity. **Slicing cucumbers** have a somewhat smooth skin surface with few warts, a thick solid dark green skin with a few white spines, and are elongated in shape—much longer than pickling types. **Beit Alpha cucumbers** are also known as the Middle Eastern slicing type of cucumber. It has smooth, solid light green, thin fruit skins, and there are no warts on the skin surface. **Greenhouse cucumbers** are solid medium-green in color, with few visible spines, and are approximately 1 to 1½ ft in length. Greenhouse cucumbers should not be grown outdoors since they are parthenocarpic and seedless. Insect pollination will cause seed formation in greenhouse types, making the fruit gourd-like at the blossom end. These are the plastic-wrapped cucumbers found in grocery stores, and marketed as bitter-free and easily digestible.

Asian trellis types of cucumbers are very long and sweet with thin skins. Their flavor is extremely mild and they are also burpless. A trellis is essential for Asian cucumber fruit to grow long and straight; otherwise they tend to curl when grown just on the soil surface.

Specialty types of cucumbers include those that do not fit into any of the other specific groupings. Many home gardeners like to grow these specialty types due to the unusual shapes and colors they can add to a garden. This includes the unusual cucumber varieties 'Lemon' and 'Armenian'. The 'Lemon' cucumber is andromonoecious, bright yellow, and has a round shape. They are excellent for salads and pickling, and have a clean, crisp taste and are never bitter. Although the 'Armenian' cucumber is actually a type of melon, *Cucumis melo* var. *flexuosus*, it produces a long, slender fruit that tastes like a cucumber and resembles a cucumber inside. It is also known by various names including the yardlong cucumber, snake cucumber, and snake melon.

Cucumbers are usually started by directly planting seeds in the garden soil. It is important to plant seed after the danger of frost has passed and the soil has warmed in the spring to at least 65° to 70°F. Warm soil is necessary for the germination of cucumber seeds and proper growth of plants. With ample soil moisture, cucumbers thrive in the warm-weather conditions of summer. A second planting for fall harvest may be made in mid- to late summer. Cucumbers can also

Figure 7.5 Four of the most common types of cucumbers: slicing cucumber (top), beit alpha (left-bottom), pickling (center-bottom), and greenhouse (right-bottom).

be transplanted into the garden to obtain early yields. It is best to sow two or three seeds in peat pots, peat pellets, or other containers about 3 to 4 weeks before the frost-free date in the spring, and thin to one plant per container at the 2-leaf stage. Do not allow transplants to get too large in containers or they will not transplant well. Seeds should be planted about 1 inch deep or transplants should be placed into the garden about 1 to 2 ft apart in rows placed 3 to 4 feet apart. To grow cucumbers in hills, grow 2 or 3 plants per hill with hills spaced 2 to 3 ft apart in rows and 4 ft between rows. The advantages of hill planting are reduced fertilizer usage and easier watering due to having less area to manage with hills compared to long row rows.

Cucumbers have a relatively short growing season, and a gardener can usually pick fruit about 40 to 45 days after seeding. So, they can be grown easily in almost every garden. Remember that by being a warm-season plant, the cucumber is very sensitive to frost. Cucumbers respond

Figure 7.6 Specialty types of cucumbers include the 'Lemon' (top left) and 'Armenian' (right), while the elongated Asian trellis types (bottom left) tend to curl if not grown on a trellis.

to mulching with soil-warming plastic in early spring or cooler organic materials in summer. Use of black plastic mulch warms the soil in the early season and can give significantly earlier yields, especially if combined with floating rowcovers.

Cucumber plants have shallow roots and require ample soil moisture at all stages of growth, and mulching helps with providing consistent moisture. During fruit set and development, adequate moisture becomes especially critical.

In cucumbers, gynoecious plant types are widely used, and these are plants that produce primarily female flowers. However, male flowers are needed to supply pollen, so usually a gynoecious variety is blended with a monoecious variety, so that pollen will be available to pollinate the female flowers. Generally, about 85% of a gynoecious variety is blended with 15% of a monoecious variety to achieve optimal fruit set. With this blend, gardeners will get more fruit production since most plants are producing only female flowers, which are the ones that will set fruit.

Determinate growth habit cucumbers are readily available to home gardeners, and are the preferred cucumber types to grow in containers where space is limited. These determinate types produce a smaller vine compared to the indeterminate (or normal cucumber) types. With the determinate types, the vine terminates with several female flowers. Most cucumbers, however, are termed indeterminate, and in small gardens, the vines may be trained on a trellis or fence to save space. When the long, burpless varieties are supported, the cucumbers hang free and develop straight fruits. However, areas with high winds can make vertical training impractical. Wire cages also can be used for supporting the plants. The longer fruited cucumber types (especially Asian slicing types such as 'Summer Top') do better when placed on a trellis. Cucumbers trained on a trellis take up very little ground space and produce more attractive fruit and fewer curled fruit (see Fig. 7.6).

Frequent harvesting of cucumbers is important. Keep the cucumber vines picked of all fruit as they reach usable size. Cucumbers suffer from **first fruit inhibition**. This is due to the partitioning of photosynthesis in the plant, as the first fruit that develops will divert most of the photosynthate for its development and leave little photosynthate for other fruits that started developing later. However, if that fruit is picked, then the photosynthate is then diverted to the next developing fruit and so on. So, even if one fruit is left on the plant to mature, the plant will stop setting new fruit. It is very important to harvest cucumbers 2 or 3 times per week to allow new fruit to develop, so higher yields can be achieved. The cucumber fruit grows rapidly to harvest size and should be picked at least every other day.

So, cucumbers can be picked really at any stage of development before the seeds become hard, since cucumbers are usually eaten when immature. The best size depends upon the use and variety. They may be picked when they are no more than 2 inches long for whole pickles, 4 to 6 inches long to make dill pickles, and 6 to 8 inches long for slicing varieties, or even 10 to 12 inches or larger for the large, burpless greenhouse or Asian trellis types. A cucumber is of highest quality when it is uniformly green, firm, and crisp. Do not allow cucumbers to turn yellow, as they will have hard seeds and oftentimes a bitter flavor as fruit become mature.

There are many important insects and diseases that are pests of cucumbers. Important insect pests include cucumber beetles, aphids, and pickleworms. Injurious insects include cucumber beetles and pickleworms. Gardeners should not handle, harvest, or work with cucumber plants when they are wet, which normally occurs in the early morning, as this will most likely spread many different foliar diseases throughout the planting. Important diseases include anthracnose, powdery mildew, downy mildew, mosaic, scab, bacterial wilt, and damping-off. Bacterial wilt is a disease transmitted by the cucumber beetle, so effective control of this insect pest prevents this disease from appearing in the garden.

There are numerous cucumber varieties to choose from. Most of the new cucumber hybrids are resistant to major diseases. A gardener must first decide on the type of cucumber to grow, and if a determinate or indeterminate variety is desired. Most home gardeners will grow either slicing or pickling types and several recommended varieties are provided next.

Slicing varieties:

'Ashley' (66 days; straight, slightly tapered fruit that are 7 to 8 inches long)

'Bush Crop' (55 days; 6- to 8-inch-long fruit on dwarf, bushy plants)

'Marketmore 76' (68 days; uniform, dark green, straight fruit, with multiple disease resistance)

'Poinsett 76' (70 days, open-pollinated; monoecious; 7 to 8 inches long; has multiple disease resistance)

'Salad Bush' (57 days; monoecious hybrid; uniform 8-inch dark green fruit on dwarf 24-inch-long vines; adapted to containers, hanging baskets, and small gardens; tolerant to a wide variety of diseases)

'Slicemaster' (61 days; gynoecious with multiple disease tolerance; dark green, and 8 to 9 inches long)

'Straight 8' (58 days; long-time gardener favorite; excellent flavor; evenly dark green fruit)

'Sweet Slice' (63 days; mild burpless and non-bitter; has multiple disease tolerance; 10 to 12 inches long).

Pickling varieties:

'Bush Pickle' (48 days; compact plant; good for containers or small-space gardens)

'Calypso' (55 days; gynoecious; multiple disease resistance; uniform dark green and blocky fruit)

'Carolina' (52 days; straight, blocky fruits with white spines; medium-sized plant with good vigor; high level of disease resistance)

MELONS (*CUCUMIS MELO*)

The term "melon" refers to cantaloupes, muskmelons, and winter melons. Melons typically require a great amount of space in a garden to grow, and are not recommended for small gardens. But melons are a welcome addition to most gardens, providing a 3 to 5 lb fruit that is both sweet and flavorful (Fig. 7.7). Melons are grouped into several categories, but the types generally grown are muskmelons, cantaloupes, and winter melons. The groupings of melons will be briefly described.

C. melo* group *agrestis are wild types having slender vines and small, inedible fruit. They are often just referred to as wild melons. These types are synonymous with *C. melo* var. *callosus* and *C. melo* var. *trigonus*.

C. melo* group *cantalupensis contain the cantaloupes or muskmelons. These are medium-size fruits that have heavy netting on the skin surface and easily separate from the vine at maturity. The fruit flesh is usually orange but can also be green, and is aromatic or has a musty smell. Melon plants in the group are usually andromonoecious. This group also includes the *C. melo* var. *reticulatus*, which was once separated out only for its thin reticulated rind and sweet green flesh.

C. melo* group *chito (Mango melon, vine peach) and ***C. melo* group *dudaim*** (Queen Anne's pocket melon) are very similar and distinguishing between these two groups is difficult. Fragrance is the key way to tell the difference between these groups, but certain accessions of *dudaim* lack fragrance, thereby eliminating the main distinction between it and *C. melo chito*. These are small, round to oval fruits, with white flesh and thin rinds.

C. melo* group *conomon contain melons known as pickling or sweet melons. They have small fruit with smooth, tender skin, white flesh, and usually with little sweetness or odor. However, some melons in this group have high sugar content when mature and are eaten like apples. Plants in this grouping are andromonoecious.

C. melo* group *flexuosus contain what is known as snake cucumber or snake melon. Fruits are long, slender, and often ribbed. They are used at an immature stage as an alternative to cucumber. Plants within this group are monoecious.

C. melo* group *inodorus is known as the winter melon group. They have smooth or wrinkled surfaces with flesh usually white or green and lack a musky odor. They are normally larger, later in maturity, and keep longer than melons in group cantalupensis, and do not separate from their stems at maturity. Melon plants in the group are usually also andromonoecious.

C. melo* group *momordica is commonly known as snap melon. The flesh is white or pale orange, low in sugar, and mealy. The smooth surface of the fruit cracks as maturity approaches and the fruit disintegrates when barely ripe. Most melons in this group are monoecious.

Although there are numerous melon groupings, the two groups that are normally grown in home gardens are *Cucumis melo* group *cantalupensis and* C. *melo* group *inodorus*.

***Cucumis melo* group *cantalupensis*.** This melon is the heavily netted cantaloupe sold in grocery stores, and is often referred to as a **muskmelon**. These melons emit a musky odor, and have a short shelf life. There are two types of muskmelons grown and they are commonly referred to as **western** and **eastern** types. **The western type** muskmelons are usually smaller in size, heavily netted on the fruit surface, do not have sutures on the melon surface, and are used primarily when shipping to distant locations (Fig. 7.8). **The eastern type** of muskmelons have some netting on the surface of the fruit, but not as much as that typically seen on western types, have **deep sutures on the melon surface**, and are sold primarily at local markets since these melons will not ship particularly well (Fig. 7.9).

Figure 7.7 Muskmelon plant vining out over the soil surface in a garden.

Muskmelons can be harvested at one of two stages; half-slip or full-slip. The major quality factor in melons, the soluble solids, is dependent on the sugar content of the fruit, which is definitely related to the harvest stage. A full-slip melon can have as high as 15% soluble solids, whereas melons harvested at the half-slip stage have soluble solids normally ranging from 8% to 12%.

Half-Slip Stage – The coloration of the melons begins to change from a greenish to yellow/orange color that covers 50% or less of the fruit. Although the fruit stem (or pedicle) will start to slip from the fruit at this stage, it does not completely detach from the fruit. A portion of the pedicle will remain attached to the fruit as it is being pulled from the vine. The sugars and flavors have not fully developed at this stage and will increase on the vine up to the full-slip stage. Fruits are often harvested at this stage for longer shelf life, as a melon picked at half-slip can be left at room temperature for several days to soften, develop more aroma, and improve flavor. However, melons will not produce more sugars once harvested, since there are no starch reserves to break down.

Figure 7.8 Western type of muskmelon.

Full-Slip Stage – Generally, the melon at this stage is completely ripe, with 90% or more of the fruit surface having a yellow/orange coloration. This is the harvest stage at which the pedicle separates cleanly from the fruit with little or no pulling. The pedicle is easily removed from the fruit when harvesting. Melons harvested at the full-slip stage have a very short shelf life.

Figure 7.9 'Superstar' muskmelon growing in a home garden on black plastic. This variety is an eastern type of muskmelon.

Many gardeners will also smell the cantaloupes at the stem to determine ripeness and sweetness. If the melons smell sweet and fruity, they are probably ripe, and a more intense smell signifies a sweeter melon.

It is important to remove field heat from muskmelons soon after

Figure 7.10 Several different types of winter melons: honeydew melon (top left), canary melon (top right), casaba melon (bottom left), and piel de sapo melon (bottom right).

harvest. The field heat can easily be removed by putting into a refrigerator or cooling by some other method. If they are not cooled soon after harvest, the melon will rapidly develop a decreased quality and reduced shelf life. A muskmelon planting can usually harvested over a period of 3 to 4 weeks. After this period of time, the vines will become non-productive as plants begin to die or become heavily diseased. After all melons have been harvested, remove vines, and compost stems, foliage, and rotten or immature melons.

***C. melo* group *inodorus*.** This is the winter melon group (Fig. 7.10). These melons ripen slowly, and have a longer shelf life than muskmelons. Winter melons include **honeydew** (skin is creamy white, smooth, hard with lime-green flesh), **Crenshaw** (football-shaped melons, with yellow and green skin having salmon-pink flesh), **canary** (melons have bright yellow smooth skin), **casaba** (skin is wrinkled, golden colored when mature, and the white flesh is sweet and juicy), **piel de sapo** (large elongated melons with a dark rind and interspersed green specks on skin surface, also known as toad or frog melon), and **Persian** (large melons that are round in shape having orange flesh, with a skin surface that is heavily netted with a green rind).

The ripeness of winter melons is difficult to determine because they do not "slip" from the vine when ripe like muskmelons. A gardener must have an idea of a winter melon's appearance when it is ripe; otherwise, it may be harvested too soon before sugar development is complete. For example, a honeydew melon skin changes from green to yellowish white as it ripens.

The flower characteristics of muskmelons and winter melons are different from most other cucurbit vegetables. Melons produce two flower types on plants: **perfect flowers** and **staminate flowers.** The perfect flowers (after pollination and fertilization) will develop into the fruit that we commonly think of as being a melon. This plant type is known as ***Andromonoecious.*** Honeybees or some other pollinating insect are required for pollination to achieve effective fruit set.

Muskmelons and winter melons are sensitive to frost conditions, and gardeners should not plant seeds until 1 to 2 weeks after the last scheduled frost date, unless they are being protected in some manner from cold-weather conditions. Melons do not tolerate cool temperatures or transplanting very well, so wait until the soil is warm before planting seeds. To warm the soil, black plastic mulch, floating rowcovers, or plastic tunnels can be used (see Chapter 12). Transplants can be started in individual containers like peat pots and moved to the garden shortly after the seeds germinate and the soil is warm. Seeds can be planted when soil temperatures are above 65°F, but the optimum temperature range for germination is between 70° and 90°F. Sow seeds at a depth of approximately 1 inch with rows spaced about 4 to 6 ft apart, having an in-row spacing of 18 to 24 inches. If plants are spaced too closely together, smaller melons will result.

Both muskmelons and winter melons require consistent moisture conditions during the growing season. At least 1 to 2 inches of water per week should be applied depending on local weather conditions, such as rainfall. However, gardeners need to reduce watering by about three-quarters of the normal amount at 2 weeks prior to harvest. Too much water will dilute the sugars in the melons at harvest, causing melons to have a bland flavor.

Muskmelons can be grown on a trellis, but the developing fruit must be supported with a sling. Otherwise, the weight of the developing fruit will cause it to remove itself from the vine and fall to the ground or pull the vine to the ground surface along with the fruit. The growth of vigorous vines can be controlled by pinching off the growing terminals once the melon crop has set.

Muskmelons do not cross-pollinate with cucumbers, squash, or watermelons, so off-flavor and poor quality are attributed to growing conditions (excess water while ripening, low soil fertility, and hot, cloudy weather).

There are several insects and diseases that are injurious to melons. Scouting and identification of important pests is necessary to obtain a high-quality crop of melons in most areas. Problematic insects in melon production include the cucumber beetle, pickleworm, squash bug, and squash vine borer. Powdery mildew, downy mildew, *Fusarium* wilt, anthracnose, gummy stem blight, and mosaic are important diseases.

The specific variety of melon chosen depends on whether muskmelons or winter melons are to be grown. Muskmelons will typically mature in 70 to 80 days from seeding, whereas winter melons will take much longer time to mature—usually 100 days or longer from seeding. Suggested varieties of muskmelon to grow in home gardens include 'Hales Best Jumbo' (83 days, older open-pollinated variety, 3 to 5 lb western type netted melons with firm flesh), 'Topmark' (90 days, tough skin, oval fruits at 3 to 4 lbs and about 5 inches in diameter, flesh is sweet, deep salmon colored with a small seed cavity), 'Superstar' (80 days, eastern type with deep sutures, will often produce 8- to 9-lb melons), and 'Athena' (85 days, 5- to 6-lb fruit, has faint sutures, and excellent foliar disease resistance). Remember that winter melons usually require an extra 2 to 3 weeks to mature compared to muskmelons. Suggested winter melon varieties include 'Honey Brew' (95 days, honey dew type with 4- to 5-lb fruit, has lime green flesh, and high disease resistance), 'Early Dew' (90 days, honeydew with lime-green flesh), 'Dorado' (92 days, canary melon with wrinkled bright yellow skin, cream-colored flesh, and high disease resistance), 'Golden Beauty' (95 days, yellow casaba melon, has wrinkled bright yellow skin, melons average 4–5 lbs), 'Canary Yellow' (95 days, produce 4- to 6-lb bright light-yellow skin that is slightly wrinkled with sweet, pale green flesh), 'Crenshaw' (115 days, fruits are pear-shaped, weigh 6 to 8 lbs, and have green to yellowish skin and salmon-pink flesh), 'Amarillo Oro' (100 days, very old casaba variety, pre-dating 1870, originated in Spain, golden yellow wrinkled skins with sweet, creamy green to slightly salmon-colored flesh), and 'Piel de Sapo' (100 days, fruits are oblong, 8 to 12 inches long and average 8 lbs, flesh is pale green to white in color and very sweet).

SUMMER SQUASH (*CUCURBITA PEPO*)

Summer squash is often grown in many home gardens since they require less growing space than most cucurbits. They do not spread their vines out over the ground like most other cucurbits, but typically grow more upright to some extent. Summer squash includes many different types, including crookneck, semi-crookneck, and straightneck yellow squash, cocozelle, scallop, vegetable marrow, and zucchini (Fig. 7.11). Summer squash is a short-season crop, and can be harvested 40 to 45 days after seeding for most varieties. Since it is a short-season vegetable, they can be grown in almost every garden, and multiple crops can be produced through the growing season in gardens located in warmer temperate climate areas.

Summer Squash Types:

Straightneck types – fruit are yellow, cylindrical with a short neck or constriction near the stem end, and are typically enlarged near the blossom end.

Crookneck types – fruit are yellow, with long, slightly to highly curved necks. Fruit of many varieties of this type tend to be warted as they reach maturity.

Scallop types – also referred to as the patty pan type. This squash has the most interesting shape among the summer squash. Fruit are flattened with scalloped margins, and can be white, green, or yellow in color.

Figure 7.11 Summer squash types: yellow crookneck (top left), yellow straightneck (top right), scallop (center left), vegetable marrow (center right), cocozelle (bottom left), and zucchini (bottom right).

Vegetable Marrow types – fruit are short, tapered, cylindrical in shape, narrow at the stem end, broad at the blossom end. Most are light green in color, although some can be dark green or striped. This squash is often prepared by slicing lengthwise, scooping out the pulp in the center, and replacing with rice or meat.

Cocozelle – fruit are long, cylindrical, slightly bulbous at the blossom end, tapered toward the stem end, light green to green in color, and often have slight ridges (or longitudinal striping) down the fruit. It is similar to zucchini except that it has striped colors or ridges on fruit.

Zucchini – fruit are uniformly cylindrical, with no taper to the stem end, and have light to dark green or yellow fruit. This is the probably most widely grown summer squash type.

Summer squash is a warm-season vegetable, and it should not be planted until the danger of frost is past. Most would recommend planting at least one week after the average last frost date, unless some types of cold protection method will be used. The optimal soil temperatures for germination is about 75° to 85°F, with little to no germination occurring below a soil temperature of 60°F. Seeds should be directly seeded into the soil about 1 to 1½ inches deep in rows placed about 4 ft wide with 2 ft spacings between plants. Plantings can be made at 2- to 3-week intervals during the spring to summer months to have a continuous supply of squash throughout the season. Summer squash plantings can be made up to 60 to 70 days before the last scheduled frost.

Figure 7.12 A summer squash plant infected with one or more cucurbit viruses. Notice the mosaic pattern on the leaves.

Most summer squash plants are monoecious, having both male and female flowers on the same plant. Similar to other cucurbits, honeybees or some other insect pollinator are required for pollination. Thus, it is important to observe flowering squash plants early in the mornings to make sure that there are sufficient pollinators visiting flowers. If bees are not readily noticeable, in the mid-morning, a gardener needs to think about alternative options to supplement the native honeybee populations.

Cucurbit viruses can be a major hindrance to squash production, especially during the late summer and fall growing season in some temperate climates. Several problematic viruses that can affect summer squash include cucumber mosaic virus (CMV), watermelon mosaic virus (WMV), papaya ringspot virus (PRSV), zucchini yellow mosaic virus (ZYMV), and squash mosaic virus (SqMV). Viruses cause what is commonly known as mosaic, which is seen as mottled, distorted foliage that results in plant stunting and loss of plant vigor (Fig. 7.12).

Fruit on plants infected by cucurbit viruses have a green flecking on the fruit surface or yellow fruit will become entirely green, and affected fruit may also develop a warty surface or knobby appearance (Fig. 7.13). This disease can be delayed by reducing aphid populations (since they vector the virus particles) by using an insect barrier cloth over the plants, or use of aluminum or white-painted plastic mulches. The bright color of the mulches tends to cause the aphids to become disoriented so that they are not able to locate squash plants. Another way to delay the onset of the mosaic disease is to use precocious yellow gene (*Py*) varieties. The *Py* gene in yellow squash masks the green symptoms on fruit associated with virus infection. Fruits from plants with the *Py* gene tend to remain yellow in color when infected with viruses compared to those without the gene, which turn green or have green flecking. However, the foliage of the *Py* varieties will still have the typical mosaic symptoms of virus infection.

Figure 7.13 Symptoms of cucurbit virus infection on summer squash.

In addition to those varieties containing the *Py* gene, improved squash varieties are available that have virus resistance. These virus-resistant transgenic varieties allow summer squash production to occur in areas that typically have severe problems with virus diseases. These are termed genetically modified organism (GMOs) and are not allowed for organic production.

Summer squash are harvested at physiological immaturity, similar to cucumbers. Summer squash should be harvested when fruit are young and tender. Seeds within fruit should be undeveloped with soft seed coats. Fruit should also be harvested continuously to prevent new flowers from aborting. Summer squash fruit are usually harvested at least every other day during the summer months. Zucchini and yellow types should be picked when about 1½ to 2 inches in diameter, and scallops when 3 to 4 inches across.

There are many insects and diseases that are problematic in the production of squash. Problem insects include the cucumber beetle, squash vine borer, squash bug, and pickleworm. Important diseases include the various mosaics caused by viruses, downy mildew, and powdery mildew. Insects and diseases are generally worse during the late summer to fall production season.

The choice of the squash variety to grow in a garden depends on the type preferred by the gardener. Most gardeners will grow yellow, scallop, or zucchini types in home gardens. Suggested yellow straightneck varieties are 'Goldbar' (47 days, very straight, uniform, golden yellow hybrid), 'Lemon Drop' (41 days, very early, creamy-yellow hybrid with smooth glossy skin), 'Multipik' (45 days, precocious yellow gene hybrid, very prolific fruit set that may reduce fruit size), 'Seneca Prolific' (45 days, creamy-yellow hybrid, shape and color are very uniform), 'Sunbar' (43 days, precocious yellow gene hybrid with cylindrical fruit), and 'Early Prolific Straightneck' (45 days, light yellow, cylindrical fruit). Recommended crookneck varieties include 'Dixie' (41 days, early variety, bright yellow fruit), 'Supersett' (45 days, precocious yellow gene hybrid), and 'Summer Crookneck' (46 days, old open-pollinated variety).

Scallop types are less commonly grown in home gardens than either yellow or zucchini types. Suggested scallop varieties to grow in home gardens include 'Early White Bush' (48 days, has flattened, uniformly white fruit with a tinge of green), 'Peter Pan' (52 days, excellent quality hybrid having light green fruit with good yield), and 'Scallopini' (52 days, compact bush grow habit, medium to dark green colored fruit, has a long continuous harvest). Zucchini squash is probably the most popular type of summer squash grown in home gardens. There are many zucchini varieties available to the home vegetable gardener. Recommended varieties include 'Elite' (52 days, dark green hybrid with uniform color, shape and length), 'Senator' (52 days, hybrid with cylindrical glossy medium green fruit that are 5 to 7 inches long), 'Spineless Beauty' (50 days, hybrid with cylindrical, medium-dark green fruit, that tend to be short on first pick, plants have spineless stems), and 'Goldrush' (53 days, fruit have golden yellow skin, with uniform shape).

WINTER SQUASH (*CUCURBITA PEPO, C. MAXIMA, C. ARGYROSPERMA, AND C. MOSCHATA*)

Winter squash is a group of vegetables that many gardeners avoid due to the excessive amounts of space that are required for plant growth and development. Although most varieties of winter squash tend to vine out all over the soil surface, there are several semi-vine and bush varieties that do work well in the small garden. Winter squashes include the commonly grown acorn, butternut, buttercup, spaghetti, and Hubbard, among several others, which are actually warm-season crops.

Winter squash are those squash that are harvested at a mature stage, whereas summer squash are those squash harvested at an immature stage. Winter squash develop a hard shell when fully mature and can be stored for several months. If winter squash is picked at an immature stage, they have a watery texture and poor taste. Flavor of winter squash is much better after cold-weather conditions since frost tends to increase the sugar content in the fruit. Most fruit turn either a brown, bronze, tan, or orange color when ripe.

Unlike their name, winter squash will not grow under cold conditions in a garden. The term "winter squash" refers not to the planting season, but to the fact that they can be stored for several months and eaten during the winter months. When fully ripened and properly stored, most varieties of winter squash will keep and still be edible at 4 to 6 months or longer after harvesting.

Winter squash is a very tender vegetable with regards to the temperatures that they can withstand. The seeds do not germinate in cold soils, and the seedlings are injured by even a very light frost. Gardeners should not plant until all danger of frost has passed and the soil is thoroughly warmed to at least 65° to 70°F.

Winter squashes and pumpkins are closely related vegetables. The common vegetables comprising winter squash and pumpkins are made of four species in the genus *Cucurbita*. These are *C. pepo*, *C. moschata*, *C. maxima*, and *C. argyrosperma*. Winter squash can be any one of these four species of *Cucurbita*, while pumpkins can be any of these, except *C. argyrosperma*. The genus *Cucurbita* is one of the most important genera in the *Cucurbitaceae* family, with the center of origin of most species in the genus in central or southern Mexico. The major exception to this is *C. maxima*, which is thought to be native to South America, most likely Peru.

The term "pumpkin" refers to edible *Cucurbita* fruit that are round or nearly round, whereas the term "squash" is normally used to describe edible *Cucurbita* fruit that deviates greatly from roundness. Pumpkins are almost always consumed mature, while squash are consumed at either an immature (summer types) or mature stage (winter types).

Winter squash is an important group and contains many different species that will be discussed next. The type of winter squash is followed by recommended varieties for the home gardener.

Cucurbita pepo includes a wide variety of many different types of winter squash (Fig. 7.14). Winter squash types in *C. pepo* are discussed next.

Acorn squash has an obovoid or conical shape with longitudinal grooves, and is pointed at the apex often resembling a spinning top. Suggested varieties are 'Table Ace' (75 days, semi-bush plant, dark black-green squash with bright orange flesh), 'Table Gold' (65 days, compact bush habit, attractive bright golden-yellow fruit), 'Table Queen' (80 days, standard dark green acorn type, flesh is a sweet golden yellow that turns more orange in storage), and 'Tay-Belle' (75 days, dark green 6- to 7-inch fruit produced on compact semi-bush plants with good resistance to powdery mildew).

Spaghetti squash derives its name from its unique flesh, as when cooked, the fibers in the cooked flesh resemble strands of spaghetti. Fruit tend to be oblong in shape and range either from ivory to yellow or orange in color. The flesh is bright yellow or orange. Suggested varieties include 'Tivoli' (90 days, compact and short vine habit, light yellow fruit color with

Figure 7.14 Various types of* Cucurbita pepo *winter squash: acorn (top left), spaghetti (top right), and delicata (bottom).

cream-colored flesh, 3 to 4 pounds), 'Orangetti' (85 days, semi-bush plant, fruit have thick, hard, golden-orange rind, flesh is also golden-orange and high in carotene), 'Stripetti' (95 days, hybrid of Spaghetti and Delicata types, produces 8- to 12-inch-long fruits with the noodle-like texture of spaghetti squash and the sweet buttery flavor of delicata, green-striped cream rinds turn orange when ripe), and 'Vegetable Spaghetti' (90 days, oblong fruits with hard, smooth, yellowish-tan rinds with string-like, golden yellow strands of flesh).

Delicata squash is a long cylindrical to oblong shaped squash with a cream-colored, green striped thick outer skin and a golden fine-textured inner flesh. The fruit size may range from 5 to 10 inches in length with an average weight of 1 to 2 lb. It is also commonly known as the sweet potato squash and really is a novelty squash. Suggested varieties include 'Delicata' (100 days, long cylindrical shape and has cream color with dark green stripes on hard rind, fine grained, light-orange flesh that tastes like sweet potatoes), 'Honey Boat' (90 days, shaped like Delicata, but with a copper-tannish colored skin with dark green stripes, very sweet flesh), and 'Sweet Dumpling' (100 days, small, 4"-diameter flattened round, fluted fruits average about 1 lb each; light cream to white background, with dark green stripes, very sweet, tender orange flesh).

Cucurbita moschata encompasses various cultivars of pumpkin and winter squash, and many common winter squash types that are grown in home gardens are in this species. *Cucurbita moschata* is better adapted to hot, humid climates than are C. *pepo* and C. *maxima*, and is also resistant to squash vine borer, an important pest of winter squashes and pumpkins. Compared to other winter squash, the foliage of C. moschata is slightly lobed, with white spots along the veins. The shape of winter squash fruits in the C. *moschata* species are usually

Figure 7.15 Butternut squash.

elongated or oblong and have a pale tan-colored skin rather than bright orange. The most common winter squash is this species is the butternut squash (Fig. 7.15). Suggested varieties include 'Waltham Butternut' (90 days, probably most widely grown butternut variety in home garden, produces uniform smooth 4- to 5-lb fruits with a light tan skin, flesh is yellowish orange with a nutty flavor and small seed cavity), 'Early Butternut' (85 days, semi-bush growth habit, 10- to 12-inch medium-sized blocky fruit, tan-colored with flesh that is dark orange, fine textured and sweet tasting), 'Butterbush' (75 days, space-saving bush-type plants only grow 3 ft long, fruit average 1 to 2 lb each and have a dark red-orange flesh), and 'Butter Boy' (80 days, hybrid variety produces 2 to 3 lb squash that are light orange in color, the tender flesh is bright orange and has a nutty and very sweet flavor).

Cucurbita maxima is a diverse species that contains many morphologically different types of winter squash (Fig. 7.16). The fruits of this species are very variable in color, shape, and texture. They can be smooth ribbed, typically round or oval, and oftentimes have a protuberance at the blossom end. Fruit colors can range from orange, green, blue, or gray. Although there are numerous types of winter squash in this group, only a few that are most often grown in home gardens will be discussed.

Ambercup squash is also commonly known as Red Kuri squash, but has other names including Uchiki Kuri, Baby Red Hubbard, and Orange Hokkaido (Fig. 7.16). It is a close relative of the buttercup squash and resembles a small pumpkin with bright orange skin. The bright orange flesh is edible and has a dry sweet taste. This winter squash has an extraordinarily long storage life. Suggested varieties include 'Red Kuri' (90 days, 5 lb bright orange teardrop shaped squash with smooth-textured flesh) and 'Sunshine' (95 days, bright orange 3 to 5 lb deep scarlet colored fruit that has a slightly flattened round shape).

Banana squash is a species of winter squash that has an elongated shape with a light blue or pink to orange skin and sweet, bright orange inner flesh (Fig. 7.16). The fruit shape is reminiscent of a large banana, and will grow up to two feet in length and about six inches in diameter. 'Pink Jumbo' and 'North Georgia Candy Roaster' are two varieties commonly grown in home gardens. 'Pink Jumbo' matures in 115 days, produces large almost cylindrical fruit (up to 50 lb or more) that have a slight taper at the blossom end, and flesh that is yellow-orange, dry, firm, not stringy, and sweet. 'North Georgia Candy Roaster' will mature in 100 days, has pink-colored banana-shaped fruits with blue tips that are up to 18 inches long and 6 inches wide, and smooth, delicious orange flesh.

Buttercup squash is a variety of winter squash with a tough roundish, greenish-colored outer skin with silvery stripes/spots and a heavy yellowish-orange inner flesh (Fig. 7.16). 'Buttercup' [95 days, one of the most commonly grown varieties and was the variety for which this winter squash type was named, has a turban shape (a flattish top and dark green skin), weighs three to five pounds, and has a heavy dense, yellow-orange flesh], and 'Sweet Mama' (85 days, semi-vining buttercup type; dark green, fruit weigh about 3 lb each, high-yielding compact vines are perfect for small gardens).

Hubbard squash is a bumpy or ribbed type of winter squash with round to oval fruits that have a blue-gray skin and taper at the ends with interior gold-colored flesh (Fig. 7.16). The extra-hard skins make them one of the best-keeping winter squashes. The fruit are very large and irregularly shaped, with a warty skin, and range from large to enormous. Like all winter squash, they have an inedible skin, large, fully developed seeds that must be scooped out, and a dense flesh. Hubbard squash can be successfully stored 6 months at 50° to 55°F and 70% relative humidity. Less rot will develop if stems are completely removed before storage. Hubbard squash and other dark-green-skinned squashes should not be stored near apples, as the ethylene from apples may cause the skin to turn orange-yellow in color. 'Blue Hubbard' is probably the most popular variety that has been grown for generations in gardens and can be found from multiple seed sources. It matures in 110 days, produces teardrop- or football-shaped fruit, weighs about 10 lbs each, and has sweet, fine-grained, golden flesh.

Kabocha Squash is also known as Ebisu, Delica, Hoka, Hokkaido, or Japanese Pumpkin (Fig. 7.16). Kabocha is the generic Japanese word for "squash," but refers most commonly to a squash of the buttercup type that produce rounded greenish, bluish, or orange-colored fruits without the button at the blossom end. Two common varieties are 'Black Forest' (95 days, smaller-sized kabocha with dark green, flat-round fruits that average 3 to 4 lb each, and has deep orange flesh that is medium dry and sweet) and 'Hokkori' (95 days, slightly flattened globe-shaped fruit have black-green skin and sweet-tasting orange-yellow flesh). Remember the primary difference between buttercup and kabocha is that the buttercup has a button at the blossom end.

Jarrahdale pumpkin is really an edible winter squash (Fig. 7.16). This Australian heirloom is grown for its unique blue-green ribbed skin and its deep, string-less orange-colored flesh, which is perfect for pies or for other culinary uses such as breads or muffins. It is also highly desired for autumn decorations due to the unusual bluish-green color and deep ribbing of fruits. 'Jarrahdale' is the primary variety available for home gardeners. This variety matures in 100 days and produces 6- to 10-lb heavily ribbed fruits that are flattened on top like pumpkins, but has sweet orange moist flesh.

Turban Squash is named for the resemblance of the fruit shape to a turban (Fig. 7.16). The turban squash fruit has a bright orange, red, green, and white rind. It has golden-yellow flesh and has a taste reminiscent of hazelnut. The fruit has a bulb-like cap swelling from its blossom end, and come in bizarre shapes with extravagant coloration that make them popular as autumn ornamentals. It is popular for centerpieces, and its top can also be sliced off so it can be hollowed and filled with soup. Its flesh and storage ability are comparable to the buttercup squash. The main variety grown by home gardeners is a pre-1800s heirloom called 'Turk's Turban' (100 days, very popular for decoration because of its unique and colorful fruit, the 5-lb fruit has a cream-colored "turban" that is colorfully striped with green, yellow, orange, and red).

Cucurbita argyrosperma, formerly known as ***Cucurbita mixta***, are winter squashes that are typically globose to ovoid fruit with whitish cream-colored skin having green-mottled striping patterns (Fig. 7.17). Unlike summer squash, the green-mottling patterns on the fruit skin surface are not caused by a mosaic virus. Fruits average about 10 to 12 lb and are about 18 inches long. It is commonly known as the cushaw squash or cushaw pumpkin. Similar to *C. moschata* in that it has high resistance to cucumber beetles and squash vine borers. The cushaw squash is not as popular as many of the other winter squash types, and it does not keep as well as many of the other winter squash. 'Green-Striped Cushaw' is the primary variety that is grown by home gardeners, and will mature in 110 days. This long, curvy-necked pumpkin is somewhat drought-tolerant.

All types of winter squash are fairly long-season crops and generally need a 100- to 120-day growing season for fruits to reach maturity before harvest. Fruit quality and storage life are

Figure 7.16 Various types of Cucurbita maxima winter squash: ambercup (top left), pink banana (top right), buttercup (top center left), Hubbard (top center right), kabocha (bottom center left), jarrahdale (bottom center right), and turban (bottom).

Figure 7.17 Green-striped cushaw squash. The brown spots on the fruit are not typical, and are most likely either the beginning of a fruit rot or attached garden soil.

reduced if the fruit are harvested before maturity, but fruit left in the field after several frosts also have a reduced quality. Winter squash are typically mature when the skin or rind resists thumbnail pressure.

Winter squash are normally direct-seeded into a garden soil at 1- to 2-inch depths. If growing in a short growing season, the gardener may need to transplant winter squash to provide enough growing days to reach maturity. Most winter squash are large spreading vines, but there are bush types that require less space (check variety description for bush, compact, dwarf, or semi-vining types if that type is desired before purchase). Large vining types need 6 to 8 ft between rows and 3 to 5 ft within rows, whereas bush or semi-vining types require only 3 to 5 ft between rows and 2 to 3 ft within rows. This means that vining plant types require between 20 and 40 ft^2 for each plant to grow properly, while bush types require about 6 to 15 ft^2 for each plant.

Pollination is also an important part of the production process with winter squash since bees or other pollinating insect activity is needed to achieve fruit set. Most winter squash have a monoecious flowering habit with male and female flowers in separate flowers on the same plant. Again, remember to look for bee activity in flowers during the mid-morning hours. Fruit abortion or misshapen fruit are more likely to occur under poor pollination conditions.

Winter squash can be harvested whenever the fruits have turned a deep, solid color and the rind is hard (resists thumbnail pressure). Most fruit will be harvested in September or October, before the occurrence of hard and heavy frosts. Try to cut squash from the vines carefully, leaving two inches of stem attached if possible, and avoid cutting or bruising fruit when harvesting or handling. Fruits that are not completely mature, have been injured in some manner, have had their stems damaged or removed, or have been subjected to heavy frosts in the garden do not keep for a long time and should be used as soon as possible. Store fruit in a dry building or basement where the temperature is between 50° and 55°F. For prolonged storage, it is best to place the fruits in a single layer so that they do not touch each other, as this arrangement minimizes the potential spread of rots.

Winter squash has similar insect and disease pests as summer squash. Important insect pests include the cucumber beetle, squash bug, squash vine borer, and pickleworm. Problematic diseases include anthracnose, downy mildew, powdery mildew, and various mosaics caused by cucurbit viruses.

PUMPKINS: (*CUCURBITA PEPO, C. MAXIMA, AND C. MOSCHATA*)

Many home vegetable gardeners are interested in pumpkins, as they are highly prized commodities during the autumn to be used as decorations, Halloween celebrations, or eaten in pies at the traditional Thanksgiving dinner. Pumpkins come in various sizes, shapes, and colors, from gigantic ones to those that are dark orange or white. Although most grow on large trailing vines, smaller pumpkin varieties can be found and trained to grow up trellises to save space in smaller gardens. Pumpkins are rarely eaten at an immature stage but are usually grown until they reach maturity, and at this time, the rinds harden, which allows them to be kept for several months.

The term "pumpkin" refers to a plant in the *Cucurbita* genus that produces round fruit, and pumpkins can be any of three *Cucurbita* species, namely *C. pepo*, *C. moschata*, and *C. maxima*.

Figure 7.18 Two Cucurbita maxima varieties of pumpkin 'Atlantic Giant' (top) and 'Prizewinner' (bottom).

Cucurbita maxima is the species that contains the largest pumpkins that have a bright orange fruit color (Fig. 7.18). These pumpkins

Figure 7.19 Ornamental* Cucurbita pepo *fruit (or jack-o'-lantern pumpkins) commonly used for autumn decorations.

are grown for large size and often reach more than 1,000 lbs. Suggested varieties are 'Atlantic Giant' (fruit can weigh 600 lbs or more), 'Big Max' (fruit in the 30- to 50-lb range), 'Mammoth Gold' (fruit average 25 lbs), 'Big Moon' (fruit can weigh 200 lbs or more), and 'Prizewinner' (newer variety that has good shape and color, fruit can weigh in the 200 lb range). All of the large pumpkin varieties listed require at least 110 to 120 days from seeding to maturity.

Cucurbita pepo is a diverse group and this species contains the jack-o'-lantern pumpkin types, as well as the miniature and other small ornamental pumpkins that are used for autumn and Halloween decorations (Fig. 7.19). Thus, most are for decoration, and should not be eaten. These types have poor flesh quality including being excessively stringy and not smooth like other pumpkin species. Suggested varieties include 'Howden' (105 days, old standard in the pumpkin industry, has nice handle, shape, and fruit color with fruit in the 20 lb range), 'Connecticut Field' (110 days, popular variety for home gardeners, 15 to 20 lb fruits are bright orange, slightly ribbed, but do vary in shape and size), 'Autumn Gold' (90 days, has precocious yellow gene to help mask virus-infected fruit, fruit size in the 10-lb range), 'Spirit' (100 days, semi-bush plant type, fruit size in the 10- to 12-lb range), 'Trick or Treat' (105 days, semi-bush plant type, produces 10- to 12-lb fruits, three way variety as it is used for pies, carving, or seeds), 'Small Sugar' (100 days, popular garden variety for pies or canning, fruit average 3 to 4 lbs), 'Baby Bear' (105 days, has good disease tolerance, miniature type with fruits averaging less than 1 lb), and 'Baby Boo' (100 days, miniature type, produces white fruit in the ¼ lb range). There are many other pumpkin cultivars that can be grown in a home garden, and this is only a suggested list from which to begin.

Figure 7.20 'Long Island Cheese' pumpkin.

Cucurbita moschata encompasses the so-called cheese pumpkins, with their flat, ribbed, buff-colored skins (Fig. 7.20). The processing pumpkins are mostly in this group, and they tend to make poor jack-o'-lanterns. These pumpkins are primarily for eating purposes, since their flesh is smooth and very amenable for cooking. Recommended varieties include 'Buckskin' (115 days, buff-colored skin, fruit average 10 to 12 lbs), 'Kentucky Field' (110 days, buff-colored skin, good for making pies, 12- to 15-lb fruit), and 'Long Island Cheese' (105 days, lightly ribbed, medium-sized, flattened 6- to 10-lb pumpkin that resembles a wheel of cheese, rinds are smooth and tan colored, with a flesh that is medium sweet and deep orange).

Being a cucurbit, pumpkins are warm-season crops and will not tolerate any frost. Pumpkins need a long frost-free growing season of approximately 100 to 120 days. Seeds are large and should be planted about 2 inches deep after the soil has warmed to at least 65° to 70°F. Pumpkins planted too early in a long growing season may cause problems for the gardener; oftentimes, pumpkins will need to be harvested too early and will not store until the autumn when they are typically used.

Spacing of pumpkins depends on the plant type grown, which relates directly to the length of the vine. There are large vining types as well as dwarf (semi-bush) vining types just like for winter squash. For the large vining types, rows should be spaced 6 to 8 ft apart and 3 to 5 ft in the row, and for the semi-bush types, space plants 3 to 5 ft between rows and 2 to 3 ft in the row. Again similar to winter squash, this means that vining plant types require between 20 and 40 ft^2 for each plant to grow properly, while bush types require about 6 to 15 ft^2 for each plant. To produce the large *C. maxima* pumpkins, a gardener must have lots of area in which to grow the plant (e.g., 70 to 150 ft^2 for each plant).

Growing a large pumpkin. First, a large-fruited *C. maxima* variety such as 'Big Max', 'Atlantic Giant', or 'Prizewinner' must be purchased and planted. It is important to continuously remove all female flowers and fruits that set except for one, as this will allow all the energy made by the plant to go into one fruit and allow it to obtain the greatest size possible. Adequate space is essential (70 to 150 ft^2 per plant) so that the plant has no competition for moisture, sunlight, or nutrients. Furthermore, sufficient nutrients and moisture must be provided to the plant to ensure that it will maximize fruit size.

Pumpkins have monoecious flowering habits and require bees for adequate pollination. If pumpkin flowers are poorly pollinated in a garden, flowers will abort or fruit will be malformed or misshapen. Proper pollination is essential to maximize fruit size.

Pumpkins should be harvested after the vines die down during the autumn months, or the fruit have developed a deep orange color. Pumpkin skins will harden with gradual cooling, but a sudden freezing will damage the skin and shorten the shelf life. Pumpkins will not become overripe, so it is best to leave them on the vine as long as possible. Avoid bruising or scratching fruit during harvest, as these areas will provide a means of entry into the fruit for organisms that will cause fruit decay. It is best to leave 4 to 6 inches of stem attached to the fruit as a handle; then store it in a cool, dry location.

Insect and disease problems on pumpkins are similar to those of winter squash. So, refer to those discussed for winter squash. Insect and disease control is a necessity when growing pumpkins. If these pests are not controlled, the vines will often not survive to produce fruit.

GOURDS: (*CUCURBITA PEPO, C. MAXIMA, LAGENARIA SPP., AND LUFFA SPP.*)

Many gardeners are interested in gourds because of their strange shapes and colors, as well as the additional benefit that some are also edible. Most gardeners will use gourds as decoration or novelty items. However, in many developing countries, some are used as household utensils.

Gourds are extremely sensitive to frost conditions, and most will not grow below 60°F. All gourds are relatively fast growers, especially during the heat of the summer. Gourds tend to thrive under conditions where temperatures average between 80° and 95°F. Gourds will grow in almost any garden that has well-drained soil with adequate levels of organic matter.

It is important that seeds are not planted until the soil has warmed to 65° to 70°F and should not be planted until after the danger of frost has passed. Seeds are typically placed 2 inches deep in the soil, with seedlings thinned to about 2 to 3 ft apart in the row and 4 to 6 ft apart between rows, providing about 8 to 20 ft^2 per plant. Many home gardeners will often train the gourd vines on some sort of trellis, and gourds really do best if supported on trellis to get the best fruit length and shape. Otherwise, gourds are similar to other cucurbits with respect to their culture, growth, and maintenance.

There are many types of gourds in several different genera. The more common gourds are found in several different species, including *Cucurbita pepo*, *C. maxima*, *Lagenaria* spp., and *Luffa* spp. The various *Cucurbita* gourds have yellow flowers, while flowers of *Lagenaria* gourds are white. Those that most home gardeners are familiar with are the various types of *C. pepo* gourds (Figs. 7.21 and 7.22). The fruit of these gourds will be ready for harvest in approximately 90 days from seeding, and will usually be brightly colored with many being multi-colored, and having hard, glossy rinds. 'Crown of Thorns', 'Small Spoon', 'Apple', 'Pear', 'Spoon', 'Nestegg', 'Small Orange', and 'Yellow-Warted' are types of *C. pepo* gourds. 'Turk's Turban' is a widely grown *C. maxima* type of gourd, and is an orange-red winter squash used mostly for decoration but is also edible. Various species of *Lagenaria* produce the 'Bottle' and 'Birdhouse' gourds that mature in approximately 125 days from seeding (Fig. 7.23). *Luffa* gourds are often referred to as 'Chinese Okra', 'Vegetable Sponge', 'Sponge Gourd', or 'Dishrag Gourd' (Fig. 7.23). Most *Luffa* gourds are grown primarily to use the fibrous interior flesh as a sponge. However, the ***Luffa*** gourds must grow to maturity to be used as sponges, and need approximately 200 days to mature from seed. *Lagenaria* and *Luffa* gourds can both be eaten when immature. *Lagenaria* fruit can be harvested about a week after they flower and can be eaten like squash. The fruit of *Luffa* gourds is eaten like a cucumber or cooked in stews when the fruits are less than four inches long.

Figure 7.21 A yellow-warted Cucurbita pepo gourd hanging from a trellis.

Cucurbita gourds tend to be fleshy at harvest and can be easily damaged when subjected to frost or freezing conditions. They should be harvested once they reach maturity and before weather conditions become cold. These gourds can be difficult to cure and must be kept in a cool, dry place for several weeks to cure.

Figure 7.22 Various Cucurbita pepo gourds found for sale in a market during the autumn.

***Figure 7.23** Lagenaria (left) and Luffa spp. (right) gourds hanging from a trellis.*

The various *Lagenaria* gourds are not as easily damaged by light frosts and can be harvested at a later date in the autumn when mature as compared to *Cucurbita* gourds. *Luffa* gourds are harvested when the fruits turn brown, which can be after the first hard frosts.

During curing of *Cucurbita* and *Lagenaria* gourds, they should be harvested only at the mature stage and washed. Once washing is complete, the gourds should be placed in a dry area with good air circulation and out of direct sunlight. Try to place gourds so that they are not touching each other. After the curing and hardening process is complete, the gourds can be waxed, painted, or decorated to an individual's taste. *Lagenaria* gourds usually take an excessive amount of time (often several months) to fully harden off and to become completely dry.

WATERMELON: (*CITRULLUS LANATUS*)

Watermelons are a traditional summertime favorite (Fig. 7.24). These juicy, crunchy, and sweet fruits are an important crop grown in many home gardens. Watermelons are a warm-season crop, and normally require 3 to 4 months of a frost-free period to develop fruit, depending upon the variety grown. Similar to other cucurbits, watermelons need warm weather, and lots of water, sunlight, and space to grow properly. However, even if a climate lacks warmth—that is, a short-summer region with fewer than 90 growing days, or if the garden

***Figure 7.24** There are not many things in life better than freshly cut watermelons in the summer.*

lacks space—delicious, sweet watermelons can still be grown successfully in a home garden. There are short-season varieties that can be grown and season-extension methods such as hot caps, floating rowcovers, and black mulches that may prove useful for earlier production. There are also a few bush or compact varieties available that are ideal for smaller gardens.

Watermelon vines require a lot of space to grow, and a gardener should allow about 7 to 10 ft between rows, with seed sown 1 inch deep in hills spaced 6 ft apart. It is best to plant seeds once all danger of frost is past. Watermelon seeds also require minimum soil temperatures of 70°F for germination, but 75° to 85°F is needed to obtain optimal germination rates. To improve earliness or reduce the amount of time required to harvest watermelons, start the seed inside about three weeks before they are to be set out in the garden. It is best to plant two or more seeds in 3-inch-deep pots or peat pots, and then thin to the best two plants. Also, make sure to not start the plants inside too early, as large watermelon plants transplant poorly in the garden. Seedless varieties must be started indoors. Planting seedless watermelon seeds inside at warm temperatures ensure their germination, since soil temperatures between 80° and 85°F are required for optimum germination rates.

In the garden, grow two or three watermelon plants per hill. Also, be aware that if seedless watermelons are grown, a few plants of a standard seeded variety must be grown in close proximity as a pollen source for the seedless melons. The seedless melon varieties do not have the fertile pollen necessary to pollinate and set fruit. It is generally recommended that one seeded watermelon plant be planted for every 4 or 5 seedless watermelon plants since this ensures that sufficient pollination and fruit set occurs. So, if a gardener has enough space for 30 watermelon plants in a garden, at least 4 or 5 must be seeded and these must be arranged so that they are in close proximity to each of the seedless watermelon plants so that they can receive pollen to achieve effective fruit set.

Figure 7.25 'Yellow Mini Tiger' seedless watermelon growing in garden and ready to harvest.

Seedless Watermelons are becoming popular with home gardeners, since most people prefer eating watermelons without seeds. Seedless watermelons are all hybrids derived from crossing a 4x (or tetraploid) female flower with a 2x (diploid) male flower and achieving seeds that are 3x (triploid). The 3x plants are usually weaker than 2x watermelons and require more attention, but the potential rewards of eating watermelons right out of the garden without seeds are well worth the effort. However, the seeds of triploids are quite expensive, usually averaging about 25¢/seed. As briefly discussed previously, a seeded (or diploid) watermelon variety must be interplanted at about one fourth to one fifth of the total planting to supply pollen necessary for pollination and fruit set, since triploid (or seedless) watermelons produce no viable pollen. There are many varieties to choose from for the home garden, including 'Solitaire', 'Harvest Moon', 'Charleston Jr.', 'Everglade', 'Orange Crisp', 'Yellow Mini Tiger' (Fig. 7.25), and 'Triple Crown'. Before choosing the variety to grow, try to find which ones do best in your area.

Poor pollination and fruit set can result in low yields. Watermelon is similar to other cucurbits, as it is monoecious, and has male and female flowers. Honeybees and other bees transfer the pollen from the male to female flowers. The health and age of the plant and the number of melons that have already set determine the number of female flowers that will set fruit later. Therefore, pruning misshapen melons off the plant will encourage additional fruit set at a later time. Cold, rainy, and windy weather reduces bee activity, which can cause poor melon set due to inadequate pollination. When flowers are open on the plants, do not use insecticides, such as carbaryl, since they are extremely toxic to bees and will reduce your insect pollinator population.

Watermelons grow best on a deep sandy loam soil high in organic matter, well drained, and slightly acidic (pH 6.2 to 6.8). Sandy loam soils are preferred for growing watermelons because sandy soils generally warm faster in the spring, are easier to cultivate plant, and allow deep root penetration (watermelon is one of the deepest rooted of all vegetables). When planted on very heavy soils, the plants develop slowly and fruit size and quality are usually less than those planted at better soil sites. Yields on clay soils can be increased significantly by mulching with black plastic film to conserve moisture. Fine sands produce the highest-quality melons when adequate fertilizer and water are provided. Frequent watering is essential to obtain the greatest possible yields, and yields will be poor if moisture is not maintained to a watermelon planting, especially during fruit developments. Irrigation will definitely improve yields.

The number of days to harvest really depends on the watermelon variety grown. So, it is best to become familiar with the variety being grown to determine when it will be ready for harvesting. The typical number of days required to maturity is between 70 and 120 days. The proper time to harvest watermelons is difficult to determine, as there is no easy way to tell when a watermelon is ripe. Ripeness in watermelon is often difficult to determine. The flesh of a typical red watermelon will change from an immature pink to red-ripe, and then to overripe within a 10- to 14-day period. Overripe watermelon fruits are a lower quality, having a watery, mushy texture and lower sugars.

There are several ways to determine if a watermelon is ripe without having to cut it open. These ripeness indicators are often used to determine if a watermelon is ripe. Sometimes these ripeness indicators work, and other times they do not. **Ripeness Indicators: Tendrils** – A dead tendril or curl near the point where the fruit is attached to the vine is used by some as an indicator that the fruit is ready for harvest; **Underside of Watermelon** – The belly (or underside) of the watermelon that rests on the ground (often referred to as the ground spot color) turns from greenish white to pale yellow and many say that this is the best ripeness indicator; **Dull Appearance** – The watermelon tends to lose the powdery or slick appearance of the top surface (referred to as the bloom) and develops a more dull appearance; **Sound** – If a watermelon is thumped with a finger, a ping or high-pitched sound indicates that the fruit is green, whereas a dull, or dead low-pitched sound is heard if the fruit is ripe. This indicator is often unreliable, since a low-pitched dead sound also occurs in fruit that are over ripe. If you have just a few fruit, it is probably wise to include all of the above when making your decision to harvest. However, the only way to make sure that a watermelon is ripe 100% of the time is to cut it open.

Many different insects and diseases will attack watermelons with certain ones more problematic than others. The cucumber beetle can be a troublesome pest of younger plants and can also scar fruit by feeding on the fruit surface. To reduce the risk of diseases, do not plant in an area of the garden where vine crops have been grown during the past four years. Also, avoid low, damp areas in a garden or pockets where cool air may collect. Many diseases are problematic to watermelons with *Fusarium* wilt, anthracnose, and gummy stem blight usually being the most troublesome.

There are numerous watermelon varieties to grow in a home garden. A few seedless varieties were already suggested, so next is a listing of seeded varieties for a home garden.

'Bush Charleston Gray' – 90 days; small, oblong fruit similar to 'Charleston Gray' except for the smaller size of fruit and vine; fruit weigh about 10 to 12 lb each, and have a gray-green rind and deep red, fine-textured flesh; ideal for small gardens.

'Bush Jubilee' – 90 days; small, oblong fruit up to 24 inches in length and weigh 25 to 35 lb; fruit have light-green rind with dark green stripes and bright red flesh with firm texture; ideal for small gardens.

'Bush Sugar Baby' – 75 to 80 days; small, round fruit; weighs 12 pounds; dark green rind with no stripes; red flesh. Vine spreads just 3 to 3½ feet, so it is ideal for small gardens.

'Charleston Gray' – 85 to 90 days; large, cylindrical fruit up to 24 inches long; fruit weigh 30 to 35 lb with tough, light greenish-gray rind and red flesh; sunburn and heat resistant.

'Crimson Sweet' – 80 to 85 days; large, rounded fruit up to12 inches in length having small seed; fruit weighs 20 to 30 lb with thick, hard rind having light-green and darker green stripes, and bright red flesh with fine texture.

'Sugar Baby' – 70 to 75 days; small, round fruit that are 6 to 8 inches long and weigh 8 to 10 lb; fruit have thick, dark green rinds that almost turn black with bright red-orange, fine textured flesh; excellent for home garden.

'Yellow Baby' – 75 days; small, round to oval fruit reaches up to 7 inches in diameter; fruit weighs 8 to 10 lb and have a thin, light-green rind with darker green stripes; bright yellow flesh.

'Yellow Doll' – 70 days. small, round to oval fruit that weighs 4 to 8 lb; thin, light-green rind with dark green stripes; yellow flesh; semi-compact vine.

'Orangeglo' – 85 days, elongated fruits having light-green rinds with dark green stripes; fruit weigh 20 to 25 lb; one of the best orange-fleshed varieties.

'Desert King' – 85 days; 20 lb fruit have a light pea-green rind that is resistant to sunburn; one of the most drought-resistant varieties of watermelon known; has sweet, yellow-orange flesh.

JERUSALEM ARTICHOKE (*HELIANTHUS TUBEROSUS*)

The Jerusalem artichoke is native to eastern North America and is a really a species of sunflower. It is not closely related to the globe artichoke, even though it belongs to the same family, *Asteraceae*. This vegetable is a single-stalked perennial that will grow about 6 to 10 ft in height and has 3-inch-diameter sunflowers at the top of the stalk. Jerusalem artichoke is grown for the round, knobby tubers that it produces, and can be used like a white potato (Fig. 7.26).

Seeds are often produced by the sunflower heads, but seeds should not be used for propagation purposes. Flowers are often removed from the plant, since seed development competes with tuber formation and enlargement. This vegetable is asexually propagated by the use of tubers, similar to the white potato. Whole small tubers or large tubers cut into pieces with 2 or 3 eyes can be used to propagate this plant. Seed pieces should be planted about 4 to 5 inches deep, and 12 inches apart in the row with 3 ft between rows. Although most often they will

Figure 7.26 A stand of Jerusalem artichokes prior to flowering that is being used as a foundation plant (on left) and harvested knobby tubers (on right).

form small colonies in a home garden (Fig. 7.26). As these plants grow, it is best to hill soil up around stems, which will allow tubers to develop.

Tubers range in color from white to yellow, and red to purple (Fig. 7.26). The fleshy tubers are variable in size and shape, and are typically highly branched and irregular in shape with thin skins. The irregular shapes of the tubers often make them difficult to harvest since they break apart in the soil. Tuber pieces left in the ground will produce plants the following spring. Tubers should be ready to harvest in about 100 to 150 days after planting. Tubers are usually harvested during the fall months after the plant tops have died due to the frost and freezing temperatures. Plant stems are usually cut to the ground to facilitate harvesting.

OKRA (*ABELMOSCHUS ESCULENTUS*)

Okra belongs to the same family of tropical plants that include plants such as hibiscus and cotton, the *Malvaceae* family. The large flowers produced by plants in this family are highly ornamental. Okra is the only vegetable that is in this family, and in many cases, okra is grown primarily as an ornamental in home gardens because of the large, attractive flowers they produce (Fig. 7.27). This plant can fit into so many different places in the home garden. Okra works well as a tall plant in the background or as a foundation plant at the base of a fence, with other small vegetables grown in front.

Okra is a warm-season crop, growing best at soil temperatures of 75° to 90°F with minimum soil temperatures of 65°F. Okra is a tropical perennial that typically grows 3 to 7 ft in height, and grows best in full-sun conditions. Once flowering has begun, a pod will develop in the axil of every subsequent leaf. The fruit pod color can range from red, green, or white, although most varieties produce green pods. There are two types of okra grown with respect to plant stature: **dwarfs**, which only grow to about 3 ft in height, and **tall types**, which can grow to 5 to 10 ft or more in height. Okra will grow on a wide variety of soils, but does best on sandy loams having high organic matter content. On heavy soils such as clay, it is best to amend the soil with compost or manure to improve the soil structure and to increase the organic matter content. Okra is also very sensitive to soils with hard pans, which are compacted or dense soils that restrict taproot development and do not readily absorb water. Although okra is often thought of as being drought tolerant, adequate moisture is required for plant growth and pod development. One inch of water per week should be sufficient, although water needs are greatest during flowering, pollination, and

Figure 7.27 The large open okra flower is a beautiful sight to see in a garden during the hot summer months.

fruit set. Moisture during pod development is essential for obtaining high-quality pods.

Fertilization is needed to obtain the highest yields possible. It is recommended to broadcast about 3 lbs of a complete fertilizer containing about 10% nitrogen per 100 ft^2 before planting. Plants should then be sidedressed with the same material at about one third the initial rate per 100 ft^2 at 4 and 8 weeks after planting. It is important to note that excess nitrogen will favor leaf production at the expense of pod production, so do not over-fertilize.

***Figure 7.28** Freshly harvested okra from a garden.*

Okra can be direct-seeded or transplanted. Transplanting is really used only in short-season areas to give plants ample time to form pods before cool weather begins. For transplants, start seeds 5 to 6 weeks before they are to be planted in the garden. However, most gardeners will direct-seed okra. Okra has thick-walled seeds, which often delays germination. So, it is generally recommended to soak okra seed in water overnight and plant the following day, as this will improve germination rates. For direct seeding into garden soil, plant seed in the spring once the soil temperatures have warmed sufficiently and air temperatures are consistently at least 70°F, which is normally about 3 weeks after the last scheduled frost date. Okra seed should be planted at ¾- to 1-inch depths. For garden spacings, okra should be thinned to 10 to 12 inches apart in the row with rows spaced 3 to 4 ft apart.

Okra will flower approximately 60 days after germination. Once flowering starts, pods must be harvested frequently. For best-quality pods, harvest the young, tender pods when they are approximately 4 to 6 inches long (Fig. 7.28). Pick at least every other day during the peak of the growing season. Over-mature pods (those that are not picked within a week after flower opening), become woody and are too tough to eat.

Insects and diseases are not really a problem on okra grown in vegetable gardens, but some can become problematic. Some of the more important insects include the corn earworm (chew on pods and leaves) and the Japanese beetle. There are several okra varieties and suggestions include 'Clemson Spineless' (56 days, continuous harvest of 7-to 9-inch slim, tapered pods on 4 to 7 ft tall spineless plants), 'Annie Oakley II' (48 days, early, good for more northerly areas of U.S.), 'Cajun Delight' (49 days, very tender, less fibrous pods), 'Burgundy' (50 days, burgundy-colored plants and pods, plants reach 4 to 6 ft in height, spineless tender pods, nice plant just for ornamental value), and 'Lee' (50 days, dwarf plant type, only reaches 3 to 4 ft in height, great for limited space gardens or containers, pods 5 to 6 inches in length).

SNAP BEANS AND OTHER MINOR LEGUMES

Beans belong to **Fabaceae** family (or the former *Leguminosae*), which is an interesting group of vegetables, since they are the only vegetables that have the ability to fix their own nitrogen with the help of soil-borne *Rhizobia* bacteria. Several warm-season vegetables in this family are widely grown including snap beans, lima beans, and cowpeas. They are relatively easy to grow, and at the least, snap beans should be grown in all vegetable gardens, since they bring nitrogen into the garden soil. There are also several unusual types of beans that can also be grown in gardens, and these will add variety to any vegetable garden. These are beans such as scarlet runner beans, yardlong beans, horticultural beans, and soybeans. However, the type of bean grown by most gardeners is definitely snap beans.

Figure 7.29 Snap beans growing in a garden.

Snap Beans: *Phaseolus vulgaris*

Snap beans (also known as garden beans, green beans, and string beans) are grown for their **immature** green pod (Fig. 7.29). There are two types of beans: **bush** and **pole**, which describes their growth habit. **Bush beans** are generally self-supportive, grow about 1 to 2 ft in height, and are usually planted in rows. **Pole beans** require trellis support, since they have long twining vines. Pole beans grow more slowly than bush beans, but typically produce more beans per plant. Snap beans are a warm-season crop.

Snap beans typically require around 60 days of moderate temperatures (70° to 75°F) to produce a crop of green pods. The optimum temperature for plant growth is 65° to 75°F. It is important to note that temperatures above 90°F will cause newly opened flowers to abort as well as more fibrous pods. The bush types of snap beans are slightly more cold-hardy than pole types are and can typically be planted as much as 2 weeks earlier. However, bush types are also somewhat less susceptible to heat and drought than pole types are.

Snap beans and other beans have the unique ability to fix their own nitrogen. Beans require that nitrogen-fixing bacteria (*Rhizobium* spp.) be present in the soil to achieve nitrogen fixation. Nitrogen fixation is performed by these bacteria, which are found in the nodules of leguminous plants, and make the resulting nitrogenous compounds available to their host plants. Most soils have these bacteria present, since at one time or another a legume such as clover or another bean was grown in the soil. However, if snap beans have been grown in the garden before and have done poorly, a soil inoculant may help. Inoculants can be purchased at garden centers, from internet gardening sites, or through mail-order catalogs.

Snap beans will grow on a wide range of soil types, but prefer well-drained, high organic matter soils. Snap beans will have problems emerging in heavy soils, such as those with high clay levels. However, snap beans will grow sufficiently in these soils after the seedlings emerge. Snap beans are light nitrogen feeders due primarily to their relationship with nitrogen-fixing bacteria. Before planting, broadcast about 2 lbs of a complete fertilizer containing about

10% nitrogen per 100 ft^2, and then sidedress once with 1 lb of the same fertilizer material per 100 ft^2 about one month after planting.

Snap beans need frequent moisture, especially during flowering and pod set. Bean plants require on average about 1 inch of water per week during the growing season, but critical times to make sure adequate water is available to plants are during flowering, pollination, and pod development. Snap beans have shallow root systems, so it is best to provide a deep watering 2 or 3 times per week rather than all at once to prevent drought stress from occurring to plants.

Since snap beans are a warm-season crop, they should be direct-seeded in gardens 1 to 2 weeks after the last expected frost. For a fall crop, plant approximately 12 weeks before first expected frost. Bush-type snap beans can be planted at 2- to 3-week intervals during the growing season to assure harvesting throughout most, except the hottest months, of the growing season. Compared to multiple plantings that are required to maintain bush-type bean productivity in a garden, just one planting of pole beans will last throughout the growing season. Snap bean seeds should be planted about 1½ inches deep, and thinned to 4 to 6 inches between plants (pole beans) or 2 to 3 inches apart (bush beans). Rows need to be placed 2 to 3 ft apart. For pole beans, trellising needs to be prepared before or soon after planting. Pole beans require more initial set-up due to trellis construction, but will supply beans throughout the season if managed properly.

Pods should be harvested at an immature stage, so snap beans should be harvested before seeds develop in the pod. The pods are most tender before seeds develop in the pods, and if left on the vine for too long, pods will become stringy and fibrous. Snap beans are ready to harvest about 2 weeks after flowering.

There are several insects and diseases that can become problems on beans produced in a home garden. Insects that often need to be controlled include corn earworms, cutworms, Mexican bean beetles, Japanese beetles, flea beetles, and thrips. Diseases that can be problematic include bean rust, damping-off of young seedlings (usually caused by various species of *Pythium* or *Rhizoctona solani*), common bean mosaic, and anthracnose.

Before selecting one or more snap bean varieties for a vegetable garden, the gardener must decide on the plant type to be grown (bush or pole) as well as the pod color, since there are many different colors, including green, yellow, white, and purple. For bush types of snap beans, suggested varieties include 'Provider' (50 days, 5½-inch-long green pods), 'Contender' (42 days, 6-inch-long green pods), 'Blue Lake Bush' (52 days, 6-inch green pods), 'White Half Runner' (60 days, 3-ft vines, no staking needed, 4-inch green pods), 'Royalty Purple Pod' (52 days, purple pods turn green when cooked), and 'Cherokee Wax' (50 days, yellow 6-inch pods). For pole types, suggested varieties include 'Kentucky Wonder' (65 days, all-time favorite pole bean, 8-inch green pods), 'McCaslan' (65 days, slightly flattened 7-inch dark-green pods), and 'Romano Pole Bean' (64 days, flat 8-inch green pods).

OTHER MINOR LEGUMES GROWN IN HOME GARDENS

There are several other warm-season vegetables in the Fabaceae family that make great garden plants besides snap beans. A few of these vegetables will now be discussed to give the gardener an idea of some other legumes to grow in the garden, as well as their cultural requirements. Pest problems found on snap beans will be somewhat similar to those found on the other bean crops discussed in this section.

Lima Beans (*Phaseolus lunatus*)

Lima beans have similar cultural requirements as snap beans. However, lima beans need warmer soils than do snap beans to germinate properly, as well as higher temperatures during the growing season and a longer season to produce a crop. Lima beans generally require 70 to 80 days after planting to mature. Lima beans grow best at soil temperatures of 70° to 80°F and air temperatures of 75° to 85°F.

As with snap beans, there are two types of lima beans, **bush** and **pole**. Pole beans need to be planted only once during the growing season if they are managed and harvested properly. However, pole types generally take another 10 to 15 days to mature compared to bush limas. To get a continuous supply of lima beans, plant bush types every 2 to 3 weeks, similar to what is done for snap beans. Again, pole types need to be planted only once during the season.

Limas prefer a sandy or clay loam soil with high organic matter. Fertilization and watering is similar to snap beans. Remember that critical periods when plants need continuous moisture are during flowering, pollination, and pod development, and a lack of water or poor moisture conditions may cause pod drop or flower abortion.

Lima beans need warm soils in which to germinate (70° to 85°F is optimum). Seeds should be sown at 1½- to 2-inch depths in soil at least 2 weeks after the last expected frost. Thin seedlings to 4 inch (bush types) or 6 inch (pole types) spacings in the row, with rows spaced 2 to 3 ft apart. Harvest fresh lima beans as soon as the pods are well filled, but while pods are still green and plump in appearance. Fresh beans can be picked for 3 to 4 weeks from one bush lima bean planting. To get dry beans, let the pods become dry and paper-like.

Insect and disease problems in lima beans are similar to those found for snap beans. Insects that can be problematic, if not controlled, include cutworms, corn earworms, flea beetles, Mexican bean beetles, Japanese beetles, wireworms, and weevils. The damage caused by weevils (seen as small holes in bean seeds) may not be noticed until after storage. Important diseases that need to be managed, if observed in the garden, include anthracnose, bean rust, common bean mosaic, damping-off, and downy mildew.

Suggested lima bean varieties are 'Fordhook 242' (72 days, bush type, 4-inch pods with 3 to 5 large lima beans), 'Thorogreen' (65 days, bush type, old-time favorite pale-green lima, excellent for canning or freezing), 'Henderson' (72 days, bush type, 2 ft tall plants bear lots of pods with baby limas), 'Speckled Calico' (80 days, pole type, grow 6 to 7 ft in height, heat tolerant), and 'King of the Garden' (88 days, pole type, grow to 9 ft in height, big pods filled with 4 to 6 large creamy-white seeds and bear heavily over entire season).

Scarlet runner bean (*Phaseolus coccineus*)

Scarlet runner beans are closely related to snap beans, but will develop larger flowers, seeds, and pods. The large attractive flowers are scarlet-colored, and many gardeners will grow them in the garden for their ornamental value due to the large clusters of bright red flowers that are produced (Fig. 7.30). The brilliant scarlet flowers are highly attractive to hummingbirds. Other

Figure 7.30 Scarlet runner bean growing up trellis in home garden.

benefits, however, are the edible immature pods or the mature dried seeds. The pods will reach a length of 8 to 12 inches and have beautiful mottled reddish-purple beans inside. Only pole types of the scarlet runner bean are available to most home gardeners, even though bush or dwarf types do exist.

The culture of this vegetable is similar to the pole type of snap beans. High temperatures above 90°F may prevent pod set, and plants are sensitive to frost and cold conditions. It is a perennial in its native Central American climate, but due to its sensitivity to the cold, it is grown as an annual in temperate climates. The immature pods should be picked once they have formed on plants, and older pods become tough and fibrous to eat, which is similar to snap beans. The seeds are often shelled (either as green shell or dried shell) and used similar to limas. Although there are varieties, such as 'Painted Lady', 'Golden Sunshine', and 'Sunset', the primary variety grown is 'Scarlet Runner' and can be found from several seed sources. It will mature in about 68 days to snap stage and 115 to dry-shell stage.

Fresh/Dry Shell Horticultural Beans (*Phaseolus vulgaris*)

The Fresh/Dry Shell Horticultural beans are in the same species as snap beans, *Phaseolus vulgaris*, but have larger seeds, and are grown primarily for the bean seed to be used either at

Figure 7.31 Pods of 'Vermont Cranberry' horticultural bean.

the green- or dry-shell stage. Their sweet, succulent flavor is wonderful in soups, for baking, and in a variety of other delicious ways. The pod fiber in most cultivars is too tough to eat, even at an immature stage. The mature pods are yellow and are brightly colored, usually striped or mottled with red. Harvest when the pods begin to change from green to yellow. The culture of this vegetable is similar to that of snap beans. Suggested varieties include 'Vermont Cranberry' (65 days fresh shell and 85 days dry shell, 6-inch-long bright-red mottled pods, with red- and pink-streaked beans; see Fig. 7.31) and 'Tongue of Fire' (70 days fresh shell and 90 days dry shell, 6- to 7-inch-long string-less, red-streaked pods can be eaten).

Green Vegetable Soybeans or Edamame (*Glycine max*)

Edible vegetable soybeans (or edamame) are eaten in the fresh-shell stage and as green-seeded soybeans, and have a firm, tender texture (Fig. 7.32). There are also black vegetable soybeans available and these types are used primarily as supplements to rice and other vegetable dishes. Edible soybeans are grown similarly to snap beans, and they are just as easy to grow as any bush bean. After the soil has warmed to 65°F, plant edible soybeans about 1 inch deep in the soil with spacings of 2 to 3 inches apart in the row with 2½ to 3 ft apart between rows. Succession planting can be used to obtain an extended harvest. The culture is similar to snap beans.

Harvest the immature green pods when they are 80% to 90% filled, that is, when the pods are plump (the seeds are full size), but while the pods are still green. Beans reach their maximum sweetness about 1 month after flowering. Most home gardeners use soybeans at

Figure 7.32 Green soybeans that are ready to steam in the shell.

the green stage and are generally served steamed in the shell, but the shell is not meant to be eaten. The beans can also be easily shelled from the pods and cooked similar to lima beans. Suggested varieties are 'Envy' (75 days, fresh, 2' tall plant, produces good yields of soybeans, can be used as fresh-shelled or dry), 'Moon Cake' (120 days, 5 to 6 ft tall plants, large-seeded pods, bred for taste, flavor, and high nutrition); 'Owen' (120 days, compact 2 ft tall plants, dark brown seeds), and 'Black Jet' (104 days, dry, black soybean, 2' plants are early maturing and high yielding).

Common Cowpea (*Vigna unguiculata*)

The common cowpea is known by more than 50 common names in the United States including black-eyed peas, field peas, crowder peas, and southern peas. There are numerous types and varieties, with many referred to as crowders, creams, black-eyes, pink-eyes, purple hulls, and silver skins. Pods are usually 6 to 12 inches in length and contain seed typically having a dark outline or eye around the hilum, and therefore some varieties are referred to as "black-eyed peas." The other major types are called "purple hull" (because of the pod color), "cream" (because of the seed color), or "crowder" (refers to the peas being crowded in the pods). The pods of this legume vegetable can be used like snap beans, or the pods can be shelled and mixed with the broken pods, or the pods can be allowed to dry on the plants and used like dry beans.

The extreme variability of *Vigna unguiculata* has led to a number of commercial cultivars grouped by their difference in seed shape, size, and color. A few of these are presented next.

Black-eye, pink-eye, or purple-eye peas have seeds that are typically white with a black eye around the hilum, although the eye can be other colors, including purple or various shades of red. The seeds are not tightly packed or crowded in the pod and tend to be oblong or kidney-shaped.

Brown-eye peas have pods that come in variety of colors from green to lavender. The immature seeds, when cooked, are medium to dark brown, very tender, and have a delicate flavor.

Crowder pea seeds are black, speckled, and brown or brown-eyed. The seeds are crowded in the pod, hence the name, and also tend to be globular shaped.

Cream type seeds are cream-colored and not crowded in the pods. This is really an intermediate type between black-eye and crowder types.

White acre type seeds are kidney shaped with a blunt end, semi-crowded, and generally tan in color. Pods are stiff and the seeds tend to be small.

Clay types are older varieties that are medium to dark brown in color and kidney shaped, and rarely grown.

Forage types are adapted primarily for use as fodder or cover crops.

Unlike garden peas, cowpeas are a warm-season vegetable and are severely damaged by even a slight frost. Do not plant this warm-season vegetable too early in cold soils. Cowpeas tolerate hot and dry conditions better than do snap or lima beans, but will not grow to any extent when soil and air temperatures are below 60°F. Their culture is similar to snap beans, and should not be planted until the chance of frost has passed in the spring.

Peas grow in all soil types but are very sensitive to high levels of nitrogen fertilizer, and respond by having high amounts of vine growth with few pods. A slightly acidic pH of 6.0 to 6.8 is most desirable for cowpea production. Inoculants of nitrogen-fixing bacteria may increase yield, especially in soils where cowpeas have not been grown before. Like other legumes, cowpeas are light nutrient feeders, and it is generally recommended to apply 2 lbs of a complete fertilizer containing at least 10% nitrogen per 100 ft^2 about 7 to 10 days before planting.

A gardener can seed cowpeas into the garden when soil temperature reaches 60°F at the 4-inch depth and continue until 80 days before the first autumn frost. If seeds are planted too early, poor stands can result and they may have to be replanted. Plant seeds about 1 inch deep in rows spaced 2 to 3 ft apart with about 3 or 4 inches between plants in the row.

Depending on variety, harvest should begin 65 to 80 days after seeding and continue for 3 to 5 weeks. A gardener should begin to harvest well-formed pods when they begin to turn yellow in color, as this is the stage that has the highest eating quality.

Cowpeas are relatively free of insects and diseases in many areas of the country, except the southernmost portions. Major disease problems that can occur include fusarium wilt, seedling diseases caused by *Rhizoctona* and *Pythium* spp., downy and powdery mildews, several viruses, root knot nematodes, and pod rots. The most serious insect problems are cowpea curculios, aphids, and stinkbugs. The Cowpea curculio is an insect that is a very serious pest of Southern peas. This insect looks like a boll weevil and punctures the pod, leaving a small scar that looks like a blister on the pod and leaves a speck on the peas. The curculio can be found in high populations especially in later plantings. It is also important to note that weevils will sometimes infest dry seed, and that these insects can be killed by freezing dried seed in sealed containers for at least 48 hours.

There are numerous cowpea varieties available, but a gardener's preference really depends on the type of plant (bush or vining), maturity date (e.g., gardening in a short growing season will require an early maturing variety), seed shape, size, and color, and disease-resistance package required. Older varieties tend to make a vine, while newer varieties have a semi-vining to bush habit. Suggested varieties to grow in a home garden are listed next, followed by an overview of each.

'Calico Crowder'—79 days, buff-colored peas with maroon splashes; great flavor; vigorous sprawling vines.

'California Black-Eye'—70 days; an old standard that produces vigorous high-yielding vines; no disease resistance.

'Iron and Clay'—110 days, vigorous drought-hardy plants make great cover crops or fodder, and late-maturing 7-inch-long pods produce light tan seeds that can also be eaten.

'Magnolia Black-Eye'—68 days; disease resistant; high yields; excellent home garden variety.

'Mississippi Silver'—61 days; widely adapted; brown crowder type with large, light brown seeds; easy to shell; semi-vining plant with multiple disease resistance.

'Pink-Eye Purple Hull'—65 days; an old southern favorite with purple hulls and pink-eye seeds; semi-bush with good disease resistance.

'Whippoorwill'—72 days, once the standard for southern peas, drought tolerant and adaptable to most soils; small light brown seeds with dark speckles produced in 7- to 9-inch-long pods.

'White Acre'—62 days, small white seeds produced in 7- to 8-inch-long pods; high yielding and superior flavor.

'Zipper Cream'—62 days; large, creamy white seeds produced in 6- to 9-inch-long pods; very easy to shell; bushy 2 to 3 ft tall plants.

Asparagus or Yardlong Bean (*Vigna unguiculata ssp. sesquipedalis*)

The asparagus or yardlong bean is closely related to the cowpea, except that the plant has a climbing habit and pods can reach lengths of around 3 feet (Fig. 7.33). This bean is easily grown, has high yields, is easily harvested, thrives under hot weather conditions, tolerates a wide range of soils, and is slightly more resistant to drought and much more resistant to bean beetles than are snap beans.

This plant is best grown the same way as pole beans. So, it is best to train this climbing plant to grow on a trellis of some sort. This will prevent the long pods from touching the ground and ensure that straight pods develop. One planting trained up a tall trellis will produce all season. Like cowpeas, this is a warm-season vegetable, and is severely damaged by even a slight frost. It is best to plant in the spring after all chances of frost have passed.

Figure 7.33 Asparagus bean or yardlong bean hanging from trellis in garden.

The seeds need warm soils to germinate; sow seeds a couple weeks after the last scheduled frost date, about 1 inch deep and thin to 8 inches apart. For other specifics on culture, refer to pole beans and cowpeas. Harvest when pods reach 12 to 15 inches in length and before the seeds fill the pods. The long slender pods are used similarly to snap beans, although fresh and dry seeds are also consumed, but to a lesser extent.

A few varieties can be found, including 'Chinese Red Noodle' (95 days, pods reach 18 inches in length, dark red seeds in purple pods), 'Liana' (72 days, 10 to 12 ft vines producing 24-inch-long pods that are best harvested at 18 inches, black seed type), 'Purple Podded' (90 days, dark purple pods that retain color when cooked, red-brown seeds, heat resistant and prolific) and 'Stickless Wonder' (54 days, dwarf type, do not bear as long as tall varieties, 12-inch light-green pods that are best harvested at 10 inches or less).

SOLANACEOUS VEGETABLES

The **Solanaceous** family is often referred to as the nightshade family, and this family contains some of the mostly widely grown vegetables in home gardens. The term "nightshade" derives from the general name for plants in the genus *Solanum*, which often produce toxic alkaloids in fruit, and includes various weeds as well as several commonly grown vegetables. Warm-season vegetables in this family include the tomato, pepper, and eggplant. In almost every home garden, one or more of these vegetables can commonly be found.

Tomatoes (*Solanum lycopersicum*)

Tomatoes are generally the most important vegetable grown in home gardens (Fig. 7.34). Year after year, it tends to be the number-one vegetable grown by home gardeners, and home gardeners account for about 5% of tomato seed sales in the U.S. The tomato is one of the most popular and versatile garden vegetables. Tomatoes are prepared and used in so many different ways in the kitchen and this definitely contributes to their popularity. They are also very easy to grow in gardens, and gardeners with limited space just love tomatoes since they adapt easily to container culture. It is one of the vegetables that a gardener can reap high rewards from when grown in containers in their garden.

Tomatoes are a warm-season vegetable and should be transplanted at least one week after the average last frost. There are numerous varieties available that produce different fruit shapes and colors, plant types, and harvest times suitable for any garden. It is best if tomato seedlings are planted deeply into highly fertile soils, and at least one set of leaf nodes is buried when it is planted. The stem will put out additional roots for a sturdier, healthier plant. Stake or tie tomato plants to keep them off the ground, and keep them well watered. It is critical to maintain a constant adequate moisture level around tomato roots, as water plays a major role in determining fruit quality; thus, it is recommended to water deeply a few times each week.

A gardener's success in growing tomatoes can be attributed to the use of a few proven techniques. These guidelines are discussed next and are offered so that anyone can easily produce tomatoes in their home garden.

Figure 7.34 A cluster of tomatoes hanging from a staked plant in a home garden.

Variety Selection. Before selecting tomato varieties to grow, it is important to note that tomato plants have two general growth habits: determinate (short-vined plants that terminate growth) and indeterminate (long-vined plants that never stop growing until frost). It is important to know what type of plant will be growing in the garden in order to space and train tomatoes properly. The choice of a variety that has proven to be widely adapted over a number of years should be at the top the list. This will often be a difficult decision for a new home gardener, since there are just so many varieties to choose from and all say that they are widely adapted and produce the highest yields. There are lots of information available from numerous sources (e.g., friends or family members, local state extension service, or internet) about what variety a new gardener should grow. A new gardener should just try to make the best-informed decision on which varieties to grow, and then use trial and error over a few years to see what varieties work best for them, but it really depends on a gardener's preferences. 'Early Girl', 'Better Boy', 'Big Boy', 'Big Girl', and 'Celebrity' have proven themselves to be good, widely adaptable medium- to large-fruited varieties. It is important to note that varieties are usually listed with their specific disease resistances. Tomato varieties can be found with the following disease resistant combinations: VFN, VF, VFFN, and VFFNT (V = *Verticillium* wilt resistant, F = *Fusarium* wilt resistant race 1, N = root-knot nematode resistant, FF = *Fusarium* wilt resistance to race 1 and 2, and T = Tobacco Mosaic resistant). Other disease resistances can also be found for certain varieties. The disease-resistance package for a particular tomato variety is very important, as it provides resistance to diseases that are troublesome without the gardener really having to do anything.

Gardeners realize that there are many different types of tomatoes that can be grown, but the choice depends on the tastes of the individual gardener. Tomato types include grape, cherry, patio, processing (or paste), and small, medium, or large-sized fresh market types. There are many colors to choose from including light-red to dark-red, purple, yellow, and orange within the different types.

A discussion of tomato varieties would not be complete without at least a brief description of heirloom tomato varieties. **Heirloom Tomatoes** are also widely grown by home gardeners, but just be aware that although they are marketed for their outstanding variety and flavors, they tend to have much lower yields than newer hybrids and other problems (e.g., green shoulders on fruit, severe cracking of fruit, and little to no disease resistance) that many of the newer varieties do not have. Many gardeners who grow tomatoes prefer to grow older tomato varieties and many of these varieties are often referred to as heirlooms. "Heirloom" is a term that means the seed of these varieties have been handed down from one generation to another. So, heirloom tomatoes are family favorites, and seeds from these open-pollinated varieties have been saved through the years by gardeners, who have loved these varieties. Heirloom tomatoes have a real old-fashioned tomato taste, and many people say that the newer varieties being produced today are having the taste and flavors bred out of them. The taste and flavors are the most important characteristics of these heirlooms. However, the yields of these older open-pollinated varieties are much lower compared to modern hybrid cultivars as well as being softer and more perishable when ripe. If growing heirloom tomatoes, mulch heavily with an organic amendment, such as straw, to provide constant moisture to the plant, which will reduce the amount of fruit cracking. There are many heirloom varieties that can be easily found in local garden centers in the spring or purchased through mail-order catalogs, and these include the varieties 'Brandywine', 'Yellow Brandywine', 'Cherokee Purple', 'German Johnson', 'Mortgage Lifter', 'Pruden's Purple', 'Arkansas Traveler', 'Ozark Pink VF', 'Green Zebra', 'Japanese Black Trifele', and 'Striped German'.

Figure 7.35 Tomato transplants as that commonly appear at local garden centers in the spring.

Soil and Fertility. The best soils for growing tomatoes are those with high organic matter content. Soil organic matter content can be improved by the addition of compost, peat, or other forms of organic matter into the top 6 to 9 inches of soil. Lime and fertilizers also need to be added to the soil, and it is recommended to apply about one cup of lime and ½ cup of a complete fertilizer containing about 10% nitrogen to each plant. A liming material helps to not only increase soil pH but also correct Ca^{2+} imbalances in soil, which helps to correct the blossom-end rot problem that often occurs on tomato fruit.

Since tomatoes are heavy feeders, they should be sidedressed with 2 to 3 tablespoons per plant of a complete fertilizer as described previously after plants have started to flower. A sidedress application should be made once per month and each month thereafter throughout the entire growing season. When fertilizing, it is best to keep the fertilizer granules about 4 to 6 inches from the stem of plants to avoid fertilizer burn.

Spacing Requirements. It is best to space tomato plants at least 1½ to 2 ft apart in the row with 3 to 4 ft between rows. A tomato plant will be more productive if it has lots of space to grow, which reduces the plant-to-plant competition.

Transplanting. Tomatoes are normally planted into the garden as **transplants**, since it is difficult to get a proper stand from direct seeding into a garden soil (Fig. 7.35). Plant transplants deep enough into the garden soil to prevent allowing the tops of peat pots from sticking out of the soil. If the peat pot is exposed to the air, it will act like a wick and rapidly dry out the roots, which can lead to stunting of the plant or plant death.

When selecting tomato transplants at a garden center, there are several things that the gardener should be aware of to select the **highest quality transplants**. Transplants used in the garden should be disease free, dark-green, stocky in growth, free of insect damage, not flowering, and not root-bound in the container in which they are presently growing. It is best not to bring new diseases and insects into a garden, since they will eventually have to be dealt with by the gardener. Flowering is often a response to stress, and oftentimes indicating that the plant is root-bound.

Figure 7.36 Staked indeterminate tomatoes growing in a home garden.

Starter Fertilizer. A starter fertilizer solution should be used at the time of tomato transplanting to ensure that there are sufficient nutrients available to the plant during the early stages of tomato plant growth. Proper root development is very important to help a tomato plant get off to a good start in a garden. A starter solution can often be purchased, mixed with water, and applied to transplants. It normally has a high percentage of phosphorus, with a common analysis being 9-45-15.

Caging or Staking Tomatoes. Caging or staking really depends on the type of tomato that is grown. It is often a cultural practice used in the garden to keep fruits from contacting the soil, which will reduce tomato fruit damage from slugs, turtles, rodents, cracking, sunscald, and/or disease.

There are two general growth habits of tomatoes: **Indeterminate** and **Determinate.**

Indeterminate Types. These are long-vined tall growing plants that can reach heights of 5 to 8 ft in just one growing season. Plant terminals continue to grow as long as the plant is healthy, and they tend to produce a fruit cluster on the stems between every third leaf. These plants produce a shoot or "sucker" at each leaf axil, and if these suckers are not removed, they will grow into larger stems and produce fruit. Tomato fruit size is reduced when a large number of suckers are left on the plant to grow. These suckers should be removed from the plant before they reach 3 to 4 inches in length. However, some suckers must be left on the plant to grow, and the rule of thumb is to leave one stem for each foot of row. So, if plants are spaced 3 ft apart in the row, leave three stems per plant, and if plants are spaced 2 ft apart in the row, leave two stems per plant. Suckers must be continuously removed throughout the growing season. There are many home garden varieties that are indeterminate types, including 'Better Boy', 'Big Boy', 'Lemon Boy', and 'Beefsteak'.

Both caging and staking will work for indeterminate types of tomatoes, and should be done shortly after transplanting (usually within 2 to 3 weeks). **Staking** will generally produce larger tomatoes, but will produce fewer tomatoes than caging (Fig. 7.36). A 5- to 6-ft-long stake should be driven into the soil about 1 ft deep, 3 to 5 inches from the base of the plant, and try to avoid driving the stake into the main root system of the tomato plant. Nylon string, cloth, nylon stocking, or other materials can then be used to tie plants loosely to the stake. **Cages** can be bought to also keep tomatoes off the ground. After transplanting, cages should then be placed directly over tomato plants. Wire cylinders can be used to serve the same purpose as cages and can be purchased from garden centers with ends tied together to form a long circular cylinder. Cages are a good system to use if a small number of plants are being grown since they are somewhat expensive. However, the yield per plant is generally higher than from staking.

Determinate Types. These are shorter-growing plants, normally growing 1 to 5 ft in height, produce a fruit cluster on stems between each leaf, and each stem terminates in a fruit cluster. These are often called "self-pruning" types even though they still need pruning. This type is easier to prune, since they are pruned only once when suckers are 2 to 4 inches in length. It is important to note that a later pruning will reduce plant size. Determinate varieties include many of the newer varieties, including 'Mountain Pride', 'Mountain Spring', 'Mountain Fresh', 'Celebrity', 'Fabulous', and 'Better Bush'.

Gardeners should also be aware that caging is not usually as efficient with determinate types compared to indeterminate varieties, since they do not grow as tall. So, individual plants can be tied to a stake as described for indeterminate plants, or a **weave system** can be used, which is a common practice to prevent determinate tomato plants from falling over. With this system, plants are supported in a weave of strings supported by stakes. With the weave system, stakes are placed between every other plant, and once plants reach 12 inches in height, a string is tied to the stake at the end of the row and is woven down the row between plants. A second string is then added when tops of tomatoes are 8 to 10 inches above the first string. Over the growing season, a determinate tomato crop can be supported by 3 or 4 stringings.

Pruning. The pruning of tomato plants is essential to obtain the best overall tomato yields. Pruning was briefly described previously for indeterminate and determinate plant types. It is important to realize that the pruning of tomatoes is an important part of tomato growing, and for tomatoes to reach their maximum yields, pruning must be done in the garden. Many gardeners will just plant tomatoes in the garden and let them grow without any prunings taking place. Pruning will definitely allow the gardener to harvest larger tomatoes, as well as reduce foliar disease problems in the garden. So, pruning should not be overlooked when growing tomatoes. However, remember that pruning differs between the indeterminate and determinate types.

Optimum Temperatures. Daytime and nighttime temperatures of 75° to 85°F and 60° to 70°F, respectively, are optimal for tomato growth and flowering. Fruit set is reduced at temperatures below 60°F and inhibited at temperatures greater than 86°F. During the peak of high summer temperatures, tomato flowers often abort; this relates to no fruit set and no tomatoes available to be harvested about 30 to 35 days later. This is the reason that many gardeners have a gap in their tomato production, with production beginning again in late summer as temperatures become cooler. Remember that when a tomato fruit first sets, it will take at least 30 days to reach maturity.

Water and mulching. Moisture is critical to optimize tomato productivity and fruit quality. Tomatoes must receive sufficient amounts of water during the season, although water requirements are greater during flowering, pollination, and fruit development. Organic mulches are often placed around plants (e.g., wheat straw or composted leaves) and will really help in conserving soil moisture during the hot summer months of July and August, although synthetic mulches can also be used to achieve similar results. Mulching will often help to prevent or reduce the incidence of fruit cracking from developing on tomato fruit, especially older heirloom varieties. A gardener should not even try to grow an heirloom variety without having a plan to provide some type of mulch around the plant.

Tomato Harvest. Tomatoes are their most flavorful when they are allowed to fully ripen on the vine. A ripe tomato will slip off the vine with a gentle tug. Fruits can be harvested at the most appealing ripeness stage that is preferred (from light red to dark red, but also remember that there are some yellow or orange varieties). The flavor is best if stored at room temperature, but ripe fruit may be held in the refrigerator at 45° to 50°F for 1 to 2 weeks. It is important to understand that the taste and flavor of a particular tomato grown in a home garden goes back to variety selection.

The time from planting until first fruit harvest is dependent on cultivar and growing conditions and may range from a few as 50 days after transplanting (DAT) to as many as 125 DAT. Early-, mid-, and late-season cultivars will mature in 50 to 70, 70 to 85, and greater than 85 DAT, respectively.

Disease and Insect Problems. Frequent scouting of tomato plants for important insect and disease damage is essential to stay ahead of the pests. Properly identifying **tomato insect pests and diseases** is the first task when implementing an effective control program. The most important insects and diseases, as well as physiological disorders, are briefly discussed.

Figure 7.37 A large hornworm feeding on a tomato plant in a home garden.

Some of the most important insect pests of tomatoes include aphids, cutworms, and hornworms. Aphids are tiny bulbous-shaped insects that may be noticed on tomato plant stems or on new growth. While small numbers are generally not a problem, large infestations can gradually injure or even kill plants. Most aphids are easily controlled by a soap spray (see Chapter 10 – Organic Gardening) or foliage can be removed where aphids are densely concentrated; remember to dispose of the discarded foliage. **Cutworms** are the brownish-colored caterpillars that feed on young plant stems at night, and cause these young seedlings to fall to the ground. This damage can be prevented by placing collars around seedlings; collars can be made of various materials (e.g., paper, cardboard, aluminum foil, or an aluminum pie plate bent to form a circle or cylinder and stapled. The collars should be sunk about an inch into the soil around individual seedlings, with at least several inches of the collar protruding above the soil. **Hornworms** (Fig. 7.37) are very large caterpillars (three inches or more in length) that are often difficult to see on tomato plants. The problem is that their pale green color provides excellent camouflage against the tomato foliage, and the nymph and larval stages are far smaller and less obvious. These insect pests can completely skeletonize a plant overnight, so their control is essential to maintain a plant's productivity potential. Many gardeners control them by just picking them off the tomato plant; if there are only a few, this is a good management strategy.

There are also numerous diseases that affect tomatoes. Many diseases such as *Verticillium* and *Fusarium* wilt, early and late blight, damping-off, and various leaf spots are all caused by fungi that prefer moist conditions for growth. It is generally recommended that gardeners water plants in the morning to avoid conditions that encourage fungal spore germination and growth during the cooler overnight hours. The use of drip watering systems or soaker hoses tends to keep leaves drier, which will reduce fungal infections. Although many of these fungi are airborne, some reside in the soil, garden debris, or solanaceous weeds. Thus, it is critical to keep garden debris piles and weeds at a far distance from where tomatoes are being grown. It is also important to keep tomato foliage off the ground and to avoid water splashing up from the ground onto foliage while watering. Some of the more common tomato diseases are discussed briefly next.

Figure 7.38 **Fusarium *wilt in commercial tomato field. Notice wilted plants in center of photo. When the stems of these plants were cut, a brownish discoloration of the vascular tissues in the stems was noticed which is typical for this fungus.***

Damping-off is a disease caused by any of several fungi, including *Rhizoctonia solani*, *Pythium* spp., and *Phytophthora capsici*. These pathogens attack young, seemingly healthy seedlings that suddenly develop a dark lesion at the soil line, and then quickly fall over and die. Cool and moist soil conditions, overwatering, and overcrowding all increase the probability of infection. It is best to use clean potting soil and germination trays to reduce incidence, avoid crowded seed beds, and monitor watering carefully during the first two weeks after germination. ***Fusarium* and *Verticillium* wilt** are diseases resulting from soil-borne fungi that target many garden plants including those in the Solanaceous family (Fig. 7.38). *Fusarium* wilt occurs in areas that have warm-temperate climates, while *Verticillium* is more restricted to cold-temperate climates. Unlike the *Fusarium* pathogens, *Verticillium* tends to only reduce plant productivity without outright killing the plant. Oftentimes, these wilts cause no symptoms until plants are mature and green fruit begins to reach its full size. At that point, foliage on one side of the plant begins to wilt and turns yellow, and a sliced stem will show brownish, discolored tissue. Once these pathogens are established in a soil, they are difficult to manage. Crop rotation is an effective way to reduce these pathogens presence in soils, since without a host, the propagules tend to die over a period of time. Many newer varieties are resistant and it is best to use these resistant varieties in soils that have traditionally had problems with these diseases. **Early blight** is a common tomato disease caused by the fungus *Alternaria solani*. This disease can affect most parts of the tomato plant, including the leaves, stems, and fruits. Affected plants may not die, but they will be weakened and will set fewer tomatoes than normal. Although early blight will generally attack older plants, it can also infect young seedlings. Those plants that are stressed, weakened, or in poor health are especially susceptible to this fungal disease. Typically, dark spots with concentric rings develop on older leaves first, and the surrounding leaf area may turn yellow. Leaves that are infected may die prematurely and fall to the ground, which then exposes fruits previously hidden by foliage to intense sunlight, resulting in sunscald. Lesions first start on the stems as being small and slightly sunken, and as they get larger, they elongate and develop concentric markings similar to the spots on the leaves. Lesions that develop near ground level can lead to some stem girdling, which can kill the plant outright or reduce a plant's overall productivity. A good way to manage this disease is to provide sufficient air circulation to plants, as air flow will help keep the plants dry, reducing the chance for infection. Many gardeners will also remove the lower infected leaves from plants to reduce the chance for subsequent infection of newer leaves. **Nematodes** are microscopic worms that infect plants and are one of the most dreaded problems to find on tomatoes in a garden. Generally on tomatoes, the root-knot nematode is the primary pathogen, and causes galls to develop on roots that interfere with the plant's ability to take up nutrients and moisture. They are most common in gardens in warmer temperate areas that have short winters. Crop rotation is critical to effectively managing this pest. So, tomatoes must be followed with crops that are not susceptible to this pest.

There are environmental conditions that can cause **physiological problems** on tomatoes. A few of these problems are discussed next.

Blossom end rot is first noticed as a dark spot at the blossom end of a tomato fruit, which gradually widens and deepens to become a sunken dry, leathery spot. This problem is caused by inconsistent water availability or by calcium deficiency. Calcium moves only in the water-conducting tissues of a plant, and when water movement into the tomato fruit is restricted, a localized calcium deficiency often develops at the blossom end. The simplest solution is to make sure that there are sufficient amounts of calcium in soil and that consistent moisture is provided to the plant. Mulches can help in maintaining consistent moisture levels.

Catfacing is readily noticeable as deformed areas on the blossom end of fruits. These appear as deep grooves or indentations over the blossom end area of the fruit. Catfacing results from cool weather or insect damage while the plant is flowering. Flower buds that have been exposed to cold temperatures very early in development have shown to have a large proportion of catfaced fruit. Insect feeding on the blossom end areas of young fruit will also develop this disorder. Tomato varieties with large fruit are also most susceptible. Variety selection is the easiest way to limit this problem.

Cracking is a major problem on most tomatoes grown in home gardens. Although several environmental factors can lead to cracking in tomatoes, the main reason is most likely related to moisture fluctuations in the soil over a growing season. Gardeners notice that small cherry tomatoes will frequently split and this disorder is more severe under hot, dry conditions followed by rainfall. The skin of a tomato becomes less resilient as it matures, so the fruit often outgrows the skin and this becomes more pronounced as excess water absorbed into the plant has nowhere to go, so the skin bursts. Other types of cracking will occur on larger-fruited varieties. Circular splitting at the stem end (concentric cracking) or cracks running toward the stem (radial cracking) usually result from a sudden increase in moisture after a dry period. Once again, the tomato fruit expands beyond the skin's ability to adapt. It is important to maintain even moisture to the plant to avoid this phenomenon.

Sunscald occurs when fruit previously hidden by foliage suddenly become exposed to intense sunlight during hot weather. The affected area turns white and does not ripen, and eventually the skin will become sunken and leathery in appearance. This problem affects primarily those tomatoes that are more aggressively pruned. Also, tomato plants that lose foliage from diseases will oftentimes develop sunscald on fruit.

Puffiness results from poorly pollinated fruit. These fruit appear to be angular on the outside and have somewhat of a pointed blossom end to give the fruit a triangular appearance. The walls of the fruit are not firm and when the fruit is cut open there are few seeds and little locular material (gel). Although the eating quality is poor because of the lack of gel, some cultivars have been bred for puffiness so that they can be stuffed with rice or meat more easily.

Peppers (*Capsicum annuum*) [habanero pepper (*C. chinense*) and tabasco pepper (*C. frutescens*)]

Peppers have always been one of the more popular vegetables in home gardens since they are relatively easy to grow (Fig. 7.39). Although peppers are grown by many home gardeners as ornamentals, they are grown primarily to be used as a flavor additive to many different foods. *Capsicum annuum* is the most widely grown pepper by home gardeners, and includes many sweet and pungent types of varying shapes, colors, and sizes. There are many different types of *C. annuum* peppers, including bell peppers, sweet peppers, hot peppers, pimiento peppers, and ornamental peppers. Some gardeners love the heat and grow the spiciest peppers they can find, while others love the flavor of a sweet red bell pepper. There are also two other species commonly grown by home gardeners: the habanero pepper (*C. chinense*) and the tabasco pepper (*C. frutescens*). Peppers are classified as a warm-season vegetable, but their heat requirements are not always as high as one might think.

Peppers are herbaceous plants, with most becoming woody at the stem base. Some peppers will also become shrub-like (e.g., tabasco). Similar to tomatoes, peppers are tropical perennials, but are typically grown as annuals in temperate climates. Cold weather conditions will cause the plant to slow down its growth or to become stunted. In the spring, use hot caps to protect tender, young seedlings, while in the autumn, cover the plants in some manner, if a frost is expected. Peppers have less cold tolerance than tomatoes, but are more tolerant than eggplant to cold conditions.

Figure 7.39 Bell pepper plants growing in home garden.

Pepper flowers will drop when night temperatures fall below 60°F or are above 75°F. So, there is a narrow range of night temperatures to have effective fruit set, and daytime temperatures above 90°F will cause some flower drop as well, but fruit setting will return with cooler temperatures. Peppers also protect themselves from **fruit overload**, and that is when the plant is loaded with fruit, the new developing

Figure 7.40 Bell pepper fruit come in a variety of colors.

flowers will drop. However, once fruit are picked, fruit will again begin setting and developing on the pepper plant.

The easiest way to start peppers is to buy transplants from a garden center or nursery. Many varieties of peppers are generally available to the home gardener at these outlets. To grow specific types or a particular variety, peppers will normally have to be started from seed. Growing peppers from seed is not difficult, but 7 to 10 weeks should be allowed to develop a transplant that is suitable to plant in the garden. When seeding indoors to develop transplants, plant seed about ¼ inch deep into a sterilized medium and maintain soil temperatures in the range of 75° to 85°F. Pepper seed will not germinate readily when planted in cold potting soils. Transplants can be planted into the garden 1 to 2 weeks after the average date of the last killing frost. However, when night temperatures fall below 55°F, transplants will not grow, possibly turn yellow, and become stunted until plants receive warmer temperatures. Also, gardeners should try to plant high-quality insect- and disease-free transplants in the garden, and it is important to remember that any problems brought into a garden will have to be dealt with at a later time. It is best to transplant late in afternoon or on a cloudy day to prevent wilting of transplants, and to use a starter fertilizer solution when transplanting.

Figure 7.41 Mature jalapeno pepper. The highest level of capsaicin is found in the placenta layer, which is the portion that holds the seeds away from the flesh.

Peppers grow best in a well-drained, high organic matter soil. Optimal day temperatures for peppers range from 75° to 80°F, and night temperatures of at least 65°F are required to maximize growth. Plants should be spaced 18 to 24 inches apart in rows with rows about 3 ft apart. Black plastic mulch is a great choice to use in the garden with peppers since they grow better in warm soils. However, if an organic mulch is used, wait until the garden soil has warmed up sufficiently before placing on the soil around pepper plants, as it will tend to reduce soil temperatures and decrease plant growth.

Fertilize peppers similarly to tomatoes and apply about one cup of lime and ½ cup of a complete fertilizer containing about 10% nitrogen for each plant. The liming material will help correct Ca^{2+} imbalances in soil, which will aid in preventing blossom-end rot from developing on fruit. Pepper plants will need to be sidedressed with ½ cup of the complete fertilizer discussed previously every 4 to 6 weeks throughout the growing season. Pepper plants are generally drought resistant, but long periods of little moisture will drastically reduce plant growth and reduce fruit size and yields. The lack of water during flowering may cause flowers and young developing fruit to abort. Moisture can be conserved by placing an organic mulch, such as wheat straw, around plants. Pepper fruits are generally ready to harvest 70 to 90 days after transplanting, but it really depends if the fruit are picked green or allowed to fully mature on the vine.

Although there are many disease and insect problems of peppers, most gardeners find peppers to be a somewhat trouble-free vegetable to grow in the garden. The pests and diseases that attack other plants seem to leave peppers alone even when grown adjacent to each other. Nonetheless, problematic diseases include anthracnose, bacterial leafspot, Fusarium wilt, Verticillium wilt, and various viral diseases. Insects that attack peppers are usually those that also damage tomatoes (refer to tomatoes). A home gardener should be prepared to control pests in peppers, and to prevent excessive pest levels, peppers should be scouted for insect pests about 2 or 3 times per week in the garden.

Variety Selection. As with most vegetables, one of the most important steps in growing peppers is variety selection. The gardener must first determine the type of pepper, and the color or shape desired. There are numerous types of peppers from which a gardener can choose to grow in a home garden. Varieties of bell, hot, and other sweet peppers are briefly discussed.

Figure 7.42 Dried 'Thai Paper Dragon' peppers.

Bell Pepper fruit are large, blocky, and have a blunt blossom end of 3 or 4 lobes. These peppers are sweet and come in many sizes and colors (Fig. 7.40). Most change from green to red when fully ripe, but there are many other mature pepper colors including yellow, orange, purple, or brown. Most varieties are resistant to tobacco mosaic virus (TMV), but a gardener should check prior to purchase if this disease resistance is desired. There are many varieties to choose from and some suggested varieties include 'Keystone Resistant Giant #3' (74 days, 4 lobes, TMV resistant, older but popular variety), 'Sweet Chocolate' (85 days to a dark chocolate brown, small-medium in size, tolerant of cool nights), 'Valencia' (90 days, produces large, blocky orange fruit, TMV resistant), 'Purple Beauty' (80 days, medium-sized bells with thick flesh, compact plants, virus resistant), 'California Wonder' (75 days, widely adapted, smooth and blocky bell with 4 lobes and thick walls), and 'Jupiter' (75 days, widely adapted, produces blocky 4-lobed fruit, TMV resistance).

Hot Peppers may be the most intensely flavored, most loved, and most confused group of vegetables. It is important to note that the compound in peppers that provides the hot flavor is **capsaicin.** It is an alkaloid and sweet peppers do not contain this chemical. The amount of capsaicin in a fruit directly relates to its level of hotness. Capsaicin is a general non-blistering irritant that attacks any fatty tissue it contacts. The highest level of capsaicin is found in the placenta layer, which is the portion that holds the seeds away from the flesh (Fig. 7.41).

Capsaicin content is primarily cultivar dependent, but temperatures below 60°F during ripening will significantly reduce capsaicin content in fruit.

The shapes of hot peppers vary from the large Anaheim types to the short, chunky jalapeno types, the small round habanero types, the short slim serrano or tabasco types, and the long thin cayenne types. Again variety selection depends on what the gardener desires to grow. There are lots of peppers to select from, and a decision on what to grow depends on the shape of the pepper, the intended use of the pepper, the capsaicin content, and level of disease resistance. Suggested hot pepper varieties are 'Anaheim Chile' (77 days, 6 to 8 inch long pendant-shaped fruits have mildly hot flavor), 'Early Jalapeno' (63 days, thick fleshed hot fruit, standard variety with green fruit changing to red), 'Tam Mild Jalapeno' (67 days, thick flesh with mild taste), 'Habanero' (90 days, thin flesh, ultra hot chili pepper, wrinkled fruits turn orange when mature), 'Thai Paper Dragon' (55 days green and 80 days to red ripe stage, easy-to-dry chili peppers, thin flesh, used in cooking; see Fig. 7.42), 'Pretty in Purple' (60 days purple and 85 days to red ripe stage, ornamental, purple foliage, mixed fruit colors on plant), 'Hungarian Hot Wax' (70 days, yellow 5–6" long fruit that turn to red, easy to stuff), 'Tabasco' (75 days, 1½" long fruit are very hot with a thin flesh), 'Numex Big Jim' (80 days, large mild Anaheim type with fruit up to 12 inches in length, short 2 to 3 ft tall plants), and 'Cayenne' (75 days, pungent fruit that are green turning red at maturity).

Other Sweet Peppers. These are other sweet pepper types besides bell types that are widely grown in home gardens. These include sweet roasting peppers like the Italian heirloom 'Marconi' to the sweet pimiento, cherry, and banana types that are commonly grown. Suggested varieties include 'Banana Supreme' (65 days, hybrid that is widely adapted, fruit are about 8 inches long and are yellow changing to red at maturity), 'Sweet Banana' (75 days, popular home garden variety, 6-inch-long fruit, very productive), 'Marconi' (80 days, 7-inch-long horn-shaped fruit, outstanding sweet flavor when allowed to mature to red), 'Ashe County Pimiento' (70 days, sweet, bright red, thick-fleshed pimiento pepper), 'Red Cherry' (70 days, bushy 2 ft tall plants bear dark crimson red, 1½ by 1½ inch fruits), and 'Sweet Cherry' (78 days, popular home garden variety, often used for pickling, green fruit turns red at maturity).

Peppers are relatively easy to grow once you get them started in the garden. It is important to harvest fruit as they mature, and if the gardener harvests fruit at regular intervals, pepper plants planted in the spring will generally produce up until the first frost.

Eggplant (*Solanum melongena*)

The eggplant is another vegetable grown in gardens that belonging to the *Solanaceae* family. One of the main issues that gardeners need to be aware of for this plant is that it is very intolerant of cold weather conditions. Eggplants are more sensitive to cool weather conditions than either tomatoes or peppers. Eggplants should not be transplanted into the garden until daily temperatures are in the 70°F range and after the chance of frost has passed. Plants will often become stunted when subjected to cool weather conditions. Plants that become stunted will remain this way for a while, and usually do not grow as well afterwards as those that have not been stunted from cold weather conditions. So, it is very important to not rush planting eggplant in the garden, since cold-shock can stunt the seedlings and reduce their productivity during the growing season. Since eggplants are more cold-sensitive than other warm-season solanaceous

Figure 7.43 Globe-type eggplant that is commonly grown in home gardens.

vegetables, they are more productive when a plastic mulch is used. These plants love the extra heat provided by the mulch, and a plastic type mulch will improve both the earliness and yield of eggplant.

Eggplants are really short-lived perennials in tropical areas, but are grown as annuals in temperate-climate home gardens. This vegetable will typically have a bushy growth habit (if left unpruned), reach a height of 2 to 4 ft, and have an indeterminate growth habit. Fruit can be round, pear shaped, oblong, or elongated. There are two types of eggplants that can generally be found to grow in home gardens, **Italian (or globe) type (Fig. 7.43)** and **Asian (or elongated) type**. The outer surfaces of eggplant fruit are usually smooth and shiny. The fruit (whether solid or streaked) can vary in color from white, yellow, green, red, purple, or black. Optimal temperatures for plant growth occurs between 70°F and 86°F (day) and 65°F and 75°F (night). Those varieties that produce elongated fruit tend to be more resistant to high temperature extremes than those that produce the oval-shaped fruit. Eggplants are day neutral in their flowering requirements.

The culture of eggplant is similar to peppers. Eggplant will grow well in most soils unless root development is restricted in some manner, but this vegetable tends to grow best in a well-drained soil with a high organic matter content. Eggplant is a heavy nutrient feeder and should be fertilized using 2 lb of a complete fertilizer containing about 10% nitrogen per 50 ft of row before planting. Plants should be sidedressed with ½ lb of a similar fertilizer material as described previously per 50 ft of row at first fruit set and then 3 and 6 weeks after fruit set.

Eggplants can often take 100 days to mature from seed, and most gardeners will use transplants. Eggplants are typically started from seed indoors or in a greenhouse setting and grown for about 8 to 10 weeks inside before being placed outdoors. Eggplants require lots of space to grow adequately, and when planting eggplant, place them about 18 to 24 inches apart in the row, and the rows need to be about 3 to 4 ft apart. Eggplant varieties that have a large bushy growth habit are generally spaced further apart than the smaller-growing varieties.

Harvesting is critical to maintain the productivity of an eggplant planting in a garden. Fruits must be picked continuously to maintain plant productivity. Fruits are ripe when the skin appears glossy and fruit resists thumbnail pressure, but it is best to harvest fruit before flesh becomes tough and woody. It is best to use a sharp knife or small pruning shears to harvest fruit. Many gardeners will harvest eggplant at about ½ normal size to get best quality fruit. If the fruit is filled with brown seeds, then the best eating quality stage has passed. It is important to note that eggplant flowers may not set fruit during hot weather conditions.

There are many insects and diseases that will attack eggplant. Many insects can defoliate eggplants, including eggplant lace bugs, flea beetles, and Colorado potato beetles. Other injurious insects include aphids, spider mites, and fruitworms. Although there are numerous insects that can cause damage to eggplant, the control of flea beetles on young seedlings is essential since this insect can completely defoliate a young plant in just a few days. Flea beetles make holes in the leaves the size of a pin and will cause serious damage unless controlled in some manner. Organic methods for control of this insect include the use of an insect barrier cloth placed over seedlings at planting or the use of an approved organic insecticide such as pyrethrum. Important diseases are anthracnose, fusarium wilt, phomopsis blight, phytophthora fruit rot, powdery mildew, and root-knot nematodes.

There is a wide range of eggplant varieties from which a home gardener can choose. There are many small-fruited and ornamental eggplant varieties that are excellent for growing in containers or to use for ornamental purposes. Many gardeners will grow these plants just for the unique aspect they can bring to a garden. Eggplants for the table also come in a large array of colors, from shades of pink, green, yellow, white, lavender, violet, purple, black, red, rose, and orange, they can be one solid color, streaked, or multicolored. Eggplant fruits can be round, oblong, or slender and elongated with a variety of sizes from tiny, marble-sized, dwarf, or large that are either ribbed or smooth. Suggested varieties include 'Black Bell' (65 days, produces medium-sized, round to oval shaped fruit that are glossy black with a green calyx, early and highly productive variety), 'Black Beauty' (75 days, forms large bush and oval-shaped dark purple fruit, widely adapted), 'Long Purple' (75 days, long, slender 10 inch long fruits that are best when 1 inch in diameter, dark purple fruit), 'Rosa Bianca' (83 days, fruit have creamy white base color with pink-purple irregular stripes radiating from top, great-tasting Italian heirloom), 'Thai Long Green' (80 days, long, light green fruit that are 1½ inch in diameter and 10 inches in length, short 2 ft tall plants), 'Fairy Tale' (60 days, fruit are short and slender with beautifully marbled purple and white tones, and produce up to 6 fruit per cluster, small compact plants—perfect for containers) and 'White Beauty' (70 days, 6 inch long white fruit, 2 to 3 inch in diameter, 3 ft tall plants).

SWEET CORN (*ZEA MAYS*)

Sweet corn is a favorite vegetable of many home vegetable gardeners. Many gardeners find that nothing out of the garden can compare to the fresh, sweet, crisp taste of home-grown sweet corn that is picked and eaten on the same day (Fig. 7.44). Sweet corn is a member of the **Poaceae** (or grass) family, which is probably the most economically important crop family in the world, since all of the grain crops (wheat, rice, etc.) that feed most of the world's population are included in this family.

Sweet corn is an annual warm-season grass and requires full-sun conditions. The optimal temperatures for corn growth range between 70° and 80°F. However, the optimum temperature varies over the corn growing season, and between daytime and nighttime temperatures.

Sweet corn requires lots of space to effectively grow and produce at high levels in a garden. However, many gardeners will find room in their gardens for sweet corn because they can grow the varieties that they like which normally cannot be found in supermarkets. Growing

Figure 7.44 Sweet corn on display at a farmers' market.

sweet corn is really not worth the effort unless you can devote at least a 15 ft x 20 ft block of space in the garden. Sweet corn is wind pollinated and must be planted in blocks to achieve effective pollination and ear development. So, just planting a few plants in the garden will most likely not produce even a few edible ears.

The first recorded sweet corn was given to European settlers in 1779 by Native Americans. Sweet corn results from a mutation at the sugary locus (*su*), which causes the endosperm of the seed to accumulate twice as much sugar as field corn. Plants from this mutant had kernels with a sugary rather than a starchy endosperm and a creamy texture. This lack of starch and increased sugar level in the seed make the sweet corn kernel wrinkled, rather than plump. Sweet corn also has a thinner pericarp (seed coat) than field corn, which makes it more tender to eat. Other mutants (sh_2 and *se*) have since been identified to improve the sweetness and eating quality of sweet corn.

Sweet corn is divided into three distinct types according to genetic background: normal sugary (*su*), sugary enhanced (*se*), and supersweet (sh_2). However, there are also varieties that contain a combination of either two or all three of these genes, exhibiting superior qualities of each type. Many new varieties being released contain two or all three types of genes. These may be called by several different terms, some of which are mixed gene, multi-gene, sweet breeds, synergists, triple sweets, and extra tender. All of these should be isolated in the same manner as the sh_2 varieties that will be discussed later.

The standard (*su*) types. Standard sweet corn varieties contain the "sugary (*su*) gene" that is responsible for the sweetness and creamy texture of the kernels compared to field corn. The sugar content in standard sweet corn types will peak at about 5% to 11%. It is important for gardeners to understand that the *su* types are best suited for being harvested and eaten within a very short time frame, since they lose their sweetness very quickly, as the sugar will rapidly convert to starch.

The sugary enhanced (*se*) types. These sweet corn types contain the sugary enhancer gene (*se*). These types have two advantages compared to standard types. First, they have a higher sugar content (12% to 20%) than the normal (*su*) sweet corn types, and secondly, these types will remain sweet for several days after harvest if refrigerated. The *se* varieties can be planted near normal (*su*) sweet corn and do not require isolation from *su* types to maintain their superior postharvest attributes. Compared to sh_2 types (which are discussed next), the *se* types have the same tender kernels and creamy texture as *su* types, but seedling emergence is better than sh_2 types and isolation requirements are less stringent than sh_2 types.

The shrunken-2 (sh_2) types. The sh_2 sweet corn types are also called supersweets. This gene in sweet corn delays the conversion of sugar to starch, allowing kernels to stay sweet longer than *su* or *se* varieties. The sh_2 sweet corn types have two main advantages over *su* or *se* types. The sh_2 types are two to three times sweeter, and the conversion of sugar to starch occurs at a much slower rate. Supersweets hold their sweetness much longer than *su* or *se* types do. Therefore, supersweets hold up well on plants and in the refrigerator. These types will remain sweet up to 2 weeks or longer after harvest, if refrigerated, compared to *su* or *se* types. The kernels of the supersweet varieties have a crispy, tough-skinned texture and contain low amounts of the water-soluble polysaccharides that impart the creamy texture and high-quality sweet corn flavor that are present in other sweet corn types. Thus, *su* or *se* types have superior eating qualities compared to sh_2 types.

Another major disadvantage of sh_2 varieties is they must be isolated from the *su* or *se* types by at least 300 feet to prevent cross-pollination because if cross-pollination of sh_2 types with *se* or *su* types occurs, the kernels of sh_2 types will not be as sweet. Their pollen is weak and easily supplanted by other types, which causes the kernel to revert to a form with the toughness and starchiness of field corn. Besides isolation, cross-pollination of sh_2 types can be prevented by planting at different timings or selecting varieties that have different tasseling and silking dates. So, to avoid pollination of sh_2 varieties with *se* or *su* varieties, they should be isolated by 500 feet or more (especially down-wind), planted at different timings or dates, and/or sweet corn varieties should be grown that have different silking/tasseling dates. Supersweets do not germinate very well in cold soils, so a warm soil temperature (>60°F) is essential for optimum germination rates. Supersweet corn seedlings are also slower to establish than the normal and sugary enhanced types.

Flowering characteristics

Sweet corn is a wind-pollinated plant, so it is best to grow in short blocks of multiple rows instead of 1 or 2 long rows. A sweet corn plant is *monoecious*, since both male and female flowers are on the same plant in different flowers. The **tassel** is the male flower (containing pollen) and is borne on a panicle at the top of the plant. The **ear** is the female flower (silks are stigma and styles, and kernels are the ovules that form seeds on the ear). Sweet corn is a short-day plant and will not flower at photoperiods longer than 13 hours.

For an ear of corn to develop properly, corn pollen from the tassel at the top of the plant must fall onto the silks of the ear located about halfway up the stalk. Thus, it is critical to plant several short rows, rather than one or two long rows, to achieve better pollination to develop more full ears. Weather conditions are especially important during pollination.

Water requirements

The moisture needs of sweet corn are high. Adequate soil moisture must be continuous throughout the growth period and is especially critical from silking through kernel development. Insufficient moisture at the time of silk emergence will result in reduced sweet corn yields and quality. Leaf rolling is a symptom of moisture stress, and water stress at silking leads to poor pollination, which shows up as missing kernels on the cob (kernels at the end of the ear may not develop or whole rows of kernels may not develop along the cob). Hot, dry, or excessively wet and rainy conditions during pollination can result in missing kernels, small ears, and poor kernel development at the tips of ears. In rainy or wet weather, pollen sticks to the tassel and poor pollination may result. Sweet corn requires about 1 to 2 inches of water per week when actively growing.

Seed Germination

Sweet corn requires high temperatures for optimum germination and growth. Soil temperatures of 70° to 80°F are needed for optimum germination, and lower temperatures will result in reduced or no germination. The seeds of supersweets (sh_2) are even more susceptible to cold soil temperatures than other types, and the lack of sufficient germination rates can often lead to reduced, uneven stands. The reduced vigor of sh_2 seed is related to the reduced starch reserves available for germination and the increased sugars, which render the seed more susceptible to diseases. Seeding depth should be 1 inch in heavy soils and 1½ inches in lighter, sandy soils. The seed of supersweet varieties should be sown at depths that are ½ inch less shallow than other sweet corn types. The planting of sh_2 varieties should be delayed until soil temperatures are above 60°F. Some gardeners will plant corn seed into black plastic mulch, which will increase soil temperatures and improve germination rates. This allows for better germination and faster plant growth. Sweet corn will not tolerate frost conditions, and a crop of sweet corn can be lost if planted too early. Normal (*su*) and sugary enhanced (*se*) sweet corn should not be planted any earlier than 7 to 10 days before the average date of the last killing frost. Again, several rows should be sown at each planting to ensure good pollination. The time to maturity after seeding ranges from 60 to 100 days depending on variety.

Cultural Practices

Sweet corn plants should be spaced 10 to 14 inches in the row and 30 to 36 inches between rows. Gardeners should be aware that close spacing of plants will result in smaller ears or nubbins.

Sweet corn can be effectively grown on many different types of soil, as long as the soils are fertile and well drained. Low amounts of soil fertility will result in low sweet corn yields. If soils are cold and wet during early planting, deficiencies of nitrogen and phosphorus on young emerged seedlings will often occur. Small ears at harvest can also result from low fertility, and poorly filled ear tips can result from inadequate amounts of several macronutrients including nitrogen, phosphorus, and/or potassium. At planting, sweet corn should be fertilized with about 2 lb of a complete fertilizer containing at least 10% nitrogen per 100 ft of row. When the corn is about 8 to 10 inches in height, sidedress with the same amount and repeat when corn reaches 18 inches.

Harvest

Typically, sweet corn will produce only two ears per stalk. As a general rule, sweet corn is ready to harvest three to four weeks after the first silks appear. Silks should be dry, brown in color, with the husks tight around the ear (Fig. 7.45). The ears should feel full and solid within the husk, and the only sure way to check for ripeness is to open the husks of an ear and press a kernel with your fingernail. Kernels should appear plump and exude a milky liquid (*sh_2* varieties will exude a clear liquid). Harvest by breaking the ear from the stalk. Hold the ear at its base and bend downwards, while twisting at the same time. It is best to try to break the ear off as close to its base without damaging either the ear or the main stalk. Sweet corn should be harvested early in the morning when the sugar content is the highest within the kernels.

For a continuous supply of sweet corn throughout the summer and into the fall, plant early, midseason, and late varieties in the spring or make successive plantings every 2 to 3 weeks. When planting for a succession of harvests, the effect of cool spring weather should be considered. Many gardeners make the next planting when the previous planting is 18 inches in height, instead of planting every 2 to 3 weeks. Later sweet corn plantings are more likely to be attacked by insects, which are a problem in sweet corn production, especially for home gardeners.

***Figure 7.45** Sweet corn ready for harvest. Notice brown silks atop ears.*

Insect and Disease Pests

Insects are a major hindrance to growing sweet corn in a garden. The most common insects are corn earworms (chew kernels near the tips of ears), European corn borers (bore into stalks and ears), corn rootworms (feed on roots), wireworms (bore into underground stems and roots), and seed-corn maggots (attack seed planted in soil primarily during early spring when soil is cool). The corn earworm is the most problematic insect and most difficult insect to control in sweet corn. Eggs of this pest are laid on the young silks where they hatch and larvae feed on the silks and the tips of the ears. Only one corn earworm larva is found at the tip of the ear since they cannibalize each other until only one larva remains. This pest becomes worse later in season. Gardeners often spray insecticides on silks at 3- to 5-day intervals, until silks are brown and dry, to have worm-free corn. Although there are transgenic Bt sweet corn varieties (which are GMOs, meaning genetically modified organisms) available to commercial growers that will kill corn earworms and limit insect feeding in ears via toxins produced by the sweet corn plant, these varieties are not available to home gardeners.

There are only a few diseases that affect sweet corn in home gardens. Important diseases include Stewart's bacterial wilt (plants wilt, become stunted, and develop long leaf lesions with wavy irregular margins), northern corn leaf blight (lesions up to 6" long and 1" wide form on leaves), southern corn leaf blight (leaf spots first appear as small yellow flecks that later become elongated between veins, and spots will reach 1 ½" x ¼" in size becoming tan to creamy white in color with reddish to purplish brown margins; husks and ears may also be affected), and common corn smut (characterized by the presence of galls on any above-ground plant part, but are more prevalent on ears; cut off and destroy galls before they burst, and destroy affected plants to keep the disease from spreading). The use of proper sanitation practices, crop rotation, and planting resistant varieties are the best non-chemical methods to prevent diseases in sweet corn.

Other Pests

Birds can be a problem at planting time and when ears are ready to harvest. Birds will oftentimes pull seedlings from the soil just to feed on the kernels, and they will also feed on the developing kernels in ears as they approach maturity. Other animals, such as raccoons and squirrels, will feed on sweet corn and become very difficult to manage. Many will try electric fences around the garden, and place one wire about 4 inches above ground and the other at about 12 inches. It is important that the electric fence be in operation well before corn approaches maturity. Once these pests get a taste of the sweet corn, they are difficult to keep out of the planting and will fight through the electricity to get to the sweet corn.

Varieties

Sweet corn kernels come in three colors: yellow, white, and bicolor (mixed yellow and white). The cross-pollination of yellow varieties with white varieties will result in bicolor corn. Also, if a bicolor is cross-pollinated with a yellow, the kernels will be predominantly yellow. Although gardeners have preferences for kernel color, there is no relationship between color and sweetness. Sweet corn is divided into varieties that mature early- (65 to 70 days), mid- (70 to 80 days), and late-season (80 or more days).

There are numerous sweet corn varieties from which to choose, but first the gardener must decide on the kernel color and sweetness desired (*su*, *se*, sh_2). The suggested sweet corn varieties listed below are those that can be found in most local garden centers or in gardening catalogs. These suggested varieties for the home gardener are separated by kernel color and the gene present for sweetness and eating quality characters. The sweet corn varieties suggested include:

White (*su* type):
'Silver Queen' (90 days, standard white sweet corn, bacterial wilt resistant)
White (*se* type):
'Seneca Snow Prince' (66 days, early, 7½-inch-long ears, good cold soil germination)
'Spring Snow' (67 days, early, 7-inch-long ears on compact plants, tender kernels)
'Silver King' (80 days, ears 8 inches in length, disease resistant)
White (sh_2 type):
'How Sweet It Is' (85 days, 8½-inch-long ears, superior eating quality, disease resistant)
'Seneca Sugarburst' (79 days, 7- to 8-inch-long blocky ears, 6" plants)
Yellow (*su* type):
'Golden Cross Bantam' (85 days, 8-inch-long ears, old home garden variety)
'Early Sunglow' (62 days, early, 7-inch-long ears, popular home garden variety)
'Golden Beauty' (70 days, 8-inch-long ears, good home garden variety)
'Iochief' (86 days, 8-inch-long ears, popular mid-season home garden variety)
Yellow (*se* type):
'Seneca Daybreak' (64 days, early, 7- to 8-inch-long, good cold soil germination)
'Kandy Kwik' (65 days, early, 5 ft tall plants with 8-inch-long ears)
'Sugar Buns' (70 days, 7-inch-long ears on 5 to 6 ft tall sturdy plant)
'Early Choice' (65 days, 7½-inch-long ears on 4 to 5 ft tall plants)
'Bodacious' (75 days, 8-inch-long ears, excellent disease resistance, gardener favorite)
Yellow (sh_2 type):
'Northern Xtra-Sweet' (71 days, earliest yellow sh_2, large 8- to 9-inch-long ears, good cold soil germination)
'Illini Xtra-Sweet' (85 days, 8-inch-long ears, good home garden variety, the original supersweet)
Bicolor (*su* type):
'Butter & Sugar' (75 days, 7-inch-long ears, good home garden variety)
'Biqueen' (92 days, 8- to 9-inch-long ears, bicolor similar to 'Silver Queen')
Bicolor (*se* type):
'Quickie' (65 days, early, 7-inch-long ears)
'Peaches & Cream' (83 days, 8-inch-long ears, popular home garden variety)
Bicolor (sh_2 type):
'Kandy Krisp' (70 days, early, 8-inch-long ears)
'Skyline' (73 days, 8-inch-long ears, 6 ft tall plants, good cold soil germination)
'Phenomenal' (85 days, 8- to 9-inch-long ears, good home garden variety)

SWEET POTATO (*IPOMOEA BATATAS*)

Sweet potato is a member of the morning glory family (***Convolvulaceae***). The sweet potato is a perennial but is grown as an annual crop in temperate climates. Sweet potatoes, however, should not be confused with yams. Although the sweet potato and yam are often used interchangeably, they are different plants, and differ in growth habit and culture. Yams belong to the genus *Dioscorea* (family *Dioscoreaceae*) and are rarely grown outside the tropics, whereas the sweet potato belongs to the genus *Ipomoea*. Yams are mostly starch and are not as nutritious as sweet potatoes.

Although sweet potato vines are also popular ornamental plants for gardens due to the numerous leaf colors that can be attractive in landscapes, most gardeners grow sweet potatoes for the edible roots produced. The varieties specifically grown as ornamentals will produce edible roots, although the roots are small and not as flavorful as the varieties grown for their large edible roots. The **storage root** is the edible plant part eaten, not tubers as with the white potato. Depending on the sweet potato variety grown, the storage root skin color ranges from white to brown to red to orange, and the storage root flesh can be red-orange, orange, yellow, or white.

The sweet potato is a tropical root crop and there are not many commonly grown vegetables that can withstand more summer heat than this plant. In contrast, this tropical plant does not like cool weather and will be damaged by cold weather conditions, as even a light frost will kill the foliage and soil temperatures below 50°F will damage the storage roots. Sweet potatoes can be grown only in areas that have a long frost-free period and warm temperatures during the growing season. Sweet potatoes require a minimum frost-free period of 100 to 130 days, with minimum average daily temperatures of 75°F.

Since sweet potatoes grow best in warm soil and at hot temperatures, the use of black plastic mulch will help warm the soil and improve their growth and productivity in areas with cooler summers. Remember to cover the beds with a layer of black plastic about three weeks before the intended planting date. The mulch will also help the soil to retain moisture and keep the sweet potato beds weed free. Once the potatoes begin to grow, the plastic can either be removed and replaced with an organic mulch, or left in place until harvest.

Sweet potatoes grow best in loose, well-drained soils having high organic matter content. Water-logged or poor drained soils often causes **souring** (roots decay in soil under anaerobic conditions) or **enlargement of lenticels** on the surface skin of roots. Soils high in clay will often produce roots that are irregular (twisted or malformed) in shape. It is recommended to broadcast apply before planting about 1 lb of a complete fertilizer containing about 10% nitrogen per 100 ft of row. Sweet potatoes are light nitrogen feeders and if plantings receive too much, they will often develop more vines than roots. It is also recommended to sidedress sweet potato plants about 30 days after transplanting with ½ lb of a similar fertilizer material, which was described previously, per 100 ft of row. Sweet potatoes are deep-rooted plants and need approximately 20 inches of water over the growing season, which amounts to about 1 inch of water/week. However, the most critical time for irrigation is during root swelling and the lack of moisture at this time will reduce root size and plant productivity. Poor yields can result if the soil has a high clay content, is overly wet or dry, or over-fertilized with nitrogen.

The sweet potato is one of the few vegetable crops that are asexually propagated. Sweet potatoes are started from "slips" or vine cuttings grown from storage roots. Disease-free slips should be used and can usually be obtained from a garden center or via a mail-order seed company. However, storage roots can be saved from the previous growing season's crop to produce slips. Three storage roots should produce about 30 slips, enough to plant a 30-ft row. The roots that are being utilized to produce slips should be planted in a well-drained soil medium (oftentimes sand or sawdust is used); roots are then covered with 3 to 4 inches of the medium in a greenhouse or coldframe. This should be done about 5 to 10 weeks before the projected outdoor planting date. When sprouts reach 9 to 12 inches in height, they can be pulled or cut an inch above the ground surface and then directly planted into a garden soil. Additional slips develop and can be used for later planting.

Sweet potatoes should be planted in a ridge of soil about 8 to 10 inches in height to allow for proper drainage and greater warming of the soil. Slips are normally planted 10 to 12 inches apart in rows spaced 3 to 5 ft apart. Maintain high levels of moisture to slips for several days after planting, so that roots can grow and establish themselves in soil. Plants should root well with sufficient moisture. The sweet potato slips should not be planted until the soil has warmed to at least 70°F, and it is generally recommended to plant sweet potatoes at least 2 weeks after the last average frost date.

Sweet potatoes need about 90 to 120 days from transplanting the plants into the garden until harvest. Harvest roots when they are large enough to use and it is important that they are harvested before the first frost. However, sweet potatoes harvested too early will have many small roots, while a late harvest results in jumbo roots and possible cold injury. It is best to dig sweet potatoes when the soil is fairly dry, and a garden fork should be used to gently dig large mature roots out of the soil (Fig. 7.46). Avoid bruising or scratching roots during harvest or handling since damaged sweet potatoes rot more easily in storage. Also, do not let freshly dug potatoes sit in the sun for too long, as their skins tend to be easily burned by intense sunlight.

A gardener should cure freshly dug roots for 7 to 10 days in a dark room at 80° to 90°F that has about 90% relative humidity. Curing will not only improve flesh sweetness, but will also toughen the skin and heal cuts/bruises, which increase storage life. These environmental conditions will stimulate the rapid synthesis of a periderm layer on the root surface. Surface wounds of roots are somewhat healed by the curing process and protected from invasion by pathogens. Once roots are cured, store at a temperature of approximately 55°F with a relative humidity of 85%. Sweet potatoes improve in flavor during storage

Figure 7.46 'Beauregard' sweet potatoes being dug and lifted from soil with a garden fork.

since starch in the roots convert to sugars. If roots are exposed to temperatures below 50°F, sweet potatoes may develop hard spots in the roots, a condition known as hardcore, or be chilled and begin to break down.

In most instances, insects and diseases are not a real problem for sweet potatoes produced in a garden. However, sweet potato yields can be reduced by certain insects and diseases. Insects that most commonly cause problems for sweet potato in the garden include flea beetles and wireworms, although the sweet potato weevil can be a devastating insect pest in southern gardens. Most of the important diseases of sweet potato are caused by root pathogens, including soil rot or pox (caused by *Streptomyces ipomoea*, which causes a scab type of lesion on root) and scurf (dark brown to black spots develop on the fleshy storage roots, and is caused by *Monilochaetes infuscans*).

There are several **sweet potato varieties** available to the home gardener and are discussed next. In cooler temperate climate summer areas, it is important to note that the short-season sweet potatoes varieties 'Centennial' and 'Georgia Jet' work well.

'Beauregard'—100 days; roots have light rose-colored skin with deep orange flesh; attractive sweet potato with a high-quality root; high yielding; low sprout production; some disease resistance but not resistant to nematodes.

'Bush Porto Rico'—110 days; short compact vines that are ideal for gardeners with limited space, has orange flesh and red skin.

'Centennial'—100 days; high yielding; leading variety for northern climates; has bright copper-colored skin and a fine-grained deep-orange flesh.

'Georgia Jet'—100 days; best for areas with short growing seasons since it will produce edible roots in 50 days and is reported to have some level of cold tolerance; has deep red skin with moist orange-colored flesh.

'O'Henry'—100 days, white-skinned and cream-fleshed version of the popular 'Beauregard'; flavor has been compared to that of nuts and honey; drier flesh than orange types.

'Vardaman'—110 days; high-yielding bush plant type that is ideal for limited-space gardens; has golden colored skin with deep-orange flesh color).

Chapter Eight

COMMON PROBLEMS OF VEGETABLES IN HOME GARDENS

There are many common problems that gardeners have to face in their gardens. In many cases, similar problems occur year after year in a garden. Gardeners need to have an understanding of what has gone wrong with a particular vegetable, so that they can avoid making the same mistake again next year. There are many problems in a garden that can be avoided or corrected within a growing season or within a year or two. However, in some cases, it may take several years to correct a problem. This section is devoted to providing an understanding of some common garden problems and the recommendations for their prevention.

GENERAL GARDEN PROBLEMS

Compost pile not heating up to cause decomposition

Slow decomposition in a compost pile may be caused by several different factors, including too small of a pile, not enough nitrogen in pile (especially with high C:N ratio materials), lack of oxygen in pile (need to turn piles to aerate), too few microbes in pile (add garden soil to pile), too much moisture in pile (turn pile to allow drying), or not enough moisture in pile (add water to pile). The composting process is essentially complete when mixing no longer produces heat in pile, and a dark, rich humus has formed.

Lots of foliage, but no fruit development

In most cases, this is most likely due to excessive amounts of nitrogen being fed to plants. However, for certain crops, bees and other pollinators are needed for fruit set, so the lack of pollination may be the problem as it can lead to no fruit set. Also, tomatoes and peppers are sensitive to both cold and very hot temperatures, which cause their flowers to drop, resulting in no fruit.

Plants in garden are sticky and are attracting ants

This problem is due to aphids. Aphids are garden insects that secrete a sticky material on plants called honeydew, and this sugary substance attracts ants. The best control measure is to spray plants (undersides of leaves have higher aphid populations) with a liquid soap that can either be purchased or made at home (see Chapter 10 – Organic Gardening). There are many insecticides labeled for aphid control, but there is often a reentry period before you can return to the garden or a harvest-delay period before fruit can be picked.

Seedlings die shortly after emergence

This problem is most likely to occur with beans, peas, or the cucurbit vegetables (cucumber, squash, etc.). It is probably due to planting in cold, wet soils in the early spring with the presence of soil-borne fungi (*Pythium* spp. or *Rhizoctonia solani*) in the soil. This disease is referred to as **damping-off.** Seedlings that have damping-off will simply fall over, and often develop a watery-soft rot on the stem near the soil line. There is no cure for this disease except to avoid planting in the area of the garden where it has been most prevalent for several years; a gardener can also try to limit this disease by trying to prevent excessive moisture conditions in soil (such as planting in raised beds), or drenching the soil with a fungicide at planting.

Seed do not emerge through the soil

This can usually be attributed to one of several different factors: older seed with low germination rates may have been used; seeds were planted in the soil too early in the season, before soil has warmed sufficiently to allow germination to occur; seeds were planted too deep; or seeds will not emerge if the soil is crusted over. Often in the early spring, soil temperatures are low and the soil is wet. Seeds planted in these conditions will not germinate and rot in soil. It is best to wait until the soil temperatures rise to at least 60°F to plant warm season vegetables.

COLE CROP PROBLEMS

Broccoli has small plants, poor heads, and/or flowers earlier than normal

This is probably due to broccoli being planted too late in the spring. Heat stress (temperatures > 85°F) will cause broccoli plants to flower early. Broccoli needs cool weather for best head formation. The best time of the year to grow broccoli is to plant in the late summer and let it mature during the cool temperatures found during the autumn growing season. Broccoli produced during the autumn months produces a higher-quality edible head than that planted in the spring. Broccoli can be direct-seeded or transplanted when targeted for autumn production. It is also important to note that placing an oversized transplant into the garden will often result in premature flowering. Other causes of early flowering include stressing plants with too little nitrogen or water.

Cabbage fails to produce a head

Cabbage will not form a head if it is planted too late in the spring and then exposed to high temperatures. Other possible reasons for no head formation include under-fertilization or drought stress.

Cabbage heads split

Splitting is caused by the pressure from high amounts of water taken up after the heads become solid. Heavy rains or heavy irrigation after the heads reach maturity will lead to splitting. Certain cabbage varieties are more prone to splitting than others. If high amounts of water are applied, then the heads will split because they can't grow fast enough, as the excessive amounts of water bursts the cells. Splitting can be reduced by cutting the roots (spading into the soil on both sides of the plant) or breaking the roots (lifting and twisting the head to one side and then back to the other direction) prior to water uptake; this will damage the plant and require that the head be harvested relatively soon thereafter. However, these suggested practices will decrease the amount of water the cabbage absorbs and thus reduce the splitting of heads.

Cabbage leaves have numerous holes

This type of injury to cabbage (Fig. 8.1) or other cole crop vegetables is due caterpillar feeding, which can be any of several insect pests, including the larval stages of the cabbage looper, diamondback moth, or imported cabbageworm. The adults of these insects are butterflies or moths that lay eggs on these plants. The eggs hatch and the larvae (or caterpillars) feed on leaves. These caterpillars will chew large ragged holes in leaves of cole crop vegetables, and they may bore into the heads, often leaving trails of dark green frass, or excrement. There are several possible ways to control this insect. A gardener can use a lightweight rowcover (insect barrier) to exclude moths and prevent them from laying eggs on plants. The use of a Bt (*Bacillus thuringiensis*) insecticide will effectively control these worms while conserving beneficial insects, and it can be applied every few days until the worms are under control. It is also best to try to prevent infestations by spraying with a Bt product every 2 weeks from transplanting until harvest.

Cauliflower or broccoli produces small heads

The development of only small heads of cauliflower and broccoli in a garden could be due to several different factors. First, these vegetables often form small heads in **compacted or heavy clay soils**, since roots can't grow down into the soil and develop properly. Clay soils can be improved by adding organic matter such as compost or peat moss, which will improve soil structure. Another reason for small heads is that transplants may be **planted into the garden**

Figure 8.1 Cabbage plant that has been severely damaged by caterpillar feeding.

too late or too early in the growing season. If transplanted too early, the cold temperatures will cause **buttoning,** which is the development of heads less than 2 inches in diameter at maturity. Another cause for small heads may be that **potbound seedlings** were transplanted. Seedlings that are transplanted should be about 4 to 6 weeks old and not really any older, since old plants can become potbound, which can lead to small heads. When purchasing transplants, check to make sure the plant roots are not potbound in the cell packs. **Seedlings that are planted too closely** in the garden will also produce smaller heads than normal. If large heads are desired, place plants about 18 inches apart in row with at least 2 ft between rows. However, more broccoli can usually be obtained per square foot of garden if they are planted about 12 inches apart in the row.

Cauliflower heads turn yellow

The discoloration of cauliflower heads is due to overexposure to sunlight. Cauliflower heads need to be blanched, which is the exclusion of sunlight. Cauliflower heads can be protected from sunlight by bringing the largest leaves over the head and tying the leaves together with a soft string (not too tight, but allow enough room for subsequent head growth), as this will prevent sunlight from turning the head yellow. There are self-blanching cauliflower varieties available and should be used if possible.

CUCURBIT CROP PROBLEMS

Poor fruit set of cucurbits or vine crops

Inadequate or poor pollination of female or perfect cucurbit flowers is more than likely the cause of the problem. Squash, cucumbers, pumpkin, watermelons, and cantaloupes as well as other cucurbit crops produce both male and female, or perfect flowers. The male flowers do not produce fruit, but produce pollen, which pollinates female or perfect flowers. Perfect flowers have both male and female parts, but the female part of the flower must be pollinated to produce a fruit. Normally, vine crops produce 10 or more male flowers before female or perfect flowers will form on the plant. Fruit set will occur only once the female or perfect flowers develop. Various types of bees are the pollinators, since they will transfer the pollen from the male flowers to female flowers through the normal process of either pollen or nectar collection for their colony. Individual cucurbit flowers are open for pollination for only one day. To solve poor fruit set in cucurbit vegetables, wait until female or perfect flowers develop (it usually takes 1 to 2 weeks after the appearance of male flowers), and observe the amount of bee activity in flowers. If there are too few bees, then the native bee population may need to be supplemented with a beehive or the gardener may choose to pollinate flowers by hand. However, do remember that bee activity will be slowed by cool, cloudy, or rainy wet weather. Drought conditions in a garden will also cause fruit to abort, and fruit abortion can be stopped by irrigating to get adequate moisture levels back into the garden soil.

Misshapen fruit of cucurbit or vine crops

Again, this problem is probably also due to poor pollination (refer to poor fruit set problem). With improper pollination or lack of soil moisture, fruit will sometimes develop but will be malformed or misshapen (see Fig. 7.4).

Bitter cucumbers

Several factors can contribute to cucumbers developing a bitter taste, but any stress during growth can lead to an increase in fruit bitterness. This includes heat, and nutrient or water stress, but usually the most important factor is moisture. Cucurbitacins become concentrated in fruit under reduced moisture conditions, and this is really the leading cause of cucumber bitterness. Other factors such as older plants, low fertility, or high temperatures can also possibly contribute to bitterness in cucumbers as well.

Cucumber or cantaloupe vines suddenly wilt and die

Cucumbers and cantaloupes are highly susceptible to bacterial wilt, which is transmitted by the striped and spotted cucumber beetles. An infected plant can go from appearing healthy to completely flaccid and dead in just a few days. The best method for control of this disease is to control the cucumber beetles by using an insecticide.

Muskmelons or watermelons have poor taste

Poor flavor in muskmelons and watermelons is usually caused by too much water provided to plants the last one to two weeks before melons are harvested (the sugars in melons become diluted). Try to avoid watering plants during the last week before harvest. However, nothing can really be done about heavy rainfalls during this period though, and poor flavor in melons will most likely result from these environmental conditions. However, poor quality may also result from wilting vines, cool weather, or late plantings.

Malformation and Discoloration of Summer Squash Fruit

Summer squash fruit color and shape malformations are caused by various cucurbit mosaic viruses. The fruit of yellow summer squash infected with any of these viruses will develop a green warty surface, and in many cases become entirely green in color (see Fig 7.13). Zucchini squash infected with these viruses will also develop shape malformations and dark green areas on fruit that make fruit less attractive. Virus problems typically show up later in growing season as aphid populations increase, since they vector most cucurbit viruses. The only viable season-long control measure is the use of summer squash varieties that are resistant to certain cucurbit viruses. Even if the plant has been sprayed with an insecticide, aphids will successfully transmit virus particles into the plant before they are killed by the pesticide.

Watermelons suddenly wilt and die

This is most likely *Fusarium* wilt of watermelon, which is a soil-borne disease. A gardener can plant resistant varieties to reduce this problem. There are several wilt-resistant varieties, but consult seed catalogs, regional extension publications, or internet gardening sites before planting specific varieties. There is no chemical control for this disease.

Watermelon fruit are rotting on the ends

This may be caused by a combination of factors, but watermelon fruit is affected by blossom-end rot in a similar manner to that observed for tomato fruit. It can be caused by an extended period of extremely dry weather when the fruit were maturing or by a lack of calcium in the soil. This condition is normally seen on watermelon fruit if the plant has lost excessive moisture during an unusually dry period. The inability of the roots to absorb sufficient water to keep up with the water loss that is occurring will result in desiccation and blackening of the blossom end of the fruit. This condition can oftentimes be prevented by maintaining adequate moisture, especially as the fruit matures. It may also be aggravated by continued deep hoeing or close cultivation to watermelon plants. The use of a black plastic film mulch can help to reduce this problem, especially on heavy and droughty soils.

SOLANACEOUS CROP PROBLEMS

Eggplant has numerous small holes in leaves

This leaf damage is caused by the flea beetle, and there are a few ways to prevent feeding. A rowcover can be placed over plants to exclude insects and will provide some protection. There are also insecticides available, such as carbaryl, which can be used to provide some control. It is important to remember that if this insect is not controlled in some manner, young eggplant seedlings will oftentimes become completely defoliated or severely stunted, which significantly reduces their overall productivity.

Peppers are not setting fruit

Pepper fruit pollination will be reduced if temperatures drop below 60°F or go above 90°F and this will lead to poor or no fruit set. It is also important to properly manage nitrogen on peppers, as too much nitrogen can also reduce pepper fruit set. Also, if pepper plants are loaded down with fruit, no more set will occur until fruits are harvested from plants.

Pepper or tomato fruit have sunburn

This condition is referred to as **sunscald** and occurs when fruit previously hidden by foliage suddenly become exposed to intense sunlight during hot weather. The affected area turns white and does not ripen, and eventually the skin will become sunken and develop a leathery appearance (Fig. 8.2). This problem affects primarily those tomatoes that are more aggressively pruned. Also, tomatoes and peppers that lose foliage in some manner, such as through foliar diseases, can also develop sunscald.

Tomato fruit develop cracks

Although rapidly growing tomato fruit and fruit exposed to the sun will crack more readily, tomato fruit crack when there is any type of fluctuation in the growth of the plant. So, anything that causes uneven, inconsistent growth of the tomato plant can induce cracking. Usually, fruit cracking is most common when dry weather conditions are followed by excessive rainfall. There are many types of cracking, including radial (deep V-shaped opening start at the stem end and often extend to the blossom end), concentric (rings of brown scar tissue

Figure 8.2 Sunscald on bell pepper fruit. Notice the bleached-out creamy, wrinkled area on fruit surface.

encircle the shoulders on the stem end of fruit), and micro-cracks (very small cracks on the epidermis). To prevent cracking, select crack-resistant varieties and use some type of inorganic or organic mulch on top of the soil surrounding the tomato plant. Mulching provides more consistent moisture conditions to tomato roots, which results in less moisture fluctuations and less cracking.

Tomatoes are misshapen

Misshapen tomato fruit are likely the result of poor pollination. Pollination is reduced by various environmental factors, including low temperatures ($<55°F$) or high temperatures ($>92°F$ during the day, or night temperatures more than 70°F), or cloudy weather at the time of flowering or fruit set.

Tomato leaves are dying from the bottom of the plant upwards

This condition is probably due to a disease known as early blight. Early blight is a fungus that attacks the tomato leaves and fruit beginning in the late spring. The most distinguishing characteristic of early blight is the bull's-eye lesions that develop on the leaves and stems. Some control measures include staking and mulching tomatoes, increasing plant spacing to reduce humidity around the plant, watering plants early in the morning and not during late afternoon or evening, and removing some of the lower leaves that are yellow or dead.

Tomatoes are wilting

There are several factors that can lead to wilting of tomato plants. Tomatoes can exhibit wilting from several factors including:

***Verticillium* or *Fusarium* wilt** – There are many heirloom tomatoes that do not have resistance, while most new hybrid tomato cultivars have resistance to these wilt pathogens. *Verticillium* is more of a problem in cooler weather, while *Fusarium* wilt is more prevalent in warmer weather and one side of the plant tends to exhibit wilting symptoms first.

Root-knot nematodes – These are soil-borne roundworms that attack the root systems of tomatoes and produce galls on the roots. They prevent nutrient and water uptake, leading to wilting. The nematodes create a sink in the roots to which a portion of the photosynthates will flow, thus reducing plant productivity.

Tomato spotted wilt virus – This disease is transmitted by thrips and has distinct leaf, stem, and fruit lesions accompanied by wilting. Tomato spotted wilt is one of only a few viruses transmitted by thrips. Besides rogueing visibly infected plants from gardens, the primary way to manage this disease is to control the thrips vector.

Walnut wilt – The chemical juglone is released by walnut trees or related species, including butternut and pecan, and can induce a wilting response in tomatoes. Gardeners should try to not plant tomatoes in close proximity to trees that release this chemical.

Water wilt – The first sign of water wilt is a slight wilting of foliage near the tops of tomato plants following excessive amounts of rainfall that occur within a short period of time. The entire plant may collapse within six to twelve hours. Plants growing in well-drained soils generally recover within a day, but those tomatoes in heavy, poorly drained soils may not survive. In poorly drained soils, roots may be exposed to waterlogged soil conditions for several days, and this often leads to death of tomato plants. Tomato wilt is especially worsened if the air and soil temperatures are high.

Tomatoes rot on the blossom end

This rot that develops on the blossom end is caused by lack of calcium in the fruit and is perpetuated by inconsistent moisture. This problem is caused by inconsistent water availability or by a calcium deficiency. Calcium moves only in the water-conducting tissues of a plant, and when water movement into the tomato fruit is restricted, a localized calcium deficiency often develops at the blossom end. A gardener needs to make sure that there are sufficient amounts of calcium in soil and that consistent moisture is provided to the plant. Mulches can help in maintaining consistent moisture levels.

Once tomato fruit are observed on the plant as having blossom-end rot, it is really too late to correct the problem for the fruit currently growing on the vine. It is important that proper amounts of calcium be added to a soil deficient in this nutrient, since this will help solve the blossom-end rot problem. A fine lime should be applied to the soil to give the quickest results. Also, a fertilizer such as calcium nitrate will help to solve the problem for the new fruit that will develop on the plant. It is best to apply different fineness sizes of a liming material in spring, so that calcium will be released over the entire season.

Flower drop on tomatoes

Flower drop on tomato plants occurs as a result of stress. It is important to realize that anything that inhibits pollination and fertilization, such as low or high temperatures, high relative humidity, and excessive wind can cause blossom drop.

Tomatoes flower without setting fruit

Tomatoes are self-pollinated and under garden conditions there is usually enough natural movement on the plant to allow pollen to deposit on the stigma. However, temperature extremes or drought conditions will keep fruit from forming on a plant. Generally, night temperatures below 60°F and day temperatures above 90°F will prevent fruit set in tomatoes. Also, excessively dry conditions will prevent fruit set, and plants can be irrigated to solve this problem.

Large green caterpillars feeding on tomato foliage and fruit

This caterpillar is most likely the tomato hornworm (or also commonly known as the tobacco hornworm) (see Fig 7.37). This insect tends to be more of a problem with late summer and fall tomatoes. This insect can be controlled by physically removing them from the plants or killing them with an insecticide.

Greening of potatoes

Green tubers are caused by exposure to light either in the field or in storage. The green portion is chlorophyll that has developed due to sunlight exposure, but this area will also contain a glycoalkaloid called "solanine" and high levels of this chemical are toxic to humans. The green areas are bitter, and peeling or cutting away these sections prior to cooking will usually eliminate the chance of sickness because most of the glycoalkaloids are located in the skin or just below the skin. It is important to keep tubers well covered with soil in the field, and stored under dark conditions. Certain potato varieties will turn green more readily than others.

Misshapen or knobby potatoes

Potatoes become misshapen or develop knobs when they are exposed to hot, dry weather conditions. The high temperatures and dry soil prevents the potato tuber from growing evenly and consistently. Mulching with an organic mulch, such as straw, to lower soil temperatures and to maintain consistent higher moisture levels can help in preventing off-shape tubers.

Potatoes with black spots or streaks inside them

There are many factors that can cause internal blackening of potatoes, but the most common are *Fusarium* wilt, viruses, and high temperatures that occur just prior to harvest.

MISCELLANEOUS VEGETABLE PROBLEMS

Beets develop tops (leaves) but not roots

The failure to thin beet seedlings will often provide too much competition among plants, which will inhibit root formation. Beets should be thinned so that plants are spaced 2 to 3 inches apart and this should be done when plants are about 2 inches in height. Also, beets that are planted too late during the spring will have restricted root formation due to hot weather conditions.

Carrot roots develop a forked shape

Carrots need a loose, friable soil to develop a good root shape. Forking is caused by anything that impedes root growth in the soil. This includes pebbles or rocks, or heavy, compacted soil.

Carrots are bitter and woody

Carrots are a cool-season vegetable and when daily high temperatures exceed 80°F, the carrot root color and taste will diminish. It is important to harvest before hot weather arrives.

Carrot roots have bitter, green shoulders

Exposure of the roots to sunlight will cause the roots to develop a green color, which is the chlorophyll pigment. A gardener should hill soil around the shoulders of the carrot roots to prevent exposure to sunlight.

Lettuce has bitter flavor

Lettuce is a cool-season vegetable and will develop a bitter flavor when grown at high temperatures. This often occurs when lettuce planted in the spring is harvested during late spring or early summer. Some bitterness can be removed if the leaves are washed and stored in the refrigerator for 48 hours or longer.

Snap bean plants are healthy but no bean pods are forming

Bean flowers will abort and drop from the plants during hot, dry weather. In most gardens, this is the primary reason that pods do not form on bean plants. However, too much nitrogen fertilizer or manure will also prevent pods from setting on plants. Remember that snap beans are a legume and require only a small amount of nitrogen fertilizer. Bean pods will also not set if the soil is overly wet or waterlogged.

Spinach has formed a flower stalk

Spinach plants will bolt or form a flower stalk when exposed to long days and warm temperatures, which occur in late spring and early summer. However, there are some varieties available that are more resistant to bolting than others.

Sweet corn ears do not fill properly

Poor pollination results in reduced kernel filling in sweet corn ears. This can result from lack of proper environmental conditions, improper planting methods, or insect damage. First, either of two environmental extremes (hot, dry weather or excessive rainfall/irrigation) can result in poor pollination of sweet corn. In hot, dry weather conditions, the addition of moisture through irrigation can help solve the drought problem and increase kernel set. Although high amounts of rainfall cannot be prevented, do not overwater (especially with overhead or sprinkler irrigation) during pollination, as this can reduce the effectiveness of pollination. Poor pollination may also result from the planting method. Sweet corn should be planted in blocks of at least 4 or 5 rows rather than one long row, since it is a wind-pollinated plant. In addition, insects that feed on silks may also contribute to or be the problem, so insect control at this time is essential.

Chapter Nine

PEST MANAGEMENT FOR VEGETABLES IN THE HOME GARDEN

Pest management is a necessary component for the development and maintenance of a productive home vegetable garden. A garden overrun with insects, diseases, and weeds is often too much trouble, and most times will be abandoned by the gardener. Some pests are more common than others and gardeners need to learn how to manage or deal with the pest problems that commonly appear in gardens. Although **weeds and insects pests** will usually cause most problems in a garden, **diseases** can also cause problems for certain vegetables, especially under the right environmental conditions.

Home gardeners will typically use **chemical**, **cultural**, **mechanical**, and **biological** methods to control pests. Oftentimes, a combination of these methods will be used to control a particular pest in the home garden. Each specific method will be discussed in further detail.

Chemical control is primarily the use of pesticides to control pests. These would be insecticides, fungicides, bactericides, and herbicides to control insects, fungal pathogens, bacterial pathogens, and weeds, respectively. If a chemical pesticide is determined to be the best option to obtain control for a specific pest, then the product must be used with care and label directions followed. The use of pesticides should be limited to those situations in which the crops would be seriously damaged by pests without their use. Pesticides can be purchased in many different formulations, including dusts, emulsible concentrates, granulars, and wettable powders. It is important to remember that pesticide sprays can often drift to non-target areas, such as other crops in the garden or outside the garden area and cause damage. Gardeners are really limited regarding the pesticides that they can use in a garden situation, since oftentimes there are no pesticides registered for control of a specific pest on a particular vegetable crop in a garden. Anyway, if pesticides are misused, applied improperly, or disposed of inappropriately, they can contaminate water supplies, and be injurious to humans, desirable plants in or surrounding a garden, domestic animals, pollinating insects, and fish or other wildlife. Thus, it is important to remember that pesticides are poisonous and must be treated with respect.

Cultural control is the use of simple modifications of a pest's environment or habitat that can provide an effective method of control. This type of control method modifies the relationship between a pest population and its natural environment. Cultural controls are generally the cheapest of all control measures since they usually require only simple, inexpensive modifications to normal production practices. Cultural controls are dependable and are usually specific for a certain pest. It is important to note that they do not provide the detrimental side effects like pesticides do, specifically the development of pesticide resistance, undesirable residues in food, feed crops and the environment, and death of non-target organisms.

There are various types of **cultural control** methods that a home gardener can use to prevent or reduce pest outbreaks, including crop rotation, intercropping, phenological asynchrony, trap cropping, and sanitation. **Crop rotation** is one of the oldest and most effective cultural control strategies. When a specific crop or plants from a certain family are grown year after year in the same field, a pest population has sufficient time to become established and build up to damaging levels. Rotating the field to a different type of crop can break this cycle by starving pests that cannot adapt to a different host plant. Crop rotation schemes work because they increase the diversity of a pest's environment and create discontinuity in its food supply. As a rule, rotations are most likely to be practical and effective when they are used against pests that attack annual or biennial crops, have a relatively narrow host range, cannot move easily from one field to another, and are present before the crop is planted. **Intercropping** (also known as mixed cropping or companion planting) is another cultural control method that can reduce pest populations by increasing environmental diversity. This method can reduce the overall attractiveness of the environment to a pest when host and non-host plants are mixed together in a single planting. **Phenological asynchrony** is an important method that is often used in vegetables to control insect pests and its focus is to create a discontinuity in the pest food supply by altering the time of year for planting or harvesting. Sweet corn can escape much injury from corn earworms by planting earlier in the growing season and harvesting before larval populations reach high levels during late summer. **Trap cropping** is a unique way to control pests, since pests are either prevented from reaching the primary crop or concentrated in areas of the field away from the main crop. The principle of trap cropping is that insects have preferences for certain host plants or stages of crop growth and will move to a preferred host if given a choice. For example, blue hubbard squash or another susceptible *Cucurbita maxima* vegetable can be used as a trap crop to attract and retain cucumber beetles, squash vine borers, and squash bugs. It is important to note that trap cropping is management intensive, as insects are usually killed with insecticides (synthetic or biological formulations) as soon as they appear. **Sanitation** is another cultural control practice that can provide highly effective control for some pests. The removal and disposal of disease-infected or insect-infested plants from a garden are effective sanitation methods to limit pest damage to vegetables in a home garden. Also, at the end of the season, the removal of crop debris from a garden will aid in preventing insects or disease from overwintering in the garden and causing damage the following growing season. Clean cultivation is another sanitation method that can eliminate overwintering sites for pest populations in gardens. Just tilling a garden soil before winter may disrupt a pest's life cycle by increasing exposure to lethal cold temperatures. The populations of corn earworms can be greatly reduced if corn stubble is plowed down after harvest.

Mechanical control of pests involves the use of barriers, traps, or physical removal to prevent or reduce pest problems. An example of a common physical barrier used in gardens is an insect barrier cloth laid over plants to prevent insects from feeding on plants. Another physical barrier that is often utilized in gardens is the use of paper collars wrapped around the stem bases of plants to prevent cutworm injury; this physical barrier should extend into the soil for about an inch and be at least 2 inches above the soil line to effectively exclude this pest from chewing on the lower stems of vegetable plants, which often results in plant death if not controlled in some manner. Reflective mulches can also be used as barriers to repel migrating aphids, and in areas that have problems with cucurbit mosaic viruses, these mulches will not only reduce aphid damage to plants but also reduce mosaic virus incidence. The mulches reflect sunlight, which prevents aphids from detecting the green color of plants, and as a consequence, continue to fly instead of landing on plants. Traps are also effective mechanical control method for certain types of pests. For example, the color yellow attracts insects, such as winged aphids and adult whiteflies, and is often used in traps designed to capture these insects. Traps made of yellow sticky cards, similar to flypaper, can be purchased and placed in a garden at plant level to trap these insects. Although these sticky cards were originally designed as a way to monitor certain insect pests in greenhouses or in field situations, they can also be used as a method of control through trapping. Insects can also be removed from plants in a physical manner. For example, some aphids and mites can be knocked off the foliage by spraying the plant with a strong stream of water.

Biological control is generally associated with the use of biological agents (parasites, predators, or pathogens) for pest suppression and is discussed in further detail in **Chapter 10 – Organic Gardening**. Although it can also include the use of specific vegetable varieties that have genetic resistance to pests, biological pest control will be focused on various biological agents. Biological control is a particularly appealing pest control alternative because, unlike most other tactics, it does not always have to be reapplied each time a pest outbreak occurs. Although in conventional usage, biological control usually refers to the practice of rearing and releasing natural enemies, such as parasites, predators, and pathogens. The hope of biological control systems is that once natural enemies are released into a new environment, they will become established and provide a self-perpetuating form of control. Biological control is also the only pest control tactic that increases, rather than decreases, the species diversity within a gardening ecosystem. This increased diversity often results in greater stability because wild fluctuations in population density are less common in communities with a diverse food web.

Biological agents are a wide variety of life forms, including vertebrates, invertebrates, fungi, and microorganisms. These beneficial species are common in most natural communities and, although their presence is often unnoticed, they help maintain the balance of nature by regulating the density of their host or prey population. Insect species often become pests when this ecological balance is disrupted by natural events or human intervention. Biological control agents work better if they have certain characteristics and these include a narrow host range, climatic adaptability, synchrony with host's (prey's) life cycle, high reproductive potential, effective ability to search for host (prey), great appetite and short consumption time, and ability to survive at low host (prey) densities. It is best if natural enemies have narrow host ranges, and although generalized predators may be good natural enemies, they tend not to kill enough pests when other types of prey are also available; thus, it is important that predators have a narrow host range to kill the pest or pests that are causing damage. Natural enemies must also be able to survive the climate extremes of temperature and humidity that they will encounter in the new habitat. It is also important that the predator or parasite be present when the pest first emerges or appears. Furthermore, good biocontrol agents produce large numbers of offspring; it is ideal if more than one generation is produced during each generation of the pest. Effective natural enemies must also be able to locate their host or prey even when it is scarce; in general, lower pest populations result if the biological control agent has superior searching characteristics. Natural enemies that consume prey rapidly or lay eggs quickly have more time to locate and attack other members of the pest population. Small populations of efficient natural enemies will probably be more effective biocontrol agents than larger populations of less-efficient species. Lastly, if a natural enemy is too efficient, it may eliminate its own food supply and then starve to death. The most effective biocontrol agents reduce a pest population below its economic threshold and then maintain it at this lower equilibrium level.

It is important to note that biological control is not a quick fix for most pest problems. Natural enemies usually take longer to suppress a pest population than other forms of pest control and gardeners often regard this as a disadvantage. It also may be difficult to integrate natural enemies into a home gardening situation when pesticides are being used. Beneficial insects are often highly sensitive to insecticides and their resurgence (recovery to their pre-spray densities) is usually much slower than that of pest populations. Rapid pest resurgence often leads to a vicious cycle of continued chemical usage that prevents natural enemies from ever becoming reestablished in the garden.

Perhaps the greatest potential for future progress in biological control lies in improving the success of microbial pathogens. Many of these organisms are highly desirable as biocontrol agents since they attack a narrow range of insect hosts, they are not hazardous to humans or domestic animals, and they do not pose a threat to the environment. As a group, though, pathogens have never proven to be quite as reliable as other forms of pest control. They are vulnerable to desiccation, ultraviolet radiation, and high temperatures. They may not survive long enough in the environment to encounter a host, and even if they survive, their virulence may be too low to overcome a particular host's defenses.

Integrated Pest Management (IPM) is a philosophy that many home gardeners try to utilize in their garden for disease, insect, and weed control. The IPM system is a sustainable approach to managing pests by combining biological, cultural, physical, and chemical tools in a way that minimizes economic, health, and environmental risks. It helps balance pest management in a home garden so that vegetables are produced in an environmentally friendly manner. Without a doubt, the best management practice for any IPM program is pest prevention and an established IPM program includes regular monitoring for insect, weed, and disease pests. It is important that early intervention is instituted as soon as pest damage is detected, rather than later when pest populations or the amount of damage is irreversible. Scouting of a garden plays a pivotal role in IPM, and the importance of regularly monitoring crops should not be undervalued. During scouting, it is important to know where to look for insects. The upper and lower surfaces of leaves, along the stem, leaf, and stem axils, stem terminals, floral structures, and other small tight places are excellent places to look for insects. Some gardeners will tap flowers and leaves on white paper to check for insects, as they easily stand out next to the white background. A gardener should not forget to look for any mold, chlorosis, lesions, stunted or distorted growth, powdery residues, mottling, or necrosis on garden plants, which are symptoms of plant diseases. It is also recommended to make use of sticky traps and/or indicator plants (plants that consistently attract certain insect pests) for monitoring insect pest populations. Remember to monitor the garden on a weekly basis and be consistent.

Good recordkeeping is also a necessity, especially when evaluating whether or not specific control measures instituted are effective. The precursor to using control methods should always be scouting in order to pinpoint exactly which plants need treatment. Avoiding the overuse of pesticides and fertilizers is not only environmentally responsible, but it also saves valuable time and money.

Disease Pests. A plant disease is usually defined as abnormal growth and/or dysfunction of a plant caused by a microbial organism. However, diseases result from some disturbance in the normal life process of the plant caused by either living and/or non-living factors. ***Biotic*** diseases are caused by living organisms, while ***abiotic*** diseases are caused by non-living or environmental factors. Abiotic disorders are caused by nonliving factors, such as drought stress, sunscald, freeze injury, wind injury, chemical drift, nutrient deficiency, or improper cultural practices, such as overwatering or planting too deep. Unfortunately, the damage caused by these various living and nonliving agents can appear very similar, and even with close observation, accurate diagnosis can be difficult.

Diseases can often cause significant losses in a home garden (Fig. 9.1). Vegetables in a home garden can be attacked by a wide range of disease-causing organisms such as fungi, bacteria, viruses, mycoplasma, and nematodes. Diseases develop only when (1) a susceptible host plant is present, (2) the disease organism is present, (3) suitable environmental conditions are present to allow disease development, and (4) there's sufficient time for disease development. Without all of these requirements, a particular disease will not properly develop. It is important that gardeners understand that there is no single control measure that is effective against all diseases caused by these microscopic pests, and that gardeners must rely on a well-balanced and planned strategy to keep diseases under control. Really the first step in setting up a disease-control program is to correctly identify what disease problems commonly occur in a garden. The earlier that disease problems are identified in a garden, the greater the chance that a gardener can provide sufficient control. With quick action, a gardener can control leaf spots, blights, and mildews within the same growing season. Other disease problems may not be treatable this season, but correct disease identification allows a gardener to take preventive disease-control measures the following year.

Many garden vegetable diseases are easy to identify. It is important that a gardener look for **symptoms** produced by diseased plants or **signs** of the disease-causing organism. A disease symptom is the reaction of a plant host to the disease-causing organism, while signs of plant disease agents are the observable evidence of the actual disease-causing agent. Signs may include the mycelia of a fungal agent, fungal spores, or spore-producing bodies.

Symptoms can often be grouped as **Underdevelopment of tissues or organs**—symptoms include stunting of plants, shortened internodes, inadequate development of roots, malformation of leaves, inadequate production of chlorophyll and other pigments, and failure of fruits and flowers to develop; **Overdevelopment of tissues or organs**—symptoms include galls on roots, stems, or leaves, witches' brooms, and profuse flowering; **Necrosis or death of plant parts**—symptoms appear as wilts or diebacks, shoot or leaf blights, leaf spots, and fruit rots; **Alteration of normal appearance**—symptoms include mosaic patterns of light and dark green on leaves, and altered coloration in leaves, flowers, and/or fruits.

Typical disease symptoms on plants include:

Blight – a rapid discoloration and death of twigs, foliage, or flowers.
Canker – dead area on bark or stem, often sunken or raised in appearance.
Chlorosis – yellowing of infected tissues.
Decline – progressive decrease in plant vigor.
Dieback – progressive death of shoot, branch, or root starting at the tip.

Distortion – malformation of plant tissue.

Gall or **gall-like** – abnormal localized swelling or enlargement of plant part, and can result from insects, mites, diseases, or abiotic disorders.

Gummosis – exudation of gum or sap from plant parts.

Leaf distortion – leaves are twisted, cupped, rolled, or otherwise deformed.

Leaf loss – premature defoliation or loss of leaves from plants.

Leaf scorch – burning along the leaf margin and into the leaf from the outside margin inward.

Leaf spot – a spot or lesion on the leaf and can often have water-soaked or greasy appearance.

Mosaic – varying patterns of light and dark plant tissue that often appears on leaves or fruit.

Necrosis – dead tissue and necrotic areas on plant parts can result from many different diseases.

Stunting – lack of growth.

Wilt – general wilting of the plant or plant part.

As stated previously, biotic diseases are caused by living organisms such as fungi, bacteria, viruses, mycoplasma, and nematodes. Each of these organisms will be briefly discussed next to provide the gardener an idea of the differences between the different organisms.

Fungi. Fungi are probably the most frequent causes of plant diseases that occur in a garden. The body of a fungus is a branched filamentous structure known as ***mycelium***, and one single thread is called a hypha (or hyphae, plural). Most fungi reproduce by spores, and these structures are the main dispersal mechanism for fungi and can remain dormant until germination conditions are optimal for their survival. Many fungi can also overwinter as fruiting structures embedded in dead plant tissue. Thus, it is important to remove dead and diseased plant materials from a garden to remove the primary inoculum source which can infect plants the following growing season.

Figure 9.1 Cucumber plants in a garden infected with anthracnose. Notice the shot-hole appearance of leaves and all the necrosis (or dead tissue).

When a spore comes in contact with a susceptible plant, it will germinate and enter the host if the proper environmental conditions are present. Hyphae develop from the germinated spore and begin to extract nutrients from host plant cells. The hyphae secrete enzymes to aid in the breakdown of organic materials that are ultimately absorbed through their cell walls.

Fungi are spread by wind, water, soil, animals, equipment, and in plant material. They can enter plants through natural openings such as stomata and lenticels and through wounds. Fungi can also produce enzymes that break down the cuticle, the outer protective covering of plants, and enter plants in this manner as well. Fungi cause a variety of symptoms, including leaf spots, leaf curling, galls, rots, wilts, cankers, and stem and root rots.

Bacteria. Bacteria are single-celled microorganisms. They contain no nucleus and reproduce by dividing in two equal parts, which is referred to as fission. As a result, they can multiply rapidly in just a short period of time. Bacteria can function as either parasites or saprophytes and can infect all plant parts. However, unlike fungi, bacteria must find a natural or wound opening to enter plants. Bacterial cells can move from one plant to another in water, soil, and plant material, similarly to fungal pathogens. However, bacterial pathogens are more dependent upon water, and conditions must be wet and/or humid for them to cause significant and widespread damage.

Bacteria move between plant cells and secrete substances that degrade plant cell walls so that the cell contents can be utilized. Some produce enzymes that break down plant tissue, creating soft rots or water-soaking. Although bacteria cause symptoms such as leaf blights and spots, galls, cankers, wilts, and stem rots similar to fungi, symptoms caused by bacteria also tend to be more water-soaked and slimy in appearance. The lesions produced also tend to be more irregular and angular compared to fungal pathogens, which tend to cause more circular lesions. Bacterial leaf spots appear different from fungal leaf spots due to their intercellular movement, as veins often limit the development of a lesion, so they appear angular or irregular and not round in shape.

Viruses. Viruses are crystalline particles composed of nucleic acid (RNA or DNA) and protein. All viruses are obligate parasites and are unable to survive outside of their host. To move from plant to plant, the particles must be transmitted by vectors. The vector is typically another organism such as an insect or human. Insects spread viruses between plants as they feed on them. The feeding injury creates the necessary wound required so that the virus can enter the plant. A specific plant virus is normally spread by only one kind of insect vector. Aphids, leafhoppers, and thrips are examples of insect virus vectors, but not all aphids, leafhoppers, or thrips can spread a specific plant pathogenic virus.

It is also important to note that gardeners may spread plant viruses as they work in the garden. Mechanical abrasion from infected tools or touching with infected hands and creating slight wounds on plants may be all that is needed for virus transmission. Viruses overwinter in infected perennial plants or overwintering insects. Some viruses can be transmitted through seeds, although most are usually transmitted through vegetative propagation structures.

Viruses can cause mottling, spots, and mosaic-like patterns, crinkling, and other malformations on leaves and fruits, and may stunt plants. Since viruses are systemic in all plant tissues, infected plants must be rogued or discarded from the garden.

Viruses are named according to the first plant on which they were found and the type of symptom they cause (e.g., cucumber mosaic virus). There are numerous viruses that can infect vegetable plants growing in a home garden and two of the more commonly found viruses are tomato spotted wilt virus of tomato and cucumber mosaic virus of vine crops.

Phytoplasmas. Phytoplasmas are any of a group of extremely small bacteria that are similar to **mycoplasmas** in that they have a cell membrane instead of cell walls and can assume a variety of shapes, but are parasitic solely in plants. They cause plant diseases by infecting phloem tissue, are transmitted especially by homopteran insect vectors (insects with piercing-sucking mouthparts), and have proved extremely difficult to grow in artificial media, also called *mycoplasma-like organism*. They are obligate parasites, meaning they can survive only within their host. Phytoplasmas live in the phloem of host plants and are vectored by certain phloem-feeding insects, such as leafhoppers. This pathogen causes distortion, yellowing, and wilting, and immature leaf veins may appear clear (called "vein-clearing"). Also, flower parts may become vegetative and oftentimes will not develop properly on infected plants.

Aster yellows is a common disease caused by a phytoplasma that damages vegetables including carrots, lettuce, and potatoes. Aster yellows is spread by the aster (or six-spotted) leafhopper. These insects are small (about ⅛ inch long), gray-green, and wedge shaped. They feed only on phloem tissue and can generally be found on the undersides of leaves. Aster leafhoppers do not overwinter in cold climates, but are blown in from warmer winter areas during the spring or early summer. Dry weather can cause increased disease occurrence in the home garden, as leafhoppers move from plants in dry prairies, woods, or pastures to irrigated gardens. Generally, aster yellows symptoms appear in mid- to late summer. Although aster leafhoppers spread the pathogen, placing infected plants in the garden is a way to spread this disease as well. Management strategies for aster yellows include planting healthy plants, controlling weeds that may harbor the insects, and removal of infected plants from a garden. Although only one part of a plant may appear to be infected, it is generally assumed the phytoplasma infestation is systemic and is in all plant parts. The pathogen can overwinter in plant crowns and roots. Leaves and stems that develop from this tissue will always be infected and provide a source of inoculum for other susceptible plants.

Nematodes. Nematodes are microscopic roundworms that live in soil, water, and plant material. They have a spear-like stylet mouthpart, require free water to move around, and reproduce by eggs. They are readily spread by water, infected plant material, soil, gardening implements, and in some cases, insects.

Nematodes cause a variety of symptoms including stunting, yellowing, and wilting of plants. Some infected plants simply appear unhealthy but will continue to survive throughout the growing season. Some nematodes (such as the root-knot nematode) will cause strange, knot-like growths to develop on plant roots. Nematodes can be a major problem in gardens that are in warm-winter temperate climate areas and that have a high proportion of sand in the soil, although in most cold-climate gardens, they tend to not be as much of a issue.

Abiotic Disorders. Abiotic plant diseases are common in many home gardens and are non-living factors that cause physiological diseases (or environmental diseases) to appear in a garden. Abiotic agents are noninfectious and non-transmissible. These physiological problems can result from environment conditions such as moisture and stress, nutrient deficiencies in soils, or chemical injury from pesticides, to name a few. A common physiological problem that occurs on tomato fruit grown in a garden is fruit cracking, which is related to inconsistent moisture provided to plants. Chemical injury is plant damage caused by pesticides, fertilizers, or other chemical products. Herbicides commonly cause injury to non-target plants in a garden and can be seen as chlorosis, necrosis, distortion, and elongated growth. Glyphosate and 2,4-D are examples of common herbicides that cause chemical injury to desirable plants when used incorrectly. Excessive concentrations of these chemicals cause twisting and curling of stems, stem swelling, weakened cell walls, rapid cell growth, cellular and vascular damage and death.

Management of Plant Diseases is best achieved through an integrated approach, which includes a combination of cultural, mechanical, biological, and chemical practices. A few examples of each are provided next.

Cultural management includes many different practices that can be used to prevent the appearance of diseases in a garden including crop rotation, mulching, and sanitation. There are also many **Mechanical management** techniques that could be used, such as using insect barrier cloth to prevent the transmission of important diseases by excluding the insect vector. **Biological controls** include the use of compost and compost teas, which may suppress pathogens by introducing beneficial microbes. In addition, the roots of several plants can release chemicals into the soil that are detrimental to plant pathogens. For example, various chemicals released by the roots of various mints destroy the disease organism that causes clubroot in the cole crop family, and French and African marigold roots release compounds that will reduce pathogenic nematode populations in soils. **Chemical control** generally refers to the use of fungicides, bactericides, or nematicides to manage a disease problem, although home gardeners are limited to the products that can be used compared to commercial growers. It is important that the disease is first accurately identified, and then an appropriate product is selected with label directions followed.

Preventing diseases in garden

Most gardeners wait until a disease problem occurs in the garden and then take steps to control the disease. However, the best approach for disease control is to prevent their occurrence in the garden. Several garden management practices will aid in disease prevention in a garden, including site selection, crop rotation, planting disease resistant varieties, use of disease-free seeds and transplants, planting date management, proper plant spacing, water management, mulching, proper plant fertilization, maintaining a weed-free garden, use of proper sanitation techniques, and use of pesticides as the last option.

Garden management practices

Site selection is an important part of disease management since poorly drained soils with excessive soil moisture can lead to the likely occurrence of root and crown rot diseases. Soil-borne diseases such as *Phytophthora* can be devastating in a poorly drained garden soil. Thus, it is important to have a well-drained soil at the site a garden will be located. Full sunlight is also important, which relates again to site selection, as full sunlight will maximize vegetable crop productivity and speed the drying of plant foliage, which will reduce the incidence of many foliar diseases.

Crop rotation is also an important part of garden disease management. The continuous planting of plants in the same vegetable family in the same garden spot year after year will provide opportunities for pathogen build-up in the soil. Thus, it is important to remember to grow the same type of vegetable or closely related vegetables in the same soil only about once every three to five years. This practice will reduce the incidence of many pathogens that cause soil-borne diseases. However, longer crop rotations and/or other management methods may be needed to obtain adequate control for soil-borne problems such as root and crown rot diseases caused by the fungal pathogens *Phytophthora*, *Rhizoctonia*, *Pythium*, and *Sclerotium*, vascular wilts caused by *Ralstonia* or *Fusarium*, and root-knot nematodes. Many of these organisms tend to be long-lived and will infect many different plant families, so a short crop rotation is not very effective.

Planting **disease-resistant varieties** is really the most efficient way to manage vegetable diseases in a home garden. Gardeners should try to make an effort to buy disease-resistant varieties when they are available. It is important to note that resistance is a relative term and resistant varieties can be either totally immune or only partially resistant, depending on the disease and the variety grown. Some disease can generally be expected on partially resistant varieties, which is normally enough to manage the disease in most years.

The use of **disease-free seeds and transplants** is also an important part of maintaining a disease-free garden. Oftentimes, infected seeds and transplants can bring new plant pathogens into a garden that can cause severe disease outbreaks. A gardener should try to limit the introduction of disease pathogens into a garden, as the issues that develop from these pathogens will have to be dealt with at a later time. When choosing or selecting seeds and transplants to be used in a garden, try to make sure that they are free of disease pests.

Planting dates can also be used as an effective disease-management tool. It is best to follow the recommended planting dates for the particular vegetable being grown, since planting vegetable seed at an improper soil temperature will not only reduce germination but can also increase seed and seedling diseases. Remember that warm-season vegetables are most susceptible to damage from cold temperatures, while cool-season vegetable seed may not germinate if soil temperatures are too high. In addition, planting-date adjustment can also be made to avoid times of year when disease pressures are at their highest levels, which is normally during the warmest summer months. So, early planting can be used to avoid certain diseases, such as powdery mildew and virus diseases on summer squash and downy mildew on cucumber.

Proper plant spacing can be used to reduce the occurrence of many vegetable foliar diseases, especially fungal and bacterial diseases that thrive in high-moisture environments. When plants are spaced properly apart, good air circulation will often prevent disease outbreaks. If plants are wet for extended periods and are crowded together, they do not dry and have a greater chance to become infected with a disease.

Proper water management in a garden is important to prevent disease outbreaks. Soils that are consistently overly wet are more prone to develop soil-borne diseases such as seed decay, damping-off, and root and crown rot. Furthermore, constant moisture on foliage will also contribute to increasing foliar disease development. It is important that gardeners water deeply (about 1 inch) and only when needed (about once or twice per week) to supplement natural rainfall. More water will be needed during hot, dry weather conditions, while very little additional water may be required during cooler months.

Drip irrigation is an easy way for a gardener to improve water-use efficiency in a garden, as water is slowly released to obtain a thorough soaking of the soil and is the most efficient irrigation method. Hand watering can be an effective irrigation method for small garden plantings as long as the water is directed at the base of the plant and not on the foliage. Sprinkler irrigation is the least desirable watering method in terms of efficiency and disease prevention, and if sprinkler irrigation is used, it is best to water during the mornings to allow foliage to dry before nighttime. Do not use sprinkler irrigation in the late afternoon or evening. It is also important to note that a garden should not be worked when plants and soil are wet. Bacterial and fungal diseases will easily spread from one plant to another on hands, gardening implements, and clothing when above-ground plant parts are wet.

Mulching is another method that can be used to prevent disease development in a garden. There are many plant pathogens that survive in the soil and overwinter in soil debris. Organic mulches such as straw, bark, leaves, and shredded paper or inorganic plastic mulches will help to prevent soil from splashing onto plants and fruit from touching the bare ground. The use of organic mulches should reduce fungal spores or bacteria from splashing onto plant foliage from the soil. Ripening fruit such as tomatoes, squash, cucumbers, and melons can easily develop fruit rots when touching bare soil, and a mulch layer over the soil can prevent or reduce fruit rot development on vegetables that would normally contact the soil. Organic mulches also help conserve soil moisture, reduce weed infestations, and improve overall soil and plant health.

Proper fertilization practices will also help to keep plants healthy and less likely to become infected by diseases. Weakened plants are typically more susceptible to pathogen attack compared to healthy plants. It is important to follow suggested nutrient recommendations for each vegetable so that appropriate nutrients are applied. It is also important that soil pH be adjusted to obtain a pH in the range of 6.5 to 7.0 for maximum nutrient uptake.

Weeds can harbor diseases and insects. Many weed species serve as reservoirs for several insect-transmitted viruses that can infect vegetables grown in a home garden. Also, do not forget that weeds will also compete with vegetables for space, nutrients, and sunlight. Weed control will most likely also increase air movement in the garden and decrease environmental conditions, such as excessive moisture, that favor disease development.

The use of various **sanitation practices** is one of the most important ways to reduce disease outbreaks in a garden. Sanitation is focused on reducing the survival of disease-causing organisms and limiting future disease outbreaks. An easy but effective sanitation practice is the removal of plant residue from gardens after harvest, as this method removes any garden refuse that might harbor pathogens over the winter. Another way to limit leaf spot outbreaks in a garden, such as early blight on tomatoes, is to remove and destroy infected leaves throughout the growing season. Another way to prevent pathogen movement from plant to plant in a garden is to sterilize pruning and cutting tools with alcohol or a diluted chlorine solution after each plant.

Pesticides should be used only after all other disease-prevention options have failed. Home gardeners have fewer pesticides available to them compared to commercial growers, and the available products may be less effective. Most types of vegetables can be grown in the home garden without the use of pesticides. The use of fungicides as a preventative measure may be an option if certain diseases have been problematic in previous years and other management options have not been effective. Maneb, mancozeb, sulfur, and copper products are home gardener options that can be used on certain vegetable crops to control certain diseases. These products suppress a range of fungal pathogens and are most effective if they are applied in a preventive manner either before or at the very onset of a disease outbreak. As with any pesticide, read the label and follow the recommended precautions to ensure the chemical is applied in a safe, effective manner.

Plant diseases can rapidly spread under certain conditions in a garden and quickly wipe out or reduce the productivity of vegetable crops. However, a gardener can use any of several different management tactics to prevent disease development in a garden without relying just on the use of chemical fungicides or bactericides.

Insect Pests. Without some type of insect-management program in place, home gardens can be devastated by various insect pests that not only reduce the productivity of a garden but also the quality of the products produced. Thus, insect pests need to be controlled in some manner in a home garden. Insects can rapidly devastate vegetable plants and/or developing or mature fruits in a garden, and are often the most important pest of vegetables in a garden. Oftentimes, unless they are controlled, there may be no vegetables that can be harvested from a garden. Several prevention and control measures will be discussed, which should give a home vegetable gardener a starting point in preventing and controlling insects in a garden.

There are really two ways of controlling insect pests in a home garden: try to prevent insects from being problems in garden (preventative measures), or control pests once they appear in garden. **Scouting the garden for insect pests and learning to identify problematic insects is an important step in insect control**. The sooner an insect pest is identified in the garden, the easier it will be to control. Thus, without proper insect identification, management or control is impossible. A higher amount of insect control can be expected if the insect is identified when the population is small in the garden. So, it is important to frequently survey and monitor insect pests in a garden to detect problems early. Once insects have appeared in the garden, there are several control techniques that home gardeners can use to reduce or eliminate them. Some of these control measures are discussed below.

Insect Control Measures for Home Gardens. There are many control measures that can be effectively used to manage insects in home gardens. A few examples are provided next.

Scout garden. Gardeners should randomly select plants and monitor or scout the garden twice weekly. This includes inspecting the plants from the bud to the soil, including upper and lower leaf surfaces. There are numerous destructive insects that are important to know by sight in a garden. A few of these are briefly described next. **Aphids** are often referred to as plant lice, develop several generations in each growing season, and with their piercing-sucking mouthparts are able to extract sugars from phloem tissues in a plant. **Cabbageworms** include larva such as the cabbage looper and imported cabbageworms and are serious threats to cabbage, cauliflower, broccoli, and other cole crops. **Cucumber beetles** are very important and highly damaging insects that feed exclusively on vegetables in the vine crop family. There are two types: striped (yellow with 3 black stripes on its back) and spotted (yellow and green with 12 black spots on its back). Although they do little damage by feeding on leaves, it is important to remember that they are vectors of bacterial wilt, which can be a devastating disease on many cucurbit crops, especially cucumber and cantaloupes. **Colorado potato beetle** is the most important insect to control when growing potatoes, as both the adult and larval forms feed on leaves (see Fig. 6.45). When in high enough numbers, this insect can completely defoliate a potato plant. The adult is yellow with 8 black stripes on the wing covers and has black dots just behind the head. **Hornworms** are devastating defoliators of tomato and other solanaceous crops. The larvae are bright green with diagonal white stripes on the side and can grow to be quite large, as they are voracious eaters (see Fig. 7.37). **Bean leaf beetle** adults are reddish to yellowish with black spots on the back and about 1 inch in length, while larvae are a slender, whitish-colored grub that is brown at each end. They tend to eat irregular holes on the bean foliage. The **Mexican bean beetle** is oval shaped, about ¼ inch long, with a copper color and has 16 black spots, while the larvae are orange to yellow, fuzzy or spiny, and about ⅓ inch in length. Both the adult and larval forms of this insect feed on the leaf undersides; and after feeding, the foliage appears lace-like. **Cutworms** can often be a devastating insect in home gardens, as they will eat through the stems of plants just at the soil surface during the nighttime, and these plants can be found cut off and wilting on the soil the following day. The larvae of this inset are plump, soft-skinned, and usually a dark brown or blackish color. **Thrips** are tiny elongated insects that eat at the surface of plant tissues by rasping away small areas, and this often causes white streaks to appear on eaten areas of leaves. They tend to be more problematic under dry conditions and it is difficult to grow onions without some level of thrips feeding on the plants. **Flea beetles** are tiny black insects and are about ⅛ inch in length. They readily jump when touched or disturbed. Although this insect feeds on the foliage of many vegetables, it will defoliate young eggplant seedlings in a garden unless controlled in some manner. **Squash vine borers** rarely cause problems in home gardens, but it should be noted that larvae bore into the stem and consume the interior stem tissues, which first appear as wilting with eventual plant death. If the stems of a dead or wilting squash plant are opened and a worm is noticed tunneling in the stem, then this insect is most likely the cause of the problem. **Squash bugs** are serious pests of squash and pumpkins, as they suck plant sugars from the plants, which often results in death. These insects lay orangish-bronze egg masses on squash and pumpkin leaves that can be readily observed by just walking through a garden. The eggs hatch into nymphs and later develop into adults. **Spider mites** are really not true insects, but are closely related to spiders since they have eight legs. They often appear on plants, especially during dry conditions, as tiny moving specks within a webbing that covers the plant in the garden (see Fig. 9.2).

Figure 9.2 Spider mite webbing on a tomato plant growing in a home garden.

Crop rotation. Rotate vegetable plantings among different plant families within the garden so that the same vegetable is not grown in the same garden location more than once every three years. Although this will not control general feeders, this will often reduce the occurrence of insects that feed on specific vegetables.

Till garden soil. When a garden soil is tilled during the late fall, winter, or early spring months, insects that overwinter in soil will be brought to the soil surface and often killed due to the cold harsh environment. Many insects cannot withstand low or freezing temperatures. This practice is especially important if an area has been in grass that will be used to grow vegetables.

Handpick insects or use sticky traps. Often large insects can be removed by hand and destroyed, which is a very effective practice to use in a garden when few insects are present.

Another way to remove insects is to spray plants with water to knock insects off foliage or fruit. The use of sticky cards (usually yellow) can be used to trap insects, which will reduce populations of several insects in a garden including aphids, thrips, and whiteflies.

Insect-free transplants. Transplants should be inspected for the presence of insects prior to purchase and before bringing into a home garden. Transplants should be purchased from a reputable seller, and a gardener should make every effort to ensure that plants are free of insects at the time of planting. Insects brought into a garden will definitely cause problems and have to be dealt with at some point in the growing season.

Plant early. Insect populations tend to increase as the growing season progresses, up until frost. So, certain insects can often be avoided if vegetables can be planted early in the growing season to avoid large populations of insects that build up throughout the season. Sweet corn is a perfect example, as early planted sweet corn often avoids damage caused by the corn earworm, which becomes more of a problem later in the growing season.

Mechanical exclusion. The use of cloth insect barrier covers has become a popular practice to exclude and prevent insects from feeding on plants in a garden. However, remember that although the barrier can be left on certain crops that do not require pollinators for the entire growing season (e.g., broccoli, cabbage, and spinach), it must be removed at flowering for some crops such as the cucurbits to achieve pollination and fruit set.

Biological control. Biological control agents should be encouraged in a garden. These typically include natural enemies and microbial agents. Insect pests frequently have natural enemies that help provide control. These insects and spiders, referred to as beneficials, often exist in the landscape naturally but can also be introduced into a garden. Beneficial insects may be predators or parasites. A common example of a beneficial predator that most gardeners are aware of is the lady beetle. Both the larval and adult stages of lady beetles eat aphids and other soft-bodied insects. There are numerous other important predators including lacewings and spiders. Parasites are organisms that live on or inside another organism. The host organism often dies after the parasite has completed its development, and the most typical parasites are small wasps that use caterpillars, whiteflies, aphids, and other pests as hosts.

Bacillus thuringiensis (Bt) is a naturally occurring microbial insecticide that is an effective biological control tactic for many problematic garden insects. This microbial biological control agent produces a toxin that is specific toward certain types of pests, such as caterpillars. This product is typically sprayed on plant foliage and when eaten by caterpillar pests, death results in only a few days. The primary benefit of this biological control agent is that it is not harmful to humans, birds, beneficial insects, or fish.

Sanitation. Many garden insects overwinter in plant debris left from the previous year's crop and in weeds near or in the garden area. Thus, plants debris and weeds should be removed and composed or tilled into the soil as soon as vegetable harvest is completed. It is also important to control weeds growing in or near garden areas. Some insects will overwinter on weeds or live on weeds for a certain period of time and then move onto garden plants. Removing weeds around and in the garden will decrease the presence of certain insects in a garden. Also, mature, unused, or unwanted vegetables should also be removed, as many insects are attracted to and feed on over-mature or decaying fruit in a garden.

Insecticides. Although the use of insecticides is really the easiest method of insect control, it should really only be used as a last resort. It should be noted that most insecticides are "broad-spectrum," meaning that they will kill a wide variety of insects, including beneficials.

Insecticides should be carefully selected to control the insect pest in question or to reduce its numbers so that a garden can remain productive. In order to reduce bee and other pollinator mortality, insecticide sprays are best applied late in the afternoon or early evening hours. Most pest control sprays, even those that are natural or organic, can also harm bees, butterflies, and other beneficial insects, so be very careful when applying any insecticide. Lastly, it is important to note that an insecticide will not provide 100% control of the targeted insect pest.

Figure 9.3 Weed control is an important part of maintaining a productive home vegetable garden.

Weed Pests. Weeds are generally defined as **plants growing out of place** and even a vegetable like a tomato can become a weed if it is growing where it is not wanted. Since any plant growing out of place can be considered a weed, a sweet corn plant (from a carelessly dropped seed) growing in a row of snap beans is technically a weed; but some of the more common garden weeds are crabgrass, yellow and purple nutsedge, morning glories, common purslane, ragweed, bermudagrass, lambsquarters, and pigweed. Thus, weed control is an important part of maintaining a productive home vegetable garden (Fig. 9.3) Weeds can become serious garden problems, as they compete with vegetable plants for moisture, nutrients, space, and sunlight, and can also harbor insects and diseases that can attack vegetable plants in a garden. Weeds are difficult to control because they grow rapidly, produce large numbers of seeds, and spread aggressively by vegetative structures and/or seeds. Yellow nutsedge is often a very difficult weed to manage in home gardens due to the large amounts of underground tubers that are produced by just one plant (Fig. 9.4). The presence of weeds in gardens can and will reduce plant growth, which affects both the quality and yield of vegetables. Weeds can also hinder the harvesting of vegetables, provide cover for animal pests (rodents, box turtles, etc.), and serve as alternate hosts for vegetable diseases and insects.

Weeds that emerge soon after planting vegetable seeds or transplants need to be removed quickly after they appear in a garden (usually before they are greater than 3 inches in height). Large weeds are difficult to remove from a garden without damaging the roots of adjacent vegetables. Early-season competition of weeds with vegetables may reduce vegetable growth, yield, and quality if weeds are allowed to get larger than 3 inches in height.

Weeds can kill a gardener's enthusiasm for maintaining a garden, which can cause the garden to be abandoned in midsummer. So, it is important to control weeds while they are small and before they get out of control. The best time to remove weeds is just as they appear on the soil surface, and if the weeds are allowed to grow to be too large, they will shade the vegetable plants, resulting in reduced productivity. Most weeds can be prevented from becoming serious problems in the garden. There are several methods that can be used in a combined approach to control weeds in home gardens. These include **mechanical, cultural,** and **chemical** methods.

Figure 9.4 Yellow nutsedge is a difficult weed to control in many home gardens. Here it is rising above cucumber foliage and providing competition.

Weed Control in Home Gardens. There are several **mechanical methods** for controlling weeds in a garden. Hand-pulling is not always an effective way to control weeds, especially in a large garden, but it can be effective under certain circumstances. Weeds that grow between closely spaced rows of vegetables, in raised beds, through planting holes of a plastic mulch, or in small gardens must generally be pulled by hand. Also, weeds growing in containers used for vegetable production must also be hand-pulled. Hand-pulling of weeds must be done to maintain a productive garden. Cultivation is the most widely used method of garden weed control, and it must be repeated over and over throughout the growing season, as new weed seedlings emerge. A variety of tools can be used for weed cultivation, but those most commonly used are the hoe and garden tiller. Cultivation for weed control in small gardens is generally achieved by shallow hoeing or using a wheel-hoe, while in larger gardens, garden tillers, rototillers, or larger garden-type tractors are often used to remove weeds between rows. A gardener should try to use shallow cultivation if possible so as not to disturb or injure vegetable plant roots. It is also important to remove perennial weeds, such as bermudagrass, from the garden following cultivation because plant propagules (stolons) that have no roots can possibly form roots and make the bermudagrass problem worse.

There a few different **cultural control** practices that can be used in a home garden. Mulching is a common method that is oftentimes used in a garden. Mulching is probably the easiest and best method to control weeds in a garden. Organic and plastic mulches are useful in preventing most weeds from becoming established in the mulched area. An organic mulch about 6 inches deep, a black plastic mulch, or a weed barrier cloth will prevent light from reaching germinating weeds and thus prevent weeds from becoming problems in gardens. Although mulching works with germinating annual weed species, it is not effective for perennial or established annual weeds.

Some weed suppression can be achieved by certain established vegetables, as some vegetables have the ability to shade out weeds to prevent them from growing and developing properly. Those that have good weed suppression ability include sweet potatoes, cucurbits, sweet corn, and snap beans, while those with poor suppressive abilities include lettuce, carrot, pepper, tomatoes, and various cole crops. Another cultural practice is the use of "smother" crops (often called living mulches), as these are cover crops that are inter-cropped along with vegetables, and can be used to reduce weed populations in current and future vegetable crops grown in the garden.

Chemical weed control (or herbicides) is another weed-control method that some gardeners utilize for weed control, although their use is generally not recommended for home gardens. Commercial vegetable growers have a wide selection of many different herbicides that can be used to provide weed control, but gardeners have a limited number of herbicides available for use in home gardens. It is important to understand that there is no one herbicide that will control all weeds in a garden and that can be used on all vegetable crops. First, it is important to remember that **no single herbicide** will control all weeds in all vegetable crops, and secondly, some vegetables are insensitive to certain herbicides, while others are sensitive to the same herbicide. If the wrong herbicide is used, the vegetable is injured along with the weeds. There is also the problem of applying small amounts of an herbicide to certain areas of a garden. An under-application will result in poor weed control, and an over-application many times will injure or kill vegetable plants in a garden. There are several chemical manufacturers that sell a wide variety of herbicides in small quantities for home gardeners. These products can often be found at a local garden center, but make sure the **vegetable or vegetables** in which weed control is needed is **listed on the container** before its use and also to **use the correct rate** that is given on the container. So, before using an herbicide in a garden, read the product label for a listing of the vegetables it can be used on, the recommended rate of application, and the method of application.

Herbicides applied to the soil before vegetables are planted and before weeds have emerged are called **pre-emergence** herbicides. These herbicides are generally applied immediately after the vegetable seeds or plants are planted but before the weed seeds germinate. **Post-emergence** herbicides are applied after weeds have emerged. For control of weed seedlings before vegetable seedlings emerge from the soil, there are a few herbicides (e.g., Dacthal and Trifluralin) listed for some, but not all vegetables—check the label. Several herbicides are listed next that can be used in home gardens.

Dacthal. There are several brand names of this herbicide product available. Dacthal can be used on a wide variety of vegetable plants, and when applied correctly, gives good control of most grasses and a few broadleaf weeds. This is a pre-emergence herbicide that controls weeds as their seeds germinate.

Trifluralin. Again, there are several different brand names available for this product. Trifluralin is a pre-emergence herbicide used to control grass problems in various vegetable crops. Some planning of the garden is required to group vegetables together that are labeled for trifluralin use, so that vegetable crop injury is minimized. This herbicide is labeled for use before planting seeds of several vegetables and before setting transplants of others. It is important to read the package label for a list of approved vegetables and to follow directions to minimize crop injury. Broadcast the recommended amount of either granules or the liquid formulation on the soil, although granules are generally easier for most gardeners. After application, mix the herbicide into the top 2 inches of the soil, and use two very shallow cultivations to provide good mixing into the soil. Trifluralin should be slightly tilled into the soil after application to obtain the best weed control.

Sethoxydim. This post-emergence herbicide selectively controls emerged grass weeds in most vegetables, except sweet corn. This herbicide can be applied post-emergence to broadleaf-type vegetable plants and should be applied to most grasses before plants reach 8 inches in height. One application of this herbicide will generally control most annual grasses, but several applications may be required to control perennial grasses like bermudagrass. It is important that a sticker-spreader material (such as a crop oil concentrate) be mixed into the spray solution before application; otherwise this post-emergence herbicide will not work properly. It is important to read the label for specific instructions and approved vegetable crops.

Glyphosate. Formulations of this popular nonselective, post-emergence herbicide are approved for limited use in the vegetable garden. Most applications are recommended for eliminating existing weed problems before vegetable seedling emergence or before vegetable plants are placed in the garden. As with any herbicide or pesticide, the label should be read for specific application instructions and limitations. This is a systemic herbicide and spray drift to non-target plants will often cause injury.

Before herbicide weed control methods are used in home gardens, it is best to try all other methods of control and to use it only as a last resort. In general, if all weeds that appear in a garden in a given year are controlled, a gardener will have only about one half as many the following year to deal with and then one half that number the year after that if the control is continued. The weed seed bank in the soil is thus being depleted over time. It will usually take about 5 to 7 years to eliminate most weeds in garden using good weed-control practices. If this protocol is followed in a home garden, weeds will eventually become a minor problem in the garden.

ANIMAL PESTS

Animal pests such as birds, deer, gophers, moles, raccoons, and rabbits can discourage gardeners to the point where they no longer want to plant a garden. Animal pest populations in some areas are so large that they can destroy a garden in one night. Unfortunately, there is no quick method to control animal pests. There are many different control methods, and most gardeners will try several different ones to see which works best.

Birds. Birds can quickly decimate a vegetable garden. Oftentimes, it is necessary to protect young seedlings as well as certain types of vegetable fruit from birds. They will scratch into soil and eat seeds before they sprout, pull up tender seedlings, or raid the plants for maturing vegetables and berries. Birds will often pull up young seedlings, especially sweet corn, to eat the kernel that is attached to the seedling underground. Many gardeners must protect plants from marauding birds and their damage from the time seeds are sown until vegetables are brought into the kitchen. A favorite method is to hang reflectors or fluttering objects over the garden area to discourage birds, but this method is oftentimes not very effective. Bird netting is probably the best method that a gardener can use to exclude birds. Various types of bird netting can be purchased from most garden stores and allows a gardener to create a shield over the plants in a garden to protect from bird damage.

Gophers. Gophers can oftentimes be very damaging to a home garden and the surrounding landscape. A single gopher moving down a garden row can inflict considerable amounts of damage to a garden in a very short period of time. Besides damaging plants in gardens, gophers also gnaw and damage plastic water lines and other components of irrigation systems. Their tunnels can divert and remove irrigation water targeted for garden plants, which can lead to soil erosion problems. Although some gardeners try to control gophers by trapping and/or using poison baits, many gardeners who live in areas where gopher populations are abundant will place a fence around a garden area that is made of at least 1-inch galvanized wire about 2 feet high and secured to the ground to prevent damage. In addition, during construction of raised beds, wire can be laid under beds prior to filling with soil to exclude gophers.

Moles and Voles. Both moles and voles are mouse-like garden pests that cause subterranean damage. Moles are carnivorous and eat worms, grubs, and adult insects, while voles are primarily vegetarians and will actively gnaw on plant roots and stems. It is also important to note that other rodents exploit mole and vole tunnels to feed on plants. There are generally two ways to solve problems with these pests in a home garden: traps and baits. Trapping is the most effective, humane, and eco-friendly solution. Repellents and other home remedies generally do not work. Many gardeners have found that mousetraps set inside an active tunnel are really the most effective way to control these pests. However, the key is learning how to set traps properly and that the two pests will not necessarily be attracted to the same baits. A vole may be attracted to peanut butter as bait, while a mole most likely will not. Moles eat primarily worms, grubs, or insects, so it is best to bait a trap with these when trying to catch these pests.

Rabbits. These pests are generally excluded from home gardens by fencing. Rabbit-proof fences should be 2 ft high and made of at least 1- to 2-inch galvanized wire to prevent their entry into garden. The bottom portion of the fence should be sunk and buried at least 6 inches into the soil to prevent rabbits from digging and crawling under to reach the garden. There are other home remedy methods such as sprinkling dried blood, black pepper, or lime around plants to prevent feeding. Some gardeners will also use mothballs extended about 5 to 6 inches off the ground at various places in a garden to prevent rabbit feeding.

Raccoons. Raccoons are really a serious problem when trying to grow sweet corn, as they tend to eat the ears a few days before they reach optimum quality. Many gardeners say that raccoon damage is minimized when pumpkins, gourds, or winter squash are planted among the sweet corn plants, since the spiny vines discourage them from entering the area of the garden where the sweet corn is growing. Some gardeners suggest that cayenne pepper sprinkled on corn silks will also act as a deterrent. Other suggestions for preventing raccoon damage include covering ears with paper bags to prevent damage or the use of an electric fence around the sweet corn plants. In this case, a low electric fence using slippery PVC pipe as posts should do the trick to prevent their entry into a sweet corn planting. The fence should be about 1 ft high with two hot wires placed 6 inches apart.

Deer. Deer can completely destroy a home garden in one night and are one of the most difficult pests to prevent from entering a garden. Some gardeners say that various deer repellants work and there are several that are commercially available, but tend to work for a only few weeks at the most, if they work at all. Some gardeners say fences work, but deer can easily jump standard-sized fences. If deer are a major problem, then the best solution would probably be to erect a tall electrified fence.

Chapter Ten

ORGANIC GARDENING

Organic gardening is a widely used gardening method that many home gardeners utilize today. Growing organically was once seen as something practiced only by hippies, but it has become more mainstream in recent years. Consumers want the food served to their families to be safe and healthy, and to be produced without destroying the environment. This desire for wanting to do no harm to our families and the world around us is the main reason many people grow in an organic manner. Organic gardening is a way of taking control at the home level to grow foods that you know have no pesticide residues. Research has shown that the primary benefit of organic foods is the reduction in pesticide residues, although they are similar to those produced in a conventional manner (using synthetic chemicals) for vitamins, phytonutrients, and nutrients contained within the product. In organic gardening, little to no pesticides are used, and synthetic fertilizers cannot be used to provide nutrients to plants. Basic information will be provided in the chapter for the development and maintenance of organic vegetable gardens.

If you ask organic vegetable growers, they will tell you that the soil is the key component to growing crops in an organic manner. Most organic gardeners consider the soil as a living system, and try to maintain as much life in the soil as possible, which includes microbes, earthworms, and insects. The use of pesticides and synthetic fertilizers will often kill or discourage beneficial soil organisms from growing in the soil.

An ideal garden soil should have high biological activity; be well drained, loose, and friable for a deep rooting zone; easily penetrated by air, water, and roots; have an adequate water-holding capacity; provide adequate nutrients for plants; and resist erosion to some extent. It is also important to note that the structure of a soil can be improved by the addition of organic matter. **Soil structure** is determined by how individual soil particles clump or bind together and aggregate, which influences the arrangement of soil pores between them. Soil structure has a major influence on water and air movement, biological activity, root growth, and seedling emergence. Soil structure improvement will lead to a greater amount of soil aeration and water-holding capacity in the soil, as well as increasing the biological activity of a soil (such as beneficial microbes and earthworms). Organic matter tends to bond sand, silt, and clay particles into soil aggregates, which will lead to a good "tilth" or the ability of a soil to be easily tilled. The long-term objective is to build up soil humus with organic matter additions. **Humus** is the more-or-less stable fraction of the soil organic matter that remains after decomposition of plant and animal residues. There are many ways to increase organic matter levels in a soil, including the use of cover crops (green manures), animal manures, compost, and organic mulches. Composting and organic mulches were considered in Chapter 2, but cover crops and animal manures will be discussed in further detail in this chapter.

COVER CROPS

Many organic gardeners will use some type of cover crop in their garden for various reasons. Cover crops will protect the soil from erosion, prevent nutrients from leaching from the soil, provide an organic mulch on the soil surface, and increase organic matter levels in the soil. The most commonly used cover crops are **annual grasses** and **legumes**. Grass cover crops germinate quickly and grow sufficiently in the autumn months to protect the soil over the winter from erosion. Grass cover crops do not fix nitrogen and will tie up soil nitrogen temporarily when tilled into the soil. Legumes tend to grow slowly when planted in the autumn, but will grow rapidly the following spring, and provide an additional source of nitrogen for the summer vegetable crop(s). Fall-planted cover crops are more competitive with spring weeds than are those planted in the spring. Cover crops often consist of mixtures of legumes and grasses due to the advantages provided by each. For example, winter ryegrass and hairy vetch are often grown as a mixture and planted in the autumn (Fig. 10.1). The winter ryegrass provides a significant amount of biomass, which can add lots of organic matter to the soil, while the hairy vetch fixes nitrogen that can be used by the subsequent crop.

Figure 10.1 A mixture of winter ryegrass and hairy vetch growing as a cover crop during spring.

Catch crops. A **catch crop** is a fast-growing plant that is grown with or between successive plantings of the primary crop. For example, turnips can be grown between the rows of other vegetables and harvested long before the primary vegetable crop matures. Also, a catch crop, such as buckwheat, can be planted between the spring harvest and autumn planting of some crops. Cover crops can also be used as catch crops to keep nutrients from leaching from the soil, and to hold nutrients so they will be available for the next vegetable crop.

Green manures. When actively growing cover crops are tilled into the soil before they mature, they are termed as a **green manure**. Thus, it is just the practice of incorporating undecomposed green plant tissue into the soil. It will provide many benefits to a soil system and will aid in maintaining or raising the crop-producing capacity of a soil. Cover crops to be used as green manures are typically sown in the spring or summer, and there are several possible legumes and non-legumes that can be used as green manures (Table 10.1). These cover crops can be incorporated into the soil before, during, or after flowering. Since the plant foliage is young, green, and succulent when the cover crop is incorporated into the soil prior to flowering, a rapid increase in soil biological activity occurs, which leads to a quick decomposition and a fast release of nutrients into soil. Incorporating at flowering is usually recommended since plant decomposition and nutrient release will be sustained over a longer period of time compared to tilling into soil at the pre-flowering stage. Also, at the flowering stage, more plant biomass has been produced compared to incorporating earlier, and thus, more organic matter can potentially be added into the soil system. Lastly, the incorporation of the cover crop after flowering can lead to several problems, such as reseeding and high C:N ratios. If allowed to reseed, the cover crop often becomes a weed in succeeding crops, which can result in lower crop productivity. A high C:N ratio will lead to slow cover crop decomposition, which can reduce the amounts of nutrients available for the following crop(s). For most vegetable crops, a cover crop that is going to be used as a green manure should be incorporated at least two weeks or more before planting (Fig. 10.2).

There are several desirable characteristics of using cover crops as green manures:

1. Easy to establish and grow quickly
2. Produce a large amount of biomass in a short period of time
3. The growth habit should form a ground cover after it becomes established, which aids in preventing soil erosion

Table 10.1 Possible green manure crops to use in home gardens.

Legumes		Non-Legumes	
For regions with warmer temperatures	*For regions having a wide temperature range*	*All are suggested for wide temperature range areas*	
Austrian Winter Pea	Alfalfa	Barley	Annual Ryegrass
Bur Clover	Crimson Clover	Buckwheat	Mustard
Crimson Clover	Cowpea	Millet	Rape
Crotalaria	Red Clover	Oats	Sudan Grass
Lespedeza	Soybean	Wheat	Weeds
Vetch	Sweet Clover	Winter Rye	
	White Clover		

4. Have the ability to grow in poor soils (high sand or high clay content soils) and these are the soils in which green manures provide the greatest benefits, especially when trying to develop a poor soil site for a vegetable garden.

The two major problems of green manures:

1. Grass or other non-legume cover crops may have high C:N ratios, which cause less nitrogen to be available for the following vegetable(s) being grown (see Chapter 2 for more details on C:N ratios). The incorporation of a non-legume into the soil, such as wheat or winter rye, can result in a nitrogen deficiency in the succeeding vegetable crop, especially if the C:N ratio of the green manure crop(s) is higher than 100:1. Soil microbes tend to tie up the soil nitrogen as they decompose the high carbon plant material. This can be somewhat avoided by adding a nitrogen fertilizer when the green manure is incorporated or when vegetable crops are later planted into this soil.

Figure 10.2 A cover crop being tilled into the soil as a green manure.

2. Green manure crops can deplete soil moisture resulting in drought-like conditions in the following vegetable crop. Under this scenario, additional soil moisture may need to be added to the garden soil prior to planting vegetables.

Cover crops can also be used as **living mulches**, in which the cover crop and vegetable are grown together in some manner. The living mulch is often grown in combination with vegetable plants to provide some sort of weed control. The cover crop to be used as a living mulch can be planted before or after the establishment of vegetables in a garden. Living mulches are typically grown between the rows of vegetables to prevent weed growth (Fig. 10.3). However, in cole crops such as broccoli or cabbage, white clover is sometimes over-seeded once those crops have been established in a garden for a few weeks, and this suppresses germinating weed growth, which will improve weed control. It is important to remember that the living mulch will compete for moisture and nutrients with the vegetables being grown, and can become worse competitors with vegetables than weeds. Gardeners should make sure that living mulches do not grow taller than the vegetable plants, as they will tend to provide excess shade and competition, which will result in decreased productivity of the vegetable crop. Thus, living mulches are not recommended in a garden for low-growing vegetables, shallow-rooted vegetables, or drought-susceptible vegetables.

There are several important features required for a living mulch to be successful. The first attribute required is that it be low-growing relative to a vegetable crop. The relatively low-growing types of cover crops (such as white clover) should be used on shorter vegetables (such as cabbage and broccoli). Secondly, they should not be susceptible to the same diseases as the vegetables being grown. Thirdly, in certain situations, they may need to germinate and grow under shade conditions. Lastly, they must not interfere with the harvest process and allow for easy harvest of the crop. Oftentimes, living mulches are mowed prior to harvest to control vegetative growth, which allows an easier harvest.

Figure 10.3 Mustard used as a living mulch to prevent weed growth in pumpkins.

Cover crops will provide several major benefits to a soil including:

1. **Add organic matter to a soil.** The organic residues from a green manure crop can improve the soil structure stability needed to maximize plant growth. Improvements of soil structure resulting from green manure incorporation into soils will persist for many years, even after the practice has been discontinued.
2. **Nitrogen addition to topsoil.** All green manure crops will add some nitrogen into the soil due to the nitrogen present in plants. However, if the green manure crop incorporated is a legume, it will add more nitrogen, since these plants have higher nitrogen levels present resulting from their nitrogen-fixation ability.
3. **Nutrient conservation.** Cover crops can be used as a catch crop to retain nutrients and to prevent many nutrients from leaching out of a soil. These nutrients can then be utilized by the next vegetable crop grown in the soil. The cover crop used as a catch crop will help conserve nutrients, which can reduce nutrient contamination of lakes, streams, and groundwater supplies. Cover crops can also supply a significant amount of nitrogen and organic matter for the next crop when managed correctly.
4. **Ground Cover.** Cover crops will also protect soils from erosion, especially during periods of the year in which erosion can be more of a problem due to the lack of a vegetative cover on the soil.

Cover Crop Selections

There are many cover crops from which a gardener can choose. An introduction to some of the more commonly grown cover crops is provided for use in developing better soils for an organic garden.

Alfalfa is a legume and annual alfalfa cultivars such as 'Nitro' were designed to maximize symbiotic fixation of atmospheric nitrogen and have more vigorous vegetative growth than perennial forms. 'Nitro' alfalfa is a non-dormant variety that establishes rapidly and produces high amounts of biomass. Alfalfa has medium to high cold tolerance, a good nitrogen-fixing ability, low to medium shade tolerance, and is drought tolerant due to its extensive root system. However, alfalfa must develop some vegetative growth prior to the onset of cold temperatures to overwinter properly.

Annual Ryegrass will establish quickly and can often become a weed. This plant does not fix nitrogen and may tie up a portion or all of the nitrogen in the soil when incorporated into the soil. It is generally recommended to seed annual ryegrass about 40 days before the first expected frost in the autumn. It has moderate cold tolerance, high shade tolerance, and moderate to high drought tolerance.

Annual Sweet Clover is a legume that establishes quickly and is most often grown as a summer annual. This plant is very sensitive to cold conditions, has high nitrogen-fixing capability, low to medium shade tolerance, and medium to high drought tolerance.

Buckwheat is typically planted in spring or early summer and incorporated into the soil about one month after the appearance of flowers. It is very sensitive to cold weather conditions, has medium shade tolerance, and low to medium drought tolerance. Buckwheat will establish quickly after planting and is a good cover crop to smother weeds or catch nutrients that would otherwise leach from the soil.

Crimson Clover is normally seeded in the autumn and is fast to establish a cover over the soil surface. It has good cold tolerance, good nitrogen-fixation capability, and moderate drought and moderate to high shade tolerance. It is best to incorporate crimson clover into the soil at early bloom before the flower color begins to fade.

Hairy Vetch is a legume that has excellent cold tolerance, moderate to high nitrogen-fixing capability, and moderate shade and drought tolerance. It will establish quickly, but should be planted at least 40 days before the first killing frost in the autumn so that it has sufficient time to produce enough biomass to allow it to overwinter properly. Hairy vetch is often grown in combination with winter rye, which is a frequently recommended cover crop mixture in a vegetable production system.

Red Clover is a legume that is normally seeded in the early fall or early spring. It has moderate to high cold tolerance, good nitrogen-fixing ability, and good shade tolerance.

White Clover is a low-growing perennial, which can be sown in the spring or autumn, but it can also be planted in vegetable gardens during the growing season to provide a living mulch, since it is relatively low-growing at maturity. White clover is a legume and also works well as a green manure.

Winter Rye is a cover crop that will establish quickly and produce a high amount of biomass, although plant residues tend to have a high C:N ratio. Thus, once it is incorporated into the soil, winter rye will oftentimes tie up most of the soil nitrogen that is present until it has completely decomposed. Winter rye has a high amount of cold tolerance, and moderate to high shade and drought tolerance. Since winter rye has high cold tolerance, it is often the best choice for late fall plantings. It will provide excellent winter soil-erosion control and remove excess soil nutrients (especially N) from the soil that would otherwise be lost through leaching.

ORGANIC FERTILIZERS

Fertilizer use in a garden is essential to maximize plant productivity. Previously (in Chapter 3), comparisons were made between organic and inorganic (synthetic) fertilizers. Gardeners need to realize that either fertilizer form can be equally effective in promoting plant growth in a garden. It just depends on a gardener's preferences and their specific needs. Microorganisms in the soil break down organic materials to form inorganic compounds identical to those in commercial synthetic fertilizers. Plants absorb nutrients only in forms available for uptake and are unable to determine if the nutrients absorbed are from organic or synthetic sources. For organic fertilizers, the extra plant growth often observed is a response to the development of better root environments, and increased populations and activities of soil organisms working on the organic matter.

While some organic materials, such as manure, add organic matter to the soil as well as essential plant nutrients, other organic fertilizers do not supply organic matter. One of the major benefits of organic fertilizers is that they break down slowly and release nutrients slowly over a longer period of time compared to synthetic forms. Organic fertilizers are also less likely to release nutrients rapidly enough to burn plant roots even if used in large amounts.

It is important for home gardeners to realize that many synthetic fertilizers are more soluble and can burn plants if used improperly. Since many organic materials break down slowly, they supply nutrients to plants for a much longer period of time without having to apply on a frequent basis. Because they are not quickly soluble, they are not leached from the soil during heavy rains and, therefore, are more effective over a longer period of time. Some organic fertilizers also contain some micronutrients that are essential for plant growth.

The availability of nutrients from organic fertilizers depends on their breakdown by soil microorganisms, which in turn depends on climatic and soil conditions. The release of nutrients is much slower when the soil is cool or heavily saturated with water. Also, nutrient release from organic fertilizer sources is slowed during drought conditions unless soil is irrigated or heavily mulched to maintain adequate soil moisture levels and moderate soil temperatures. When a quick fertilizer response is needed, inorganic fertilizers tend to provide higher amounts of plant-available nutrients than organic fertilizer sources and are better suited for this purpose. Remember that nutrients in many organic fertilizers are really not available to plants until the organic matter has decomposed.

Animal Manures

Many gardeners will often use animal manures as a general-purpose fertilizer, especially when trying to grow the plants in their garden in an organic manner. Animal manures will supply a wide variety of nutrients along with organic matter for improving the physical characteristics of soils and their water-holding capacities. Manures provide an effective source of nutrients for most vegetables grown in home gardens. About one ton (2,000 lb) of an average cow manure will supply as much N, P_2O_5, and K_2O as 100 lb of a 10-5-10 inorganic synthetic fertilizer (Table 10.2). An average manure composition is 0.5% N, 0.5% P_2O_5, and 0.5% K_2O. So, compared to synthetic fertilizers, manures have a lower nutrient analysis. Another problem is that all the nutrients in manure will not be available for the vegetables being currently grown in a garden soil. Only about one half of the N, one fifth of the P_2O_5, and one half of the K_2O in most manures will be available for the current plants being grown, since most nutrients are in the organic form and must be broken down by microbes to be released for plant uptake. On the basis of readily available nutrients, one ton of an average manure supplies about 5 lb N, 1 lb P_2O_5, and 5 lb of K_2O.

Manures when used in gardens do have some problems, such as possibly contributing to increasing weed problems. Weed seeds will often pass through an animal's digestive tract and are added back into the soil with the manure. This can become an enormous problem, since home gardeners will use manures at rates far exceeding those in commercial agriculture. Also, salt build-up over time can be a problem with manures. Salinization is the accumulation of soluble salts in soil to the extent that soil fertility is severely reduced; the ions responsible for salinization are Na^+ (sodium), K^+ (potassium), Ca^{2+} (calcium), Mg^{2+} (magnesium), and Cl^- (chlorine). Anyway, most home gardeners find that manures will provide a great source of organic matter, which will create a soil that is more friable and becomes easier to till over time.

Some manures are hotter or have higher amounts of nitrogen than other animal manures, and gardeners need to be aware of the average nutrient contents of the different animal manures. These manures can easily burn plants and hot manures are oftentimes composted first to prevent plants from being burned easily if overused. Overall they are much richer in many nutrients, especially nitrogen, which most plants require in their vegetative phase but not the flower phase. Table 10.2 provides the average content of N, P_20_5, and K_2O in various animal manures.

Next, various types of animal manures will be discussed in further detail. It is important for gardeners to understand the differences between the different types of animal manures and how they can benefit home gardens. Animal manures vary by species, and nutrient content also depends on how the animals are kept and the way that manures are collected. Urine contains a large percentage of nitrogen and potassium. This means that animals boarded in a fashion where urine is absorbed with their feces (by straw or other similar bedding) can produce organic compost that is richer in nutrients.

Cow manure is widely used in organic gardens, but it is really not the best for home gardens since it often contains unwanted salts and weed seeds. It is considered a cold manure

Table 10.2 The Nutrient Content of Some Common Animal Manures.

Animal	Manures (lbs./ton)		
	N	P_20_5	K_2O
Cattle	12	5	10
Poultry	30	25	12
Pig	10	6	10
Sheep	28	10	24
Horse	14	6	12
Goat	14	6	18
Rabbit	45	25	12

since it is has a lower nutrient content than many other animal manures. Cow manures tend to break down at a slow rate, and thus will provide plant available nutrients at a fairly slow pace as well. It can be an especially good source of beneficial microbes for the soil. However, the recent increase in the use of bovine growth hormones to increase milk production certainly could become a concern for organic farmers trying to source safe cattle manures. The healthier the cow, and the more healthful a cow's diet, the more nutrients its manure will carry.

Poultry manures are generally thought of as simply just chicken droppings, but the manure of other domesticated birds, including duck, pigeon, and turkey are also included in this grouping. Poultry manures are hotter than most animal manures, as it is high in nitrogen and can easily burn plants unless composted first. A small amount of dried chicken manure should be mixed in small concentrations directly into soil to improve fertility. One potential concern for the organic gardener is the large amount of antibiotics fed to domestic poultry in large production facilities, which may render the product as non-organic. It is also suggested that some caution be used when handling chicken droppings, whether fresh or dried, due to the potential for harboring the bird flu virus.

Pig Manure is a highly concentrated or hot manure. It has less nitrogen than poultry manures, but has more than many of the other animal manures. Pig manure is best used when mixed and composted with other manures and/or large quantities of plant matter. Pig manure is an excellent fertilizer after composting when you consider the nutrients and the ability to add organic material to the soil. However, pigs share many similar intestinal parasites with humans and can pose a health risk to those exposed to manure, whether fresh or composted. Thus, it is often strongly advised not to use pig manure as a fertilizer for home gardens.

Sheep Manure is another hot manure that provides high amounts of nitrogen. It also contains phosphorus (P) and potassium (K), and provides several essential micronutrients such as boron, calcium, copper, iron, manganese, molybdenum, sulfur, and zinc, depending on the source. Because of its low odor, sheep manure can easily be used to top-dress garden beds. Manures from sheep that are fed hay and grain will have greater nutrient content than manure from animals that just feed on pasture forages.

Horse Manure has about half the nitrogen content as chicken manure, but has more than cow manure. Horse manure often contains high amounts of weed seeds, which means it should be composted before use. The heat produced from composting can effectively kill most of these seeds, as well as any harmful bacteria that may be present. Composted horse manure can also be used in the garden any time of the year and it also provides a significant source of new organic matter to the soil. Simply spread the composted horse manure across the garden area and till into the soil.

Goat Manure is similar to sheep or horse manure, although the form of the manure is in neater pelletized droppings. In addition, it doesn't typically attract insects or burn plants like cow or horse manures. It usually has a fairly high nitrogen content and is best if composted before use. Goat manure use in a garden will significantly enrich the soil for better plant growth. Its pelleted form makes it suitable for direct applications to flower and vegetable gardens, without the worry of burning plants. In addition, the pellets are easy to spread and till into the garden soil.

Rabbit Manure is the hottest of the animal manures. It is even higher in nitrogen than most poultry manures and it also contains fairly high amounts of phosphates. Due to the high nitrogen content, rabbit manure is best when used in small quantities lightly mixed into soil or composted before use.

It is generally thought that the best types of manures for the garden are cattle, horse, and swine due to the lower nitrogen content and that these will often contain straw as well. Cattle and pig manures normally contain about 10 to 12 pounds of nitrogen per ton (Table 10.2). Poultry and sheep manure have higher levels of nitrogen and may supply more nitrogen than needed early in the growing season. Whatever manure is used, try not to apply fresh manure to the garden before planting, and it is preferable to compost the manure for about a year prior to use. These manures can also be plowed under in late summer for garden planting the following spring. It is generally safe to use composted manures at a rate in the garden of 50 to 100 pounds per 100 ft^2.

Once manure is applied to a soil, it should be tilled into the soil as soon as possible after spreading to retain most of its nutrient value. Manure lying on top of a soil can lose as much as 30% of its N content in a week, depending on climatic conditions and manure age. In a new area being used for a garden, it is generally recommended to apply about 200 to 400 lb per 100 ft^2. This is equivalent to about 1 inch of manure over the entire soil surface of the garden. On established gardens, apply 50 lb of cow manure to each 100 ft^2 or about one-quarter of an inch over the entire surface of a garden.

Guidelines for applying animal manure:

1. Apply most fertilizer to plants in the early spring (also applies for green manure).
2. During the growing season, sidedress with extra animal manure as needed to increase fertility levels in soil.

Typical organic garden schedule: Grow a cover crop during winter, add 1/4 to 1 inch of compost and/or manure over garden soil surface in spring, and then till into soil. The green manure crop and animal manure should add sufficient nutrients back into the soil to allow sufficient amounts of plant growth. Furthermore, if an organic mulch is later used over the soil surface in the garden, the gardener may not need to provide a sidedress animal manure application to vegetables especially if the plants are light nutrient feeders.

Lastly, it is important for gardeners to understand that manures are relatively low in nutrients and it is best to compost them for later use in the garden. To prevent the possibility of burning plant roots with hot manures, fresh manure should not be used in the garden. However, well-aged manure, or that which has been allowed to dry over the winter months, can be worked into the soil without the worry of burning plant roots.

ORGANIC METHODS FOR PEST CONTROL

Synthetic chemical pesticides provide many benefits to food production and human health, but they also pose some hazards. Laboratory research studies have shown that pesticides can cause health problems, such as birth defects, nerve damage, cancer, and other detrimental health effects that can occur over a long period of time. However, these effects depend on how toxic the pesticide is and how much of it is consumed. Some pesticides also pose unique health risks to children. Therefore to avoid these risks, many gardeners refuse to use pesticides in their garden. Anyway, many gardeners grow vegetables in their gardens to have fresh produce that has been grown without pesticides. Many consumers today are concerned about the high levels of pesticide residues that can be found on our food, and unfortunately, pesticides have a reputation for doing more harm than good. However, insects and other pests can easily destroy vegetables in gardens, and must be often controlled in some manner. There are many organic and safe pest-control options available to home gardeners and, when handled properly, organic pesticides can be very safe to use. However, the home organic gardener must know what is available for organic use and which products provide the best option for specific pest problems.

To improve pest control in a home garden, a gardener must **monitor** insect, disease, and weed pest levels in the garden several times per week. The garden should really be monitored on a daily basis, but this is not feasible for most. Gardeners also need to realize that many different weeds will harbor insect pests that can later injure vegetables (e.g., solanaceous weeds such as black nightshade or groundcherry are related to tomatoes, peppers, potatoes, and eggplants; and the morning glory is related to the sweet potato). Weedy areas need to be destroyed well before vegetables are planted to reduce insect population levels. However, destroying weedy areas by mowing after vegetables have emerged from soil will encourage migration onto the vegetable plants.

In many instances, alternative methods of disease and insect management offer adequate levels of pest control and will pose fewer hazards. A few of these methods are outlined next, and include insect exclusion using lightweight rowcovers, the use of companion plantings, beneficial insects, microbial insecticides, natural plant products, and mined materials for disease control.

Insect Barrier Cloth

The commonly used lightweight garden rowcover is often utilized as an insect barrier cloth. It will prevent most insect pests from reaching vegetable plants in a garden, as long as plants are completely covered and the edges securely pinned to the ground. This cloth will effectively prevent various insects, including aphids, flea beetles, Colorado potato beetles, cucumber beetles, grasshoppers, leaf miners, cabbage worms, and whiteflies from feeding on plants. However, insect pest problems can still occur if insect eggs were laid on plants before the fabric was applied, if insects emerge from the soil, or if insects were introduced when seedlings were transplanted. Insect barrier cloths are porous to both air and water, and some will allow up to about 90% light transmission.

Companion Planting for Insect Control

Companion planting is a gardening technique that involves planting two or more plants in close proximity to derive some type of benefit for one or all plants. The benefit provided to vegetable plants growing in the garden can be more vigorous growth or higher yields, and/or repelling insect pests or attracting predators of specific insect pests. Companion vegetable planting has been used for many years and is generally used to either attract beneficial insects or to deter insect pests. Several examples are provided next.

Sage is a widely grown culinary herb, and will repel carrot flies, slugs, bean beetles, and cabbage moths and should be planted near vegetable plants in which these insects can be a problem. Also, nasturtiums planted around broccoli plants will repel aphids, and chives planted among sunflowers and tomatoes will discourage aphid infestations in these crops.

Furthermore, beans and potatoes also work together in plantings for organic insect pest control, and planting bush beans with potatoes in alternating rows will protect potatoes from the Colorado potato beetle and beans from the Mexican bean beetle. Broccoli, cabbage, and other cruciferous vegetables, as well as potatoes, onions, and beets do well when companion planted with dill, peppermint, sage, and rosemary. Tomatoes are a useful companion plant for asparagus, as the tomato plant repels the asparagus beetle.

Companion planting is often done in home gardens to reduce pesticide sprays on plants growing in the garden, which contributes to organic sustainability. Thus, it can become an important part of many organic gardens. Although fully understanding the interactions among plants can take a lifetime to understand, there are a few easy things a home gardener can do in their garden right now to make use of this intriguing method of gardening. Table 10.3

Table 10.3 Repellant plants recommended for use in vegetable plantings and insect pests controlled.

Insect Pests Controlled	Repellant plants
Aphids	Catnip, Chives, Coriander, Fennel, Garlic, Marigold, Mint, Mustard, Nasturtium, Onion, Oregano, Petunia, Sunflower
Asparagus Beetle	Basil, Calendula, Nasturtium, Parsley, Petunia, Tansy, Tomato
Bean Beetle	Rosemary, Savory
Cabbage Looper	Dill, Garlic, Hyssop, Peppermint, Nasturtium, Onion, Sage, Spearmint, Thyme
Cabbage Maggot	Garlic, Marigold, Radish, Sage
Cabbage Moth	Hyssop, Mint, Rosemary, Sage, Savory, Thyme
Cabbageworm	Celery, Thyme, Tomato
Carrot Fly	Leeks, Lettuce, Onion, Rosemary, Sage, Tobacco
Colorado Potato Beetle	Catnip, Coriander, Eucalyptus, Green Bean, Marigold, Nasturtium, Onion
Corn Earworm	Cosmos, Geranium, Marigold
Cucumber Beetle	Catnip, Marigold, Nasturtium, Radish
Cutworm	Spiny Amaranth
Flea Beetle	Catnip, Garlic, Mint, Sage, Tobacco
Imported Cabbageworm	Dill, Garlic, Geranium, Hyssop, Mint, Nasturtium, Onion, Sage, Thyme
Japanese Beetle	Ageratum, Chives, Garlic, Hydrangea, Juniper, Pansy
Leafhopper	Geranium, Petunia
Mexican Bean Beetle	Marigold, Petunia, Rosemary, Savory
Nematodes	Calendula, Marigold
Spider Mite	Coriander
Squash Bug	Catnip, Mint, Nasturtium, Petunia, Radish, Tansy
Squash Vine Borer	Radish
Tomato Hornworms	Dill, Borage, Calendula, Marigold, Petunia, Opal Basil
Whitefly	Basil, Marigold, Oregano, Peppermint, Thyme, Wormwood

provides an overview of many important vegetable insect pests that can be repelled through companion planting of specific plants with the vegetable crop in which the pest is a problem.

Beneficial Insects

Beneficial insects are an important component of vegetable gardens. As home gardeners, we usually do not even recognize that these insects are required to maintain an ecological balance in gardens. Garden plants attract not only insect pests by the hundreds, but also beneficial insects that feed on these pests, which can reduce their populations in gardens significantly. So, from an insect-pest standpoint, beneficial insects can be used in a garden to reduce their populations which can relate to reduced amounts of damage to vegetables.

Beneficial insects provide a safe, alternative way to using insecticides in a vegetable garden. Predator insects can control various insect pests, such as aphids, caterpillars, grubs, and slugs, which can do significant damage to vegetables growing in a garden. The use of synthetic chemical insecticides to control insect pests can place toxic residues on or in vegetables, so using predator insects is a much more natural way to provide control.

Beneficial insects are more likely to remain in a garden if there is a readily available food supply. While many of these predators can be purchased, it is probably cheaper and more effective to encourage the ones already in the garden. To encourage these beneficials, they must be provided with an alternative food source when meals of pest insects are scarce. Many beneficial insects need to sip flower nectar to survive, so choose a variety of plants that will provide flowers throughout the year to feed beneficial insects. Flowers produce nectar and pollen that are used as food by the adults of many beneficial insects. Many gardeners will often set aside an area of their garden for perennial flowers and wildflowers. Asteraceae flowers (sunflowers, marigolds, etc.) and Apiaceae (carrots, celery, coriander, dill, fennel, and parsley) are favorites with insects. Many **composites** (or plants in the Asteraceae family) will attract beneficial insects, due to the large attractive flower that most produce. Many commonly grown flowers are composites including *Chrysanthemum*, *Rudbeckia* (Black-Eyed Susan) *Coreopsis*, *Asters*, *Helianthus* (Sunflower), *Tagetes* (Marigolds), *Heliopsis* (Ox-Eye Daisy), *Solidago* (Goldenrod), and *Echinacea* (Purple Cone-Flower). Allowing these plants to flower in a garden will provide a good food source for beneficial insects.

Many **cover crops** can also provide a habitat for beneficial insects, and leaving strips of a cover crop in a garden will allow populations of beneficial insects to increase throughout the season (the living-mulch concept). Vegetables must be grown before the cover crop dries down, as beneficial insects will leave once the cover crop dies. Several cover crops have been reported to attract beneficial insects. **Crimson clover** attracts minute pirate bugs, big-eyed bugs, and lady beetles; **buckwheat** in flower will attract tachinid flies, minute pirate bugs, predatory wasps, lacewings, and lady beetles; **hairy vetch** attracts lady beetles, minute pirate bugs, and predatory wasps; and **white clover** will attract tachinid flies, bees, and predatory wasps.

The first step to enticing predators to colonize a garden is to stop using pesticides. Beneficial insects are much less resistant to pesticides than are pest insects. Even botanical pesticides, such as rotenone and pyrethrum (which will be discussed later), will kill pests and beneficial insects. Another essential part for developing a habitat for beneficials in a garden is to design and plant the garden so that it contains perennial beds or areas that remain undisturbed each year. Uncultivated areas, such as grass-covered or mulched pathways through the garden, are needed to give refuge to ground-dwelling predators and provide a stable habitat from year to year so their populations can build up.

A source of water will also help attract insects, and this becomes especially important during dry weather conditions. Bird baths and other small, shallow containers are best suited for this purpose. Sticks or rocks placed in the water serve as perches for insects to access the water. Also, remember that standing water will also attract mosquitoes, and the water should either be changed twice a week or a microbial mosquito control product (Bt—*Bacillus thuringiensis*) placed into the water to prevent mosquito breeding.

Beneficial Insects for Home Gardens

A listing of some of the more common beneficial insects is provided next, followed by a short description.

Assassin Bug. The long and narrow-body assassin bug is usually brightly colored, but can also be light brown (Fig. 10.4). Assassin bugs feed on the body fluids of many different insects, including caterpillars, Mexican bean beetles, aphids, leafhoppers, Japanese beetles, and Colorado potato beetles. These predators can be attracted by flowers such as marigolds,

Figure 10.4 Assassin bug feeding on a bumble bee.

chrysanthemums, black-eyed Susan and purple cone-flowers, which are rich in pollen and nectar. Assassin bugs also like some type of cover to hide, so place mulch around these plants to help them attack their prey.

Lady Beetle. Lady beetles tend to be reddish in color with black spots although they can also be black with reddish spots. The lady beetle has a hemispherical-shaped body and some colors vary by species. These insects are usually less than ¼ inch in length and are very active on plant surfaces. The larvae are soft-bodied, elongated and taper to a point at the rear of the insect; additionally, they are gray- or black-colored with distinct orange or yellow markings (Fig. 10.5). Both adults and larvae of lady beetles feed primarily on aphids, but will also feed on mealybugs, spider mites, and other soft-bodied insects. Gardeners can attract lady beetles by planting goldenrod and dill near the vegetable garden. Also, if the insects on which lady beetles feed are not present or are in low numbers in a garden, they will fly to new areas to search for a food source, and they may also fly away even if prey is available. If lady beetles are purchased, release during the late evening, as this will encourage them to stay in the field at least temporarily since they don't fly at night.

Ground Beetle. Most ground beetles are black in color (Fig. 10.6), although some have a brown or greenish tint. These beetles have long, thin antennas and legs, and most species are carnivorous and actively hunt for any invertebrate prey that they can overpower. The feeding habits of this beetle are particularly beneficial to home gardeners, as they feed on many insects

Figure 10.5 Lady beetle larva.

Figure 10.6 A common ground beetle.

that can be harmful to gardens including cutworms, fly maggots, caterpillars, aphids, weevils, slugs, snails, and many other soft-bodied insects. The beetle's ground habitat oftentimes makes it difficult for them to find a source of food, as they cannot fly to hunt their prey. Most larvae are predatory like the mature adult beetles, as they feed on worms and other insects that live beneath the soil. To attract ground beetles, place a shallow pan of water near the vegetable garden and fill with gravel. The gravel prevents the insects from drowning.

Praying Mantis. The praying mantises have stick-like bodies with long legs and can grow up to 6 inches in length (Fig. 10.7). Adult praying mantises have wings and are capable of flying. These insects are well camouflaged and oftentimes difficult to see, and are usually brown or green in color. Praying mantises feed on grasshoppers, bees, flies, caterpillars, and other soft-bodied insects. It is recommended to plant bee balm near your garden to attract this insect.

Green lacewings. Lacewings are typically green in color with their long wings tapered together over their bodies (Fig. 10.8). The adult forms can grow up to 1 inch in length and are widest at the middle part of their bodies. Although the green lacewing's preferred foods include aphids, mealybugs, spider mites, leafhopper nymphs, moth eggs, scales, thrips, and whiteflies, they will also feed on numerous other insect pests. Green lacewing larvae are voracious predators of soft-bodied pest insects, including aphids, whiteflies, and the caterpillars of many moths. However, this beneficial insect feeds mainly on aphids. One lacewing larva—also

Figure 10.7 A praying mantis.

Figure 10.8 A green lacewing.

known as an aphid lion—can devour 100 aphids in a week. One or two or three lacewing larvae in each square foot of a garden can rid the garden of almost all soft-bodied pests in two to three weeks. Lacewings can be attracted to a garden by nectar-producing flowers such as Queen Anne's lace, coreopsis, cosmos, fennel, yarrow, sunflowers, dill, and angelica. Lacewings and several other beneficial insects are also attracted by a yeast-sugar water mix called Wheast. Green lacewings are commercially available through several sources and are among the most commonly released beneficial insect predators.

Big-Eyed Bug. Big-eyed bugs are generally black and white, and feed on several different insect eggs. These bugs get their name from their large eyes (Fig. 10.9) and feed on leafhoppers, mites, immature insects, caterpillar eggs, and aphids. A shallow dish with water and gravel can be placed in a garden to attract these insects. They are also attracted to light in the evening, so place solar lights around the garden to attract big-eyed bugs.

Figure 10.9 A big-eyed bug feeding on aphids.

Figure 10.10 A tomato hornworm with parasitic wasp eggs (left) and a close-up view of a parasitic wasp and eggs on a tomato hornworm larva (right).

Parasitic Wasps. Parasitic wasps are often used for control of *Lepidopteran* insects in organic gardens (Fig. 10.10). *Lepidopteran* insects are moths and butterflies, and the larvae (often referred to as caterpillars) of these insects often consume vegetable plant foliage and/or fruit in a garden. Parasitic wasps vary in color, but the most common colors are black and yellow (Fig. 10.10). These insects can grow up to one inch in length. Parasitic wasps tend to feed on grubs, caterpillars, and other soft-bodied insects. *Trichogramma* is a common parasitic wasp, and lays its eggs into the eggs of moths and butterflies; after hatching, the developing wasp larvae will destroy the host egg by consuming its contents. *Trichogramma pretiosum* is the species that is most often sold for the control of caterpillar pests of vegetables.

ORGANIC INSECT AND DISEASE CONTROL FOR VEGETABLES IN HOME GARDENS

It would be much easier to grow vegetables in a garden if no pesticides were necessary, but pest problems often occur that require some type of control measure(s). If left unattended, pest populations can rapidly increase to damaging levels, and garden pests such as insects can ravage leaves, flowers, and fruits in a short period of time, and diseases can spread quickly throughout a garden. This makes organic gardening even more difficult since most pesticides are not allowed. When a problem is severe enough to apply a pesticide for control, it is best if that the organic gardener evaluates his/her options and uses the least-toxic options available. Furthermore, just because a pesticide is organic doesn't mean it does not do any harm to human health or the environment. When applying pesticides, remember to always follow the label directions, and even with organic products, more is not better.

The home vegetable gardener should always keep a close eye on their garden and try to identify pest problems early, before they become major outbreaks. If integrated pest-management principles are practiced (see Chapter 9), using good crop growing conditions, natural predators (e.g., beneficial insects), and healthy soil, plants should not require pesticide applications as frequently.

There are many organic insect- and disease-control products that can be used in home gardening. Organic products use for insect control are (1) derived from natural substances, (2) tend to be less toxic to humans than synthetic pesticides, and (3) tend to break down quickly in the environment to harmless substances. There are many organic products available; insect-control products will be discussed first, followed by disease-control products.

Microbial Insecticides

Synthetic chemical insecticides provide many benefits to improve food production, but they can also be hazardous to the environment and human health. In many instances, alternative methods of insect management provide adequate levels of pest control and pose fewer hazards. One such alternative is the use of microbial insecticides.

Microbial insecticides contain microorganisms or their byproducts to kill insects. Microbial insecticides are a form of a pesticide that works by infecting selected insect populations with bacteria, viruses, nematodes, protozoa, or fungi. Although this sounds potentially dangerous, many argue that it is actually quite safe, since the insecticide is specific only to the species being targeted. These pesticides usually have no effect on animal populations, unless reducing a certain insect in a specific area disrupts the food chain. Remember that each microbial insecticide typically works against only one type of insect. They are especially valuable when growing vegetables because their toxicity to non-target animals and humans is extremely low. Compared to most other commonly used synthetic insecticides, they are safe for both the pesticide user and consumers of the treated vegetable crops. Because their residues present no hazards to humans or other animals, microbial insecticides can often be applied when a crop is almost ready for harvest. Microbial insecticides also are known as biological pathogens or biological control agents. They are formulated to be applied as conventional insecticidal sprays, dusts, liquid drenches, liquid concentrates, wettable powders, or granules. Each product's specific properties determine the ways in which it can most effectively be applied.

Bacterial insecticides are often used to control certain types of caterpillars that feed on vegetable crops. They will kill caterpillars of both moths and butterflies that often cause damage to vegetable crops. Normally, microbial insecticides are dusted or sprayed directly on the foliage of various vegetable crops to control these insects.

There are some limitations or disadvantages that often prevent the successful use of microbial insecticides. It is important to understand these limitations so that home gardeners can choose the most effective products for a given situation and take the necessary steps to achieve the most successful results in their garden. Since a single microbial insecticide is toxic to only a specific insect species or group, each application may control only a portion of the pests that are present in a home garden situation. If other types of garden insect pests are present in the treated area, they will survive and will continue to cause damage. Also, the effectiveness of several microbial insecticides can be decreased if exposed to heat or ultraviolet radiation, or if the material is allowed to sit in storage somewhere for several years prior to application. Consequently, proper timing and correct application procedures are especially important for some products to maximize their efficacy. Two of these products will be discussed next in further detail.

***Bacillus thuringiensis* (*Bt*)** is probably the most widely used microbial insecticide in home gardens. It is a type of bacterium that kills leaf-eating caterpillars by invading the caterpillar's digestive system. These bacteria produce crystalline spores and proteins. Caterpillars that eat these crystalline spores and proteins tend to die within a few days because of toxins released when the spores and protein crystals are dissolved by digestion in the insect's midgut. This microbial insecticide provides effective control for many types of *Lepidopteran* larvae, including cabbage loopers, which are hard to control in many instances. This bacterium does not reproduce or overwinter in nature, so it must be applied each year as the caterpillars are observed feeding in the garden. Remember that Bt breaks down in sunlight, so spraying during the late afternoon or on cloudy days will be the most effective use of the product. The timing of Bt applications is critical, since it has little residual effect. For Bt to provide the most effective control, it should be applied when the foliage-feeding larvae are actually present so that it can be eaten. Both powdered and liquid forms of Bt are available. The powdered form will remain viable for 7 days after application and the liquid spray for only 24 hours. The foliage of plants to be treated with the powdered form of Bt should be sprayed with water prior to application since this will allow the pesticide to stick to the foliage better.

The most commonly used strain of Bt is the *kurstaki* strain, which kills only leaf-feeding caterpillars. The *kurstaki* strain is sold under various trade names including Dipel, Biobit, MVP, and Thuricide. The *kurstaki* strain of Bt is widely used as a spray to control caterpillars on vegetable crops. Besides controlling cabbage looper, imported cabbageworm, and diamond-back moth, which are all important pests of cabbage, broccoli, and other Brassicaceae-related vegetables, this strain provides control of tomato hornworm, which can be a major problem on tomatoes. In addition, Bt is used in agriculture as a liquid applied through overhead irrigation systems or in a granular form for control of European corn borers. These treatments funnel down into the corn whorl to kill the larvae where they are generally feeding.

There are also Bt strains have been developed that control certain types of fly larvae (*israelensis* strain, or Bti). These are widely used against larvae of mosquitoes, black flies, and fungus gnats. The *israelensis* strain is often marketed by the trade names of Vectobac, Mosquito Dunks, Gnatrol, or Bactimos. For effective control of mosquito larvae, formulations containing the *israelensis* strain are often placed into standing water near mosquito breeding sites. For these applications, Bt is usually formulated as granules or solid, slow-release rings or briquettes to increase persistence. The rates used are determined by the size of the water body. It is best if applications are made shortly after insect eggs are expected to hatch, especially after rain events or irrigation applications. The persistence of Bt in water is longer than on sun-exposed leaf surfaces, but reapply if favorable mosquito breeding conditions last for several weeks. The *israelensis* strain of Bt is often applied as a soil drench to provide control of fungus gnat larvae.

More recently, some strains (*san diego/tenebrionis*) have been developed with activity against some leaf beetles, such as the Colorado potato beetle, which can rapidly defoliate a potato plant in a short period of time. When Bt is applied for Colorado potato beetle control, the *san diego/tenebrionis* strain is sprayed onto leaves in a manner similar to the formulations used for caterpillars. The *san diego/tenebrionis* strains can be found marketed through various trade names including Trident, M-One, M-Trak, Foil, and Novodor. It is important for the home gardener to remember that among the various Bt strains, insecticidal activity is specific, for example, Bt strains developed for mosquito larvae will not control caterpillars.

Spinosad is a natural insecticide and is certified for use on organic farms. Spinosad is a nerve and stomach poison derived via fermentation from the naturally occurring bacterium *Saccharopolyspora spinosa*. It is a broad-spectrum organic pesticide but is active only if ingested or contacted while in liquid form, so has little residual effect on most beneficial species. It is also relatively non-toxic to mammals and beneficial insects. The active ingredients in spinosad, spinosyn A and spinosyn D, are complex organic compounds made by soil microbes. This chemistry represents an alternative to Bt for worm control and offers excellent control of targeted pests and low toxicity to non-target organisms, including most beneficial insects.

If this natural insecticide product is carefully used, only insects that actually eat something that has been treated, such as a leaf, will be affected. This is different than a lot of other broad-spectrum insecticides that are toxic if the insect merely comes in contact with dry insecticide residues. Depending on the specific product, spinosad insecticides can be used in **some garden vegetables** and **fruit trees** to control a variety of important pests. This product works as a **biological** control for the following insects: cabbage looper, imported cabbageworm, Colorado potato beetle, Diamondback moth, and European corn borer. There are a number of products currently on the market that contain spinosad, such as Entrust Naturalyte for the commercial organic grower market. Entrust Naturalyte is non-persistent in the environment and is non-toxic to birds and earthworms, and provides a reduced risk to aquatic organisms. The mammalian toxicity is also low compared to many synthetic insecticides, and is safe for most beneficial insects, although honeybees and parasitic wasps can be killed by the product.

Other Organic Insect-Control Products

There are several other organically certified insect-control products that the home gardener can use. The specific product chosen depends on the particular situation and the insect that requires control.

Insecticidal Soap. Insecticidal soaps are any of the potassium fatty-acid soaps used to control many plant pests. The active ingredient of all insecticidal soaps is potassium salt of fatty acids. An insecticidal soap is typically sprayed on plants in a similar manner as other insecticides. Insecticidal soaps are effective only while the soapy liquid is in direct contact with the waxy cuticle of soft-bodied insects such as aphids, mealybugs, scales, thrips, and white flies. Soaps and detergents act strictly as contact insecticides, with no residual effect. To be effective, insecticidal soap sprays must be applied directly to the insect pest and must also thoroughly cover the insect.

The fatty acids in insecticidal soaps control insect pests by disrupting the structure and permeability of the insect cell membranes. This causes their cell membranes to collapse and leak, resulting in dehydration, and the insect quickly dies. Soaps and detergents may also remove the protective waxes that cover the insect, causing death through excess loss of water.

Several insecticidal soaps are available for control of insects and mites, and can be found under various trade names. Soaps are chemically similar to liquid hand soaps. Some household soaps and detergents also make effective insecticides. In particular, certain brands of hand soaps and liquid dishwashing detergents can be effective for this purpose. They are also substantially less expensive. However, there are many features of commercial insecticidal soap products that distinguish them from the dishwashing liquids or soaps that are sometimes substituted for controlling insects. There is increased risk of plant injury with these homemade products, and they are not specifically designed for use on plants to control insects. Dishwashing soaps and detergents (such as Lemon Joy) can be effective against soft-bodied insects, but have not been formally evaluated as pesticides and may damage the plant by breaking down the waxy layer protecting the leaf surface. Regardless of what product is used, soap-detergent sprays are always applied diluted with water, typically at a concentration of around 2 to 3 percent.

One of the most serious drawbacks to the use of soap sprays is their potential to cause plant injury or what is known as their **phytotoxicity**. Certain plants are sensitive to these sprays and may be seriously injured. Certain tomato varieties can be damaged by insecticidal soaps. The risk of plant damage is greater with homemade preparations of household soaps than with commercial formulations. Since soaps can injure plants, treat a small number of leaves and observe if injury occurs; if no plant injury is seen after a few days, the entire plant can then be treated. Commercially available soaps are formulated to reduce damage to the plant, but in some cases plant injury can still occur. To reduce leaf injury, wash plants within a couple of hours after the application. Limiting the number of soap applications can also be important, as leaf damage can accumulate with repeated exposure. However, because of the short residual action, repeated applications may be needed at relatively short intervals (4 to 7 days) to control certain pests, such as spider mites and scales. Also, it is important for the gardener to understand that applications must completely cover the insect pest. This usually means spraying undersides of leaves and other protected sites. Insects that cannot be completely wetted by soap sprays, such as aphids within curled leaves, will not be controlled.

Soaps and detergents can offer a relatively safe and easy means to control many insect pests. Soaps are safe, effective, and non-toxic. Insecticidal soaps are non-toxic to humans and other mammals, degrade rapidly in the soil, and are effective in controlling some insects. However, as with all pesticides, there are limitations and hazards associated with their use. Gardeners should understand these limitations, and carefully follow all label instructions.

Oil Products. Various oils have been used for centuries to control insect and mite pests. Oils remain an important tool to manage certain pest problems (e.g., scales, aphids, mites) on fruit trees, shade trees, and woody ornamental plants. Several recently developed oils extend this usefulness to flowers, vegetables, and other herbaceous plants. Furthermore, oils can also be used to control some plant diseases, such as powdery mildew. Diluted horticultural oils, often mixed with a small amount of baking soda, can be an effective control for this common plant disease.

Historically, oils became popular for their effectiveness on hard-to-control pest problems on fruit trees. They were used as a dormant-season application (before bud swelling and bud break) to kill mites and insects, such as scales and aphids, that spent the winter on the plant. Thus, oils were first used to manage insects before the growing season even began. This would prevent the use of toxic insecticides on the edible fruit product. This "dormant oil" killed overwintering insects such as aphids, mites, and scales. When the term "dormant oil" is used now, it generally refers to the application timing, during the dormant season, rather than a type of oil.

The primary limitation of spray oils is their potential to cause plant injury in some situations. However, many of the newer spray oils can largely eliminate phytotoxicity problems if they are properly applied. The recent improvements in refining have allowed the development of oils with increased safety to plants and thus expanded their potential uses. Summer or foliar treatments are now possible for a variety of pests during the growing season. Oils also can be mixed with other insecticides, providing a broader spectrum and greater persistence of control.

Spider mites, whiteflies, and young stages of scales are common pests that can be controlled by oils during the growing season.

Oils will definitely have different effects on pest insects, although the most important is that they block the air holes (spiracles) through which insects breathe, causing them to die from asphyxiation. In some cases, oils can also act as poisons, interacting with the fatty acids of the insect and interfering with normal metabolism. Oils also may disrupt how an insect feeds, a feature that is particularly important in the transmission of some plant viruses by aphids. Thus, oils are sometimes applied to prevent transmission of viruses. Many viruses are spread by aphids (non-persistent viruses), as well as some that are mechanically transmitted and these can be somewhat inhibited by oil applications. These oils that interfere with virus transmission are sometimes called "stylet oils," which is a reference to the piercing and sucking mouthparts (stylets) of aphids that transmit these viruses.

The toxicity of oils is minimal, at least compared to other insecticides, and oils quickly dissipate through evaporation, leaving little residue. Oils pose few risks to people or to most beneficial insect species that are natural enemies of many insect pests. This allows oils to integrate well with various biological controls. Oils are also easy to apply with existing spray equipment and can be mixed with many other pesticides to extend their performance.

There are many types of oils available that an organic gardener can use, but most think of horticultural oils as it is widely marketed for this purpose. The majority of pest control oils are some type of mineral oil, a refined petroleum product. There are a few vegetable oils that are also effective pesticides, such as cottonseed oil and soybean oil. Other oils that can be used to control insects include canola, citrus, and various vegetable oils. These oils are usually combined with some type of emulsifying agent so that it can be mixed with water and used as a spray.

Horticultural Oils. Horticultural oils act by suffocating insects and eggs, disrupting insect metabolism, and repelling insects. The primary way horticultural oil kills insects is by suffocating them, since the oil blocks the spiracles through which insects breathe. These oils can also disrupt the metabolism of insect eggs and reduce insect feeding, causing them to starve to death. Some insects are also repelled by the oil.

Horticultural oils can be used on vegetables during the growing season to control certain garden pests. However, it is important to note that oils can damage plants. Plants that are flowering or water-stressed are more susceptible to oil damage than those that are not flowering or growing under nonstressful conditions. These oils should not be applied if daytime temperatures are above 85°F, or if night-time temperatures are expected to go below freezing. Horticultural oils are concentrated and must be mixed with water before applying. An added benefit is that oils have the potential to control virus spread in vegetable crops, as oils slow the spread of viruses by delaying primary infection.

Essentially all commercially available horticultural oils (e.g., Sunspray®, Scalecide®, Volck®) are highly refined petroleum products also known as mineral oils. The impurities in these products most associated with plant injury—such as aromatic compounds and compounds containing sulfur, nitrogen, or oxygen—have been removed. Filtration, distillation, and dewaxing complete the production of the finished base oil. Final formulations of horticultural oils are normally combined with an emulsifying agent that allows the oil to mix with water. This mixture is normally used at about a 2% dilution and then sprayed onto foliage.

Horticultural oils should be sprayed directly on the pests, to be the most effective. The excess oil evaporates and dissipates quickly, so there is no toxic residue and horticultural oil is considered safe to use around both humans and pets. However, horticultural oil is a labeled pesticide and should always be used according to the label's instructions.

Citrus Oils. The peels of different citrus fruits (such as lemons and oranges) contain oils that have insecticidal properties. Citrus oils can be used as an organic insect killer and repellant that is completely safe to use around the home. The active ingredient in this product is d-Limonene, a steam-distilled byproduct of citrus peels that destroys the waxy coating of an insect's respiratory system. Citrus oils cause suffocation, when applied directly to insects. This natural broad-spectrum insecticide product will cause mortality in most insects within 15 minutes. The citrus fragrance of d-Limonene also acts as an insect repellent. Citrus oils will control many insect pests in gardens, including aphids, leaf-eating caterpillars, whiteflies, and spider mites.

Citrus oils are also easy for the home gardener to make and use in their garden. First, soak the rinds of two oranges in a closed 1-gallon container of water for about 10 days, and then strain out the rinds/peels and store the solution until ready for use. If spraying on plants, dilute 2 ounces of the citrus oil solution with one gallon of water to avoid burning plants.

Canola, Cottonseed, Soybean, and Related Vegetable Oils. There are many different vegetable oils that can be used to control insects in home gardens. Canola, cottonseed, soybean, and other vegetable oils can be used as insecticides for many vegetable crops. These oils work well because they smother insects, which cause suffocation and death over time, and prevent future generations from developing. It is important to use a low oil concentration in the pesticide spray, since overuse can severely injure a plant or cause plant death.

Vegetable oils also can be used as insecticides, although the type of oil can greatly affect insecticidal activity. Canola oil is non-toxic and an environmentally safe pesticide that is effective at killing a variety of insect pests. Canola oil can be used to control many common garden insects, such as aphids, mites, and whiteflies. Cottonseed oil is generally considered the most insecticidal of the vegetable oils. Soybean oil, the most commonly available vegetable oil used in cooking, has often provided fair to good control of some insects and mites. It is important for the organic gardener to remember that the oil should be carefully sprayed onto the leaves and stems of plants, and to spray a light misting of the oil mixture onto the plants, as too much oil can actually damage the plants.

Repellants. Although some oils, such as citrus oils, are effectively used as insect repellants, there are a few other materials that are available that can be used for this purpose. The most widely marketed organic repellant is capsaicin.

Capsaicin is the chemical compound in hot peppers that provides their characteristic hot taste. The chemical capsaicin has been shown to repel insects from plants. Both dusts and liquid capsaicin sprays are available. Capsaicin dusts have been used to repel onion maggots and other root maggots of vegetable seedlings, when sprinkled around the base of plants. Liquid sprays can also be applied to plants, and to areas around plants to prevent insect damage. When applying, avoid inhaling or getting on skin or in eyes since it is an irritant.

Botanical insecticides are naturally occurring chemicals extracted from plants that are used as toxins to provide insect control. Several botanical insecticides have been formulated for insect management in the home garden. They tend to act quickly, causing immediate paralysis, death, or cessation of feeding. Many are broad-spectrum insecticides and will harm beneficial insects. Although most have low mammalian toxicity and are safe for the environment, some are very toxic and should be used as a last resort for insect control. Several botanical insecticides are harmful to fish or other wildlife. When some are used inappropriately (e.g., pyrethrin), they can cause more damage by disrupting the biological balance in a garden and trigger secondary pest outbreaks. Since many will break down rapidly, they often must be applied frequently.

Commercial preparations of some botanical insecticides include a synergist such as PBO (piperonyl butoxide) or MGK 264 (N-octyl bicycloheptene dicarboximide) to increase the insecticide's effectiveness. Both of these compounds have low mammalian toxicity, but are not accepted by some organic certification programs. Without the addition of a synergist, the activity of some active ingredients (e.g., pyrethrin) is too low for effective insect control. It is important to note that gardeners can be poisoned by careless handling or misuse of botanical insecticides. Always follow the label directions and wear recommended protective clothing.

Although many different botanical insecticides are available, **Pyrethrum**, **Neem**, **Rotenone**, and **Sabadilla** will be discussed next in further detail.

Pyrethrins are common botanical insecticides derived from the dried flower heads of the daisy, *Chrysanthemum cinerariaefolium*. The dried flower head is called pyrethrum and the extracted active ingredients that have insecticidal properties are called pyrethrins. This broad-spectrum, contact insecticide is an extract of these flowers and is organically certified for use in certain situations. The crude flower dust has six related insecticidal compounds that occur naturally. Pyrethrins act as stomach poisons and will kill insects immediately after application. Thus, pyrethrins are known for their unique "knock-down" properties that cause flying insects to quickly fall to the ground. Pyrethrins are broad-spectrum insecticides and are very toxic to honeybees.

The addition of PBO (as described previously) greatly increases the effectiveness of pyrethrins and insect mortality. This botanical insecticide degrades rapidly (usually within a day), leaving little or no residual activity. Pyrethrins are moderately toxic to humans, and can be used to control many different insect pests of vegetables, including aphids, cabbage loopers, Colorado potato beetles, spider mites, thrips, and whiteflies.

Pyrethroids, such as resmethrin and permethrin, are synthetic compounds that are closely related to natural pyrethrins. Pyrethroids tend to be more toxic to insects and will persist in the environment longer than pyrethrins. Due to their effectiveness, pyrethroids are applied at lower rates compared to many other synthetic insecticides.

Neem is a botanical insecticide that is extracted from the seeds of the tropical neem tree (*Azadirachta indica*), which commonly grows in Africa and India. A number of compounds found in neem seeds, notably azadirachtin, have proven useful as insecticides. The primary active ingredient, Azadirachtin acts as an insecticide in many different ways, including a broad-spectrum repellent, a growth regulator (prevents molting), a suppressor of feeding, and as a poison. It is typically just sprayed on leaves using water as a carrier, and is non-toxic to humans, birds, animals, plants, or earthworms. It tends to be slow-acting and is biodegradable as it will break down in sunlight. Neem has been used for centuries against aphids, whiteflies, thrips, hornworms, leafminers, gypsy moths, weevils, webworms, loopers, sawflies, and many more insects. Neem is most effective against insects that pass through all stages of metamorphosis. It is a broad-spectrum insecticide and will also kill beneficial insects that may be residing in the garden as well.

Neem Oil has unusual properties compared to the seed extracts that were just previously discussed. The oil fraction of neem seed extracts, which is mostly free of azadirachtin and related terpenoid compounds, has demonstrated efficacy as a broad-spectrum insecticide, fungicide and miticide. As a fungicide material, neem oil products have shown effectiveness against powdery mildew and rust, and they tend to be most effective when used as a preventative or at low disease pressures. It works primarily by preventing fungal spore germination and killing fungi on plant surfaces, such as leaves or fruit. Thus, thorough coverage on the leaf surface is important. As an insecticide/miticide, it is most effective when insect/mite pressure is light to moderate. The oil suffocates insects and mites on contact and is especially effective on whiteflies, aphids, and other soft-bodied insects. However, since this product is an oil, it will also kill bees, so it is recommended to apply this product at a time when bee activity is low.

Rotenone is a natural insecticide that can be extracted from more than 65 different species of tropical plants. It is a broad-spectrum insecticide used mainly to control leaf-eating beetles and caterpillars, and this organic insecticide breaks down quickly in sunlight or when mixed with soaps or lime. Thus, it must be reapplied at relatively short intervals to obtain optimum control. Since rotenone breaks down rapidly in sunlight, it is generally less damaging to beneficial insect species than are synthetic pesticides that tend to be more persistent. Rotenone is moderately toxic to mammals, but the chemical can remain active on sprayed plant parts for several weeks. However, it is highly toxic to fish and when used by home gardeners, they should try to avoid application to bodies of water such as ponds, fountains, or streams that contain fish. The effectiveness of rotenone is usually enhanced by mixing with PBO. Remember that PBO is a synergist and does not have insecticidal properties, but it increases the effectiveness of pyrethrum and rotenone.

Sabadilla. Sabadilla is made from the seeds of the South American sabadilla lily (*Schoenocaulon officinale*), a tropical small perennial bulb. This insecticide is a contact and stomach poison that can provide control of various vegetable pests, including caterpillars, leafhoppers, thrips, stink bugs, and squash bugs. It is an organic insecticide material that is very effective against the true bugs (members of the Hemiptera order), although it should be used as a last resort for control. Sabadilla breaks down rapidly in sunlight, leaving very little residual toxicity. Synergists are also often added to sabadilla to increase insect mortality rates. Sabadilla is extremely toxic to honeybees, and gardeners should avoid applying sabadilla whenever bees are present.

Nicotine Sulfate. Nicotine is extracted from tobacco or related *Nicotiana* species and is one of the oldest botanical insecticides in use today. Nicotine sulfate is sold as a 40% nicotine sulfate concentrate and is very effective for the control of a wide range of insect pests affecting vegetables, flowers, and fruits in the home garden, including aphids, caterpillars, thrips, and leaf-hoppers. However, nicotine is highly toxic to warm-blooded animals and readily absorbs through the skin. Due to its high toxicity to humans and other animals, nicotine sulfate must be used with caution and only as directed.

It is important to note that nicotine is a prohibited pesticide in organic farming operations, and it is not certified for use on organic farms.

Organic Disease-Control Products

Although there are many organic disease control products, remember that with an organic garden synthetic fungicides aren't typically used to eliminate disease problems and these products are definitely not allowed for certified organic farms. However, there are a few organically certified products that can be used by home gardeners to maintain healthy, productive plants in a home garden. The old saying "an ounce of prevention is worth a pound of cure" is right on target in the world of organic disease control in a home garden. This concept is of the utmost importance in organic gardening, as many of the disease-control methods focus on healthy plants and avoiding diseases. A few of the organic products that are available to provide organic disease control are described next.

There are several organic materials that can be used to control vegetable diseases, including the mined products of **copper, sulfur, lime-sulfur, Bordeaux mixture**, and a beneficial bacteria product called **serenade.**

Copper

Copper is a broad-spectrum protectant fungicide and bactericide. It will control a wide range of disease-causing organisms in the garden, including anthracnose, bacterial blights, bacterial and fungal leaf spots, and downy mildew. Copper products are toxic to humans and other mammals. However, the toxicity of copper can be reduced when mixed with lime (which is known as the Bordeaux mixture). Repeated applications over one growing season can oftentimes hinder plant growth, as copper levels increase in the soil. Fixed copper products can be purchased at most home garden stores, and can be applied as a dust or liquid spray.

Sulfur Sprays

Sulfur is one of the oldest known pesticides. Sulfur can be used in gardens to control many different disease-causing organisms. Several formulations are available, including dry powder, liquid, and wettable powder forms. Sulfur is used primarily for the control of powdery mildew and the various rusts that appear on many vegetables. Spider mites, psyllids, and thrips also are susceptible to sulfur applications. Although sulfur is non-toxic to humans, be careful during application, as it may irritate eyes and exposed skin as well as damage lungs if not used properly.

Sulfur has the potential to cause plant injury in dry weather where temperatures reach 90°F or more. Also, remember that it's also incompatible with many other pesticides, and do not apply sulfur onto plants within 20 to 30 days of applying spray oils, as sulfur reacts with the oil to cause plant phytotoxicity.

Lime Sulfur

Lime sulfur is made by boiling lime and sulfur together. The mixture is used as a dormant spray on fruit trees to control diseases such as fire-blight, anthracnose, powdery mildew, and some insects including scales, thrips, and eriophyid mites. Its drawbacks include its potential to burn exposed skin and eyes, and to injure plants if applied when temperatures exceed 80° to 90°F.

Bordeaux Mixture

The Bordeaux mixture is a combination of copper sulfate and hydrated lime. It was used first in Bordeaux, France, to control downy mildew on grapes, hence its name. Bordeaux mixture can be dusted or sprayed onto vegetable plants. It will aid in the control of several diseases commonly found in gardens, including anthracnose, bacterial leaf spots and wilts, and downy and powdery mildew. Although it is a natural-based pesticide, it is not labeled for organic use.

Serenade

Serenade is a microbial pesticide that can provide control of numerous plant diseases. The active ingredient of serenade is a *Bacillus subtilis* (strain QST 713), which is a bacterium that provides broad-spectrum plant-disease control. Since this product is made from a bacterium that occurs naturally, it is safe to use and provides no adverse effects on humans or the environment. It can be used to control powdery mildew, bacterial spot, and various other diseases on vegetables. Serenade biofungicide is approved for organic farming in numerous countries.

OTHER TECHNIQUES USED IN ORGANIC GARDENING

Compost

Composting (or creating compost) for gardens is an integral part of organic gardening. Compost provides an excellent source of organic matter to any soil. Organic matter will improve soil structure, soil drainage, add some nutrients to soil, and encourage earthworms and other biological activity in the soil. Several procedures were discussed in compost development and maintenance (refer to section on composting in Chapter 2).

Mulches

Many organic growers will use organic mulches in their gardens, and have found improved vegetable crop productivity when they are utilized. Organic mulches provide several benefits to plants, and were described previously (refer to section on mulching in Chapter 2).

Chapter Eleven

GARDENING AIDS

Experienced home gardeners will use many different methods to help them grow vegetables better and more efficiently. Three such methods are coldframes, high tunnels, and raised beds, which are discussed next.

COLDFRAMES

A coldframe is a **bottomless glass-covered box** that is heated only by the sun (Figs. 11.1 and 11.2). It is termed a coldframe since there is no bottom heat provided, and if bottom heat is provided (by a heat source such as fresh animal manure or heating cables), then it is termed a hotframe (Fig. 11.3). Coldframes are an important resource for many gardeners and some gardeners in colder climates would not be able to garden without them. The coldframe should be oriented toward the south in order to receive maximum winter sunlight; painting the inside walls white or silver will help to reflect more light to the plants, which can increase plant growth. Also, the success of a coldframe requires placing it against an existing structure, such as your home, garage, storage shed, or fence. A heated home will give added warmth to the coldframe, but other buildings will also work. If the coldframe is placed next to a building of some sort, then it is receiving added protection from winds and other weather conditions that can potentially damage your plants.

The coldframe has many different uses. It can be utilized to grow transplants for later planting into a garden, or growing cool-season vegetable plants over the winter months to supply a source of fresh vegetables during this time. Coldframes can be modified to have exchangeable covers, typically glass (or Plexiglas), but the glass covering could be removed and either a lath covering (which allows more air exchange) or shade cloth (reduces light and wind to protect plants) can be placed on top of the coldframe during the summer months.

Several suggested coldframe uses include:

Starting seeds earlier than they can be seeded into a garden soil, and in areas with short growing seasons, those few extra weeks can really help in getting a jumpstart on the growing season. This can allow certain vegetables to be grown in a particular climate that would not normally produce a crop due to the shortness of the growing season.

Hardening-off seedlings grown indoors before planting into a garden.

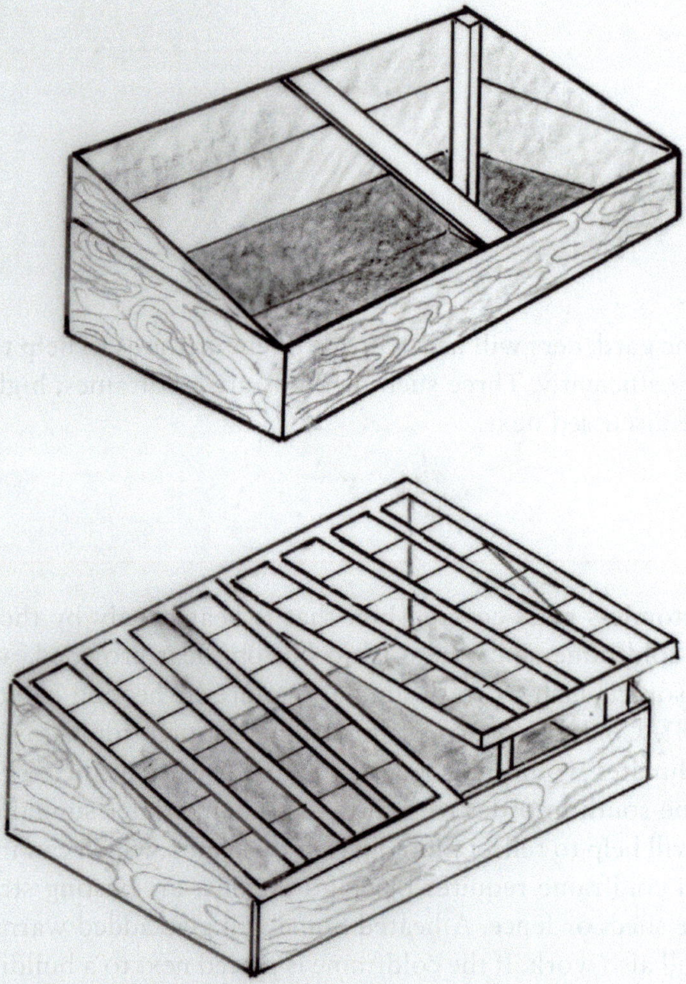

Figure 11.1 Structures for a simple coldframe design.

Forcing flower bulbs to have blooms during a time of the year other than their normal time to produce flowers.
General plant propagation of many different ornamental or fruiting plants.
Out-of-season vegetable production for many cool-season vegetables. Many vegetables (e.g., lettuce, carrots, spinach, and cruciferous crops) can be grown into and during the winter months to provide a fresh source of home-grown vegetables at this time of the year.

Coldframes have so many uses in a garden and they just rely solely on the heat and warmth of the sun. A coldframe is a great asset to a home gardener as it costs very little to use and to build. A coldframe is an easy way to force nature to work in various ways to benefit a home garden.

Figure 11.2 A simple coldframe with an easy-to-open-and-close lid.

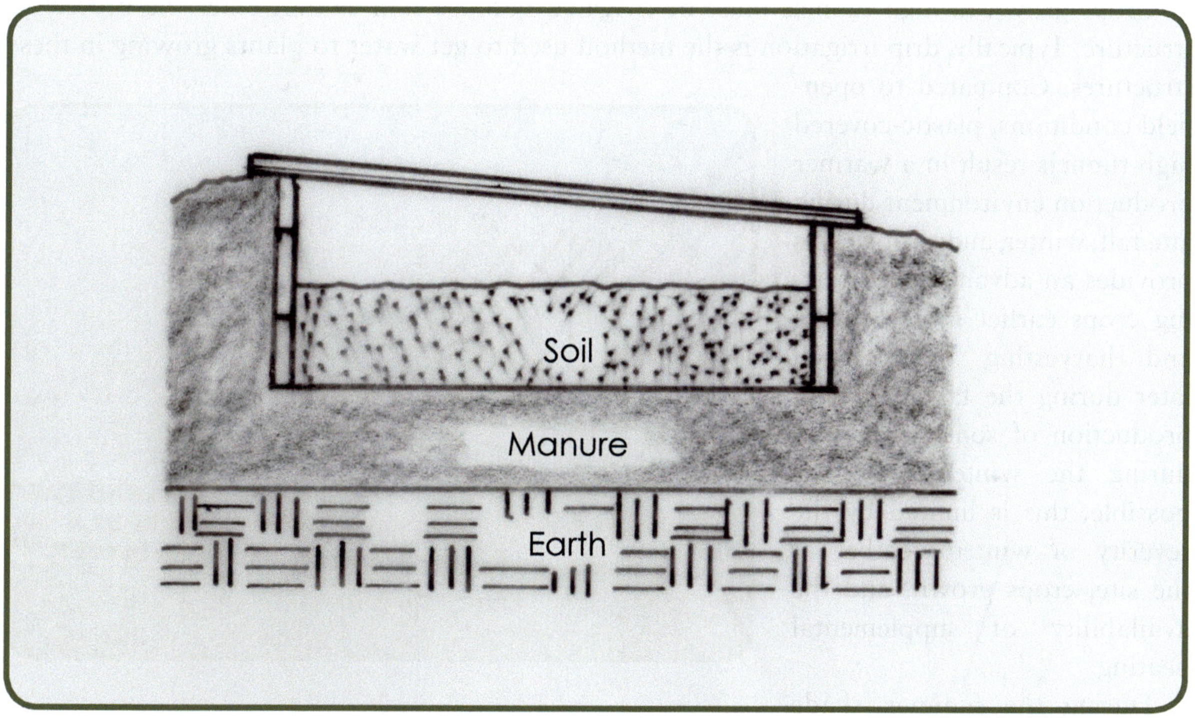

Figure 11.3 A cross-section view of simple hotframe.

HIGH TUNNELS

The terms "Hoophouse" and "High Tunnel" are often used interchangeably and refer to unheated, plastic-covered structures that provide an intermediate level of environmental protection for plants compared to open field conditions and heated greenhouses (see Fig. 11.4). Although greenhouse structures are typically covered with glass, rigid panels, or double layers of plastic and heated in some manner, high tunnels are generally covered with a single or double layer of plastic and heated only by the sun. High tunnels are easy to walk into comfortably and allow trellised crops such as tomatoes or cucumbers to easily be grown.

In contrast, low tunnels are much smaller, very susceptible to wind damage, and cannot be used to produce some crops. There are no standard dimensions for high tunnel sizes, but they typically range 16 to 30 feet in width by 30 to 96 feet in length. Most high tunnels are passively ventilated by roll-up sidewalls or end walls that can be opened or removed whenever ventilation is necessary due to high temperatures. Crops grown in high tunnels are not generally provided extra heat besides solar radiation, although supplemental heat can be provided for protection on cold nights or used to extend the growing season even longer. Generally, a single layer of 6 mil polyethylene plastic over a high tunnel will provide one hardiness zone of protection, and a second layer of plastic will provide another zone of protection. For example, vegetable crops grown in zone 6 with two layers of protection are approximately equivalent to zone 8 as far as winter hardiness is concerned. Crops are typically grown in ground beds within the high tunnel rather than in some type of container. Tomatoes are often grown in these high tunnels (Fig. 11.5), although peppers (Fig. 11.6) and cucumbers (Fig. 11.7) are sometimes also grown in these structures.

Crops grown in high tunnels must be irrigated because rain cannot enter the protective structure. Typically, drip irrigation is the method used to get water to plants growing in these structures. Compared to open-field conditions, plastic-covered high tunnels result in a warmer production environment during late fall, winter, and spring. This provides an advantage of starting crops earlier in the spring and harvesting them much later during the fall. Although production of some vegetables during the winter season is possible, this is limited by the severity of winter weather at the site, crops grown, and the availability of supplemental heating.

***Figure 11.4** A high tunnel structure near Columbia, Missouri.*

During the summer, shade cloths can be placed over the high tunnel structure to reduce

Figure 11.5 Tomatoes grown inside a high tunnel near Cobden, Illinois.

Figure 11.6 Peppers grown inside a high tunnel in southeastern Austria.

Figure 11.7 Trellised cucumbers growing in high tunnel in southeastern Austria.

air temperatures inside, and temperatures can be reduced by 4°F with 30% shade covering. Shading will also reduce light intensity inside the high tunnel, which is suggested for some crops, like lettuce. The use of sprinkler irrigation or a misting system of some sort during the summer months, in combination with shade cloth, will significantly reduce temperatures in high tunnels. It is also important to note that high tunnels also provide protection from wind and rain, which can result in increased crop yields and improved product quality, as well as decreased incidence of pest and disease problems.

RAISED BEDS

Many gardeners have learned the advantages of using raised beds in their garden and would not garden any other way (Figs. 11.8 and 11.9). When selecting materials to use for raised beds, it is best to choose rot-resistant lumber such as cedar or redwood. Other materials, such

Figure 11.8 A well-constructed wooden raised bed as it appears prior to planting.

as brick, stone, concrete, or composite materials also work well and will create attractive raised beds you won't need to rebuild.

In your raised-bed gardens, try to leave enough space between the beds to easily maneuver a wheelbarrow for adding soil, harvesting, spreading mulch, or other activities. It is best to spread mulch over the paths between your raised beds to prevent weed growth and this will also allow a gardener to have access to a garden even in wet conditions. Similarly, if you have grass paths between your raised beds, make sure there's enough space to comfortably run a lawn mower.

There are some advantages to using raised beds, including:

1. Providing for a quicker soil warm-up in the spring compared to non-raised areas. The soil in raised beds typically warms earlier in spring than do the surrounding soils. The soil also tends to dry faster, so you can get cool-season crops planted sooner, extending the growing season and your vegetable crop choices. A raised bed also allows the gardener to plant seeds and seedlings earlier compared to non-raised areas in the garden.
2. The gardener can choose the soil in which the raised beds are filled. Thus, a gardener can easily develop soils with high organic matter content that provides many advantages to the garden soil. Raised beds can make good gardens in areas where poor soils are prevalent, such as those with high-clay soils due to the improvement through the additions of high amounts of organic matter.
3. Some lawn grasses will often infiltrate a vegetable and flower garden and create a major weed problem. But in many cases, the nearby turf won't spread into your vegetable crops planted into raised beds.

Figure 11.9 Raised beds provide numerous advantages to a home gardener. This is just a beautiful raised-bed garden!

4. Raised beds also provide a healthier soil environment for beneficial microorganisms and earthworms due to less soil compaction. Additionally, soils in raised beds oftentimes do not require tilling, which can save a gardener time and money.
5. Greater amounts of weed reduction are possible in raised beds. Once raised beds are filled with fresh soil, an inch or two of some type of mulch can be used to cover soil. This reduces weed competition and preserves soil moisture.
6. Raised beds also allow for avoidance of water-logged soils. Water will drain much quicker in a raised bed and plant roots will avoid water-logged conditions that can be detrimental to a plant's growth and development.

GARDEN TOOLS

Gardeners need to have an understanding of the tools that will be needed for developing and maintaining a home garden. Tools needed to garden successfully include spade or shovel, garden rake, garden fork, hoe, trowel, irrigation equipment (garden hoses, watering can, sprinklers, drip irrigation, etc.), compressed air sprayer, wheelbarrow or garden cart, fertilizer spreader, planter/seeder of some sort, and tiller. With these necessary tools, most people can garden a small area successfully. However, many gardeners will garden successfully without all the tools listed, but some tools are needed more than others, such as a hoe and watering can, to produce a quality garden. It really just depends on an individual gardener's needs, as every gardener is different. Gardeners will also oftentimes place a shed in an out-of-the-way place in the landscape to keep all tools that are required for their gardening needs (Fig. 11.10).

Figure 11.10 A tool shed is commonplace in many home garden landscapes.

Chapter Twelve

SEASON-EXTENSION TECHNIQUES FOR VEGETABLES GROWN IN HOME GARDENS

Gardeners can't control the weather, but they can work with the weather conditions that occur in their area. Almost all types of vegetables benefit from increased early- and late-season warmth, so many gardeners have developed ways to obtain longer gardening seasons just by planting seeds indoors or by using various materials that will increase the temperatures around plants. Starting seeds indoors is a way to extend the growing season that any gardener can do with an indoor grow light or sunny window. Starting seeds indoors will allow quicker production of vegetables, and a gardener can choose specific vegetable cultivars for the garden instead of depending on the local garden center for the few cultivars that they have available to purchase as plants in cell-paks or other containers.

SEASON EXTENDERS

Season extenders are materials or structures that are used to keep the air and soil around plants warmer. There are various types of season extenders, depending on the specific needs of a gardener, and include coldframes, high tunnels, hotcaps, plastic materials to cover plants, plastic mulches, and rowcovers. Coldframes are probably the best-known season extender with home gardeners, while high tunnels are another popular season extender with commercial vegetable growers, and both were discussed previously (see Chapter 11 – Gardening Aids). However, other season extenders warrant discussion and will be briefly covered next.

Hotcaps are paper or plastic structures that are placed over plants for frost protection. They generally have a hole in the top of the structure to allow for ventilation due to excessive heat that can buildup within the hot cap, especially during sunny days. Hotcaps act as mini-greenhouses for individual plants, and will provide cold and frost protection that allows plants to be placed into a garden earlier in the season than normal. They trap heat from the sun and protect plants from frost. Hotcaps also keep valuable soil moisture from evaporating, as well as maintaining soil and air warmth around young seedlings. When frost or low temperature protection is no longer needed, they can be removed and stored for use in future crops.

Various types of **plastic protectors** can be used to cover plants for frost protection and many serve the same purpose as hotcaps. These coverings trap the heat of the sun and raise the air temperature around individual or small groups of plants. These coverings can be made at home or purchased. A plastic milk jug with the bottom cut out works well when placed over an individual plant. There are also various water-filled plant protectors to use on small plants that can be purchased. These are made of clear plastic tubes joined together that are filled with water and placed around plants. The warming principle is that the sun heats the water during the day, and at night the water releases this heat slowly to protect the plant as the air temperature decreases.

Plastic mulches are often utilized in gardening to increase or moderate soil temperatures, conserve moisture, and suppress weed growth. They can be used as a season extender since they can provide a quicker soil warm-up in the spring, which relates to extra plant growth and development during this time than what would have normally occurred under bare-soil conditions. Plastic mulches were discussed in greater detail in Chapter 2 – Compost, Soil Modification, and Mulching for the Home Garden.

Rowcovers are season extenders that can be easily used to protect rows, small areas in a garden, or an entire garden from frost or cold temperatures. Rowcovers have added a new dimension to the home gardener's management strategy to overcome frosts, a short growing season, excessive sunlight, and insect damage. Rowcovers are available in a variety of materials, including both plastics and fabrics. They can enhance yields of some crops, while producing an earlier-than-normal harvest on others. Although they offer only a few degrees of frost protection, this is very important during the slightly below freezing cold periods in the spring or autumn to protect crops, as just a few degrees will often make a big difference in the survival of a crop. Also, when properly used, they can effectively block the feeding activity of some insects, which can save time and money through reduced insecticide applications.

Rowcovers are sheets of transparent plastic or fabric, and are made of permeable and impermeable plastic, slitted plastic, or spun-bonded or woven fabrics. Rowcovers can offer a few degrees of frost protection to a crop growing in a garden; however, their major benefit is associated with more rapid growth by increasing daytime temperatures, not with frost protection. Cooler rowcover materials can be selected so that the risk of damaging high temperatures is minimized, but cooler materials will not increase growth and yield as quickly.

Figure 12.1 Slitted plastic rowcovers used over cantaloupes early in the growing season.

Rowcovers tend to provide about 2° to 8°F of frost protection during cold nights, and during the day, temperatures under rowcovers can be 5° to 30°F higher than the surrounding air. That means that your plants will be protected from frost damage down to about 24° to 28°F, depending on the fabric. For additional protection when a frost is forecast, the fabric can be double layered. If rowcovers are effectively utilized in the spring and fall, about 4 to 6 weeks of extra growing time can be added to many garden plants (Fig. 12.1). When using a rowcover during the spring, plants should be covered soon after planting or transplanting to prevent insects from gaining entry and to maintain warmth around plants. The rowcover should be applied loosely, so that plants are not restricted as they grow, and the edges of the rowcover should be secured with heavy materials, such as soil, bricks, rocks, soil, etc., to prevent displacement from plants by blowing winds. Rowcovers are often placed over cool-season vegetables such as broccoli and cabbage in a fall garden to extend the growing season of these vegetables into the early winter.

Floating rowcovers are made of lightweight spun-bonded or woven fabrics that can be placed loosely over the plants without wire supports. Rowcover fabric comes in several weights, and the heavyweight material is best used to extend the growing season, while lighter weights can be used for protection against insect damage and to provide shade. The thin, strong fabric is permeable to light, water, and air, but excludes insect pests. The edges must be held down with soil or stones to prevent displacement by blowing winds. These are very easy to apply to plants in a garden and a great choice for small gardeners. Fabric materials can be used several times (usually 3 or 4 or more) unless they have been damaged or ripped in some manner. Fabric rowcovers are lightweight blankets made of spun-bonded synthetic materials that are easy to use and have become quite popular. They provide several benefits to a plants, including maintaining a warmer temperature around plants to obtain quicker growth and earlier yields, insulating plants from damaging winds, providing a highly effective form of insect control, and protecting plants from excessive sunlight or frost conditions.

Plastic rowcovers will keep air temperatures around plants warmer during the night and day than rowcover fabrics. However, due to excessive heat build-up from solar radiation that can occur during the day, plastic rowcovers generally require removal or ventilation during the day. Most plastic rowcovers are supported by wire hoops or some other materials to provide structural support (also referred to as "low tunnels"). A low tunnel is a kind of mini-greenhouse for your garden and will also protect plants from cold weather, wind, hail, rain, disease, and flying insects. They are often used to overwinter cool-season vegetable crops and allow a gardener to get an early start on the spring and summer growing seasons. They can also be used to provide shade (with the use of fabric rowcovers) to cool-weather crops in early summer. A disadvantage is that a gardener cannot stand up in them to care for the crop and to harvest vegetables.

Low tunnels can be made from a variety of materials, including galvanized wire pieces, PVC, or conduit pipes for support. These materials are usually formed in the shape of an arch and placed in the soil at 3- to 4-foot intervals. Pre-made hoops can also be purchased for this purpose. To use PVC pipe, just bend and slip it over rebar stakes driven in the ground. The height of the low tunnel is determined by the height of the crop to be grown. For coverings, slitted plastic rowcovers are available and will not require removal due to their ventilation. Plastic rowcovers can also be purchased that can be opened and closed by a pull string at either end of the row. This makes ventilation or closing the low tunnel so much easier than removing the rowcover by hand. Normally, rowcovers are permanently removed from vegetable crops about 3 to 4 weeks after transplanting, and possibly sooner (at the beginning of the flowering period) for those crops that require bees to pollinate flowers.

Figure 12.2 Rowcovers used as an insect barrier over summer squash during early-summer at the Southern Illinois University student organic farm in Carbondale.

Besides the typical plastics and fabrics used, several other coverings can be used for these structures, with each having a different purpose. **Bird netting** can be placed over the hoops to protect young plants and fruits from bird damage. Also, **shade cloth** can be used during the heat of the summer to protect cool-season vegetable crops, such as lettuce, from the intense sunlight that can make them bitter or bolt (and go to seed). Both bird netting and shade cloth are easily supported by low-tunnel hoops.

Insect control with rowcovers

Most all weights of fabric rowcovers will provide some insect protection for vegetable crops (Fig. 12.2). To achieve the highest amount of insect control possible, rowcovers should be placed over vegetable crops before insects are present in the garden. It is also important to note that for these rowcovers to be effective barriers to insects, they must be sealed at the sides and ends of the rows, so that insects cannot enter them at any point. It is important for the gardener to be especially careful that insect pests are not trapped inside the cover, or excessive crop injury may result.

SPECIFIC VEGETABLE CROP RESPONSES TO ROWCOVERS

Rowcovers are widely used on vegetable crops to provide not only frost protection, but also to improve yields and protect from insect feeding. **Cantaloupe** (or muskmelon) is a warm-season vegetable crop that is highly responsive to the use of plastic-covered low tunnels. Ventilated clear plastic is most often used, which will result in a 1- to 3-week head start on the growing season (see Fig. 12.1). Cantaloupe plants are not as sensitive to the high temperatures that build up under the clear plastic tunnel compared to many other vegetable crops. Tunnels should be removed once plants begin to flower since bees are required for pollination. Low tunnels can also be used to exclude cucumber beetles, which vector an important disease called bacterial wilt that affects cantaloupes. If cucumber beetles are prevented from feeding on plants, then this disease can be effectively managed. Similar to cantaloupes, **summer squash** is one of the most responsive vegetable crops to grow under low tunnels. Remember that tunnels must be removed from squash plants prior to flowering since bees must pollinate flowers for adequate fruit set to occur. **Cucumbers** are more sensitive to high temperatures than melons, but this vegetable crop will show an improved yield response to the use of clear plastic ventilated rowcovers. Low tunnels are also an important mechanical method to exclude cucumber beetles from feeding on young plants, which will prevent the development of bacterial wilt, an important disease of cucumber. Again, remember to remove tunnels when the plants begin to flower, since cucumbers require bees to set fruit. For **peppers**, flower abortion and loss of early yields can occur under high temperatures in clear plastic low tunnels, so adequate ventilation is critical. However, research has shown that the use of ventilated clear plastic low tunnels for peppers can improve both early growth and productivity. Clear plastic rowcovers are generally not recommended for **tomatoes**, since early yield loss often occurs due to high temperatures that build-up underneath the cover. Although tomato vegetative growth and fruit yields can be enhanced by clear plastic rowcovers, fruit produced early in the season will often abort due to the high temperatures contained within the tunnel relating to early yield loss. **Leafy vegetables**, such as **lettuce, spinach** and **Swiss chard**, as well as the various **cole crops** (**cabbage, broccoli, cauliflower**, etc.) are all vegetable crops that have a great growth response to the use of rowcovers, with most best-suited to fabric or spun-bonded floating rowcovers. Floating rowcovers can be used to not only increase earlier production in the spring for these cool-season crops, but to also extend their production into the late autumn or early winter. Another advantage of using rowcovers for these crops is they can provide protection from various insect pests, including several different types of caterpillars and leafminers. When used for insect exclusion, place the fabric rowcover over the newly seeded or transplanted crop and remove at harvest or when plants outgrow them.

SUMMARY

Since warm-season vegetables are very sensitive to cold weather and cool-season vegetables succumb to repeated freezing events, there are various ways that home gardeners can extend the outdoor growing season for these vegetable crops. Most vegetables are grown primarily during the frost-free periods, so cold-weather conditions pose a great challenge to home gardeners trying to grow vegetables outdoors at this time. However, gardeners have ways to protect plants from frost and cold conditions by using various season extenders. For home gardeners who want to start growing vegetables early or continue growing vegetables into the late autumn or early winter, it is essential to use techniques that will extend the growing season. In this section, several types of season extenders were discussed to provide a better understanding of how to protect plants during cold weather conditions. If a gardener is growing at times of the year in which frost conditions are likely, then season extenders, such as rowcovers, can quickly become a garden necessity and a worthy investment.

Chapter Thirteen

AUTUMN VEGETABLE GARDENING

VEGETABLE GARDENING IN AUTUMN

The autumn is a great time to produce vegetables from a garden. However, this time of the year is often overlooked by many home gardeners who normally just plant in the spring and finish with gardening activities in late summer or early autumn. Vegetables planted in July, August, or September for an autumn garden is a great way to obtain fresh produce until Thanksgiving or even Christmas in some cases. Some of the best-quality garden vegetables are produced and harvested during the fall season when warm, sunny days are followed by cool and humid nights. Under these climatic conditions, plant metabolism is low, and therefore, more of the food manufactured by the plant is focused into developing a high-quality and very tasty vegetable product.

Late plantings of cool- and warm-season vegetables extend the harvest season into the autumn months once spring-planted vegetables have ceased production. Most cool-season vegetables grow much better in the autumn than those planted in the spring, as they prefer the weather conditions produced during the shorter, cooler days. The flavor of many cool-season vegetables that are allowed to mature during the cool autumn weather are sweeter and milder (less pungent) than those grown during hot summer weather. This is especially true for cole crops such as broccoli, cabbage, cauliflower, kale, and mustards. For example, cauliflower planted in late summer produces a high-quality curd that can be harvested usually in early to mid-November in temperate climates (Fig. 13.1). Cauliflower is a plant that performs much better in an autumn garden compared to when it is planted during the spring. Furthermore, vegetables, such as lettuce and spinach stay productive longer during the autumn months (compared to spring plantings that bolt rapidly at high summer temperatures) due to the shorter day lengths and cooler weather produced at this time of year. Extending the harvest as long as possible into the autumn season is a great practice in which many home gardeners do not take advantage. Gardeners can easily produce fresh, high-quality vegetables during the autumn, as well as have vegetables for storage and use during the late autumn and early winter months.

Warm-season crops planted in the spring, such as tomatoes, peppers, eggplant, sweet potatoes, and okra will continue to produce until frost, while other warm-season vegetables, such as snap beans, cucumbers, and summer squash, will slow down, cease their production, or even die in late summer. Mid-summer plantings of these vegetables will provide vigorous plants from which to harvest until or slightly beyond the first frost (with protection). Generally, the first frost is

Figure 13.1 This beautiful cauliflower curd was harvested from an autumn garden.

followed by a few weeks of warm weather, providing an even longer time to harvest with minimal plant protection from cold weather.

Although many tomatoes, such as 'Early Girl', will produce until it is killed by frost, midsummer plantings of short-season tomato cultivars can provide vigorous, robust plants from which to harvest high-quality fruit into the late autumn. Tomatoes harvested just before frost from healthy plants tend to keep better in long-term storage than fruits salvaged from older, diseased plants. While semi-hardy vegetable plants can tolerate light frosts, hardy vegetables will continue to grow until temperatures drop to approximately 20°F. Although the tops of semi-hardy root crops may be killed by hard freezes, the edible portion will store well in the ground if mulched with a heavy layer of straw. Cool-season crops are excellent for autumn gardening because they continue to grow well into the cooler months and can be harvested after many tender plants have been killed by frost. **Semi-hardy vegetables** are those that can withstand a few light frosts (28° to 32°F) and include beets, lettuce, mustard, potatoes, Swiss chard, and green onions. **Hardy vegetables** can withstand several frosts, but are killed when temperatures drop near or below 20°F, and include vegetables such as cabbage, broccoli, cauliflower, Brussels sprouts, carrots, collards, spinach, turnips, rutabagas, and kale.

Home gardeners should select the shortest-season vegetable cultivars available for autumn gardening to make sure that the harvest occurs before the first killing frost. Those cultivars listed as "early season" or those cultivars with the fewest days to maturity should be selected, and this information is usually printed on the seed packet or in the cultivar description in seed catalogs. This is especially important for those warm-season crops such as cucumbers, squash, and tomatoes planted in the late summer for an autumn garden. A gardener should try to obtain seed for the fall garden when they are planning their spring garden, as seed is often not available during the late summer or autumn months because seed sources have been depleted.

Remember that quality seed disappears from garden store shelves in mid-summer when seed companies remove their displays. To determine when to plant a particular vegetable for a late harvest, the average date of the first killing frost in your area must be known, as well as a particular vegetable cultivar's number of days to maturity.

PREPARING FOR AUTUMN GARDEN PLANTING

It is important to first remove and compost any remaining dead vegetable crops or weeds. Any diseased and/or insect infested plants should be buried deeply, burned, or removed from the garden site in some manner to reduce the spread of pathogens. Weeds with seedheads should be pulled from the garden soil and disposed of to prevent repopulating the garden soil with fresh weed seed. Tomato plants that were killed with a disease such as Fusarium or Verticillium wilt should be pulled and removed from the garden. Cucumber and melon plants infected with or killed by bacterial wilt should also be pulled and possibly placed in a closed garbage container for disposal. Cucumber beetles are attracted to the dying plants and if allowed to continue to feed, these insect pests may spread the disease to new crops of cucumbers or melons that are being planted for an autumn harvest. Most other plant residues can be composted; that is, if the gardener has a well-maintained compost pile that generates enough heat to kill disease pathogens that are possibly present.

Climatic conditions in late summer produce high soil temperatures, high light intensity, and rapid drying of the soil, resulting in an increase in problems asociated with obtaining a uniform stand of plants due to poor germination. Although most seeds will germinate quickly in the warm summer soil, some, such as lettuce, peas, and spinach, will not germinate well if the soil temperature is above 85°F, and fall-planted potatoes have the same problems with poor emergence when soil temperatures are high. Achieving sufficient seed germination in the heat of summer often requires special production practices. This might include placing a shade covering over rows at seeding and/or supplemental watering to reduce soil temperatures. Shading the soil with a shadecloth or an organic mulch will keep the soil cooler, and improve germination. It is best to remove the temporary shade once the young seedlings emerge. Light coverings of an organic mulch, such as straw or dry grass clippings (½ inch deep or less) can be placed around plants and young seedlings to help suppress weeds and conserve moisture. Oftentimes, a thickly placed organic mulch over the soil will need to be pulled back from around young seedlings, so the seedlings can receive full sunlight. Once the seedlings are 1 to 2 inches in height, apply additional mulch to help retain soil moisture and insulate the soil from the sun.

Soil moisture is also critical for proper seed germination. Oftentimes in the late summer, the soil moisture is depleted due to the lack of rainfall. So, if the soil is dry at planting, irrigation is essential to promote rapid seed germination and plant growth. Many crops such as carrots, spinach and beets do not germinate well during hot, dry weather, and soil moisture is of the utmost importance to obtain adequate germination rates for these crops. Remember that seed germination will not occur unless the soil remains moist at the depth where the seeds have been planted for an extended period of time. It is easy to just use drip irrigation over the area prior to planting, and then again, once the seeds have been planted. This will significantly improve the germination rates for these vegetable crops planted in the late summer.

If summer temperatures are excessively hot, cool-season vegetable seeds can be pre-sprouted inside at lower temperatures. To sprout seeds, sprinkle them on a moist paper towel, roll up the towel and seeds, and place in a plastic zip-lock bag. Check the seeds daily, and once most have sprouted, plant the pre-germinated seeds into a moist garden soil as soon as possible. Also, try not to damage the small roots when planting the sprouted seeds. Pre-germinated seeds must be kept moist until they emerge from the soil, so they will generally have to be irrigated in some manner for a few days.

FERTILIZER FOR AUTUMN GARDENS

Vegetable plants for the autumn garden require fertility applications for their growth and development just like those planted at any other time of the year. Since the soil nutrient levels are often depleted by spring and summer gardens, it is important to replenish these nutrients to have a successful autumn crop of vegetables. Although other nutrients can be diminished in the soil, nitrogen is an important nutrient that definitely needs to be applied to an autumn garden soil. The gardener needs to make sure that nutrients are applied at planting as well as later in a sidedress application to ensure that the soil has adequate fertility to achieve optimal plant growth, unless some type of slow-release fertilizer is used.

INSECT AND DISEASE MANAGEMENT

Insect pests tend to be worse on fall crops compared to those planted in the spring. By mid-summer into fall, general feeders, such as corn earworms and aphids, have high populations that have built up and can be heavy feeders on a variety of vegetable crops. Other insect populations, including cucumber beetles and squash bugs, generally peak in late summer and can quickly destroy newly emerging cucumber and squash seedlings. Squash bugs are troublesome pests that often occur in high populations in the late summer and early autumn and wreak havoc on squash and pumpkins growing in fall gardens.

Other insect pests, such as imported cabbage worms and cabbage loopers, can severely defoliate both young and mature cole crop plants if not controlled to some extent. Regular sprays of the bacterium, *Bacillus thuringiensis*, commonly called *Bt* and marketed under different trade names, will kill the larvae of many butterflies and moths that tend to cause the most damage to cole crop plants. This product was discussed in detail in the organic gardening section (Chapter 10).

If a gardener's goal is to reduce pesticide use, then lightweight fabric floating rowcovers is a method that can be used to exclude insects. As discussed in Chapter 12, floating rowcovers can be placed over new transplants or newly seeded areas following planting to prevent insects from feeding on emerging plants. Both light and moisture will pass through fabric rowcovers, and removal for sprinkler irrigation or rain events is not necessary to ensure that plants receive water. However, remember that floating rowcovers should be removed from insect-pollinated crops when plants begin to flower. Rowcovers will also help provide shade to the soil and may enhance germination of cool-season crops. Thus, fabric rowcovers can be an important part of autumn gardening.

During the late summer and early autumn, plant diseases, such as powdery mildew, downy mildew, as well as several other fungal diseases, will often become worse and are most prevalent during this time of the year. There are various fungicides available (both organic and non-organic) that can be applied when disease symptoms first appear to reduce their spread. Otherwise, visibly infected plants can be removed from the garden and disposed of in some manner to reduce the spread of various diseases.

WEED CONTROL

It is important to remove all weeds from the garden prior to preparing the soil for an autumn garden. Also, remember that tilling and preparing the soil for new plantings will bring weed seeds near the soil surface where they can germinate. It is best to remove young weeds as your new autumn crops grow. Mulching will help reduce weed germination by both preventing light from reaching the soil and smothering the tiny germinating weed seedlings.

TRANSPLANTS

As with spring plantings, some crops are more easily established when started early and transplanted into the garden. Garden centers or other stores that offer garden products rarely stock vegetable transplants for autumn plantings, but seedlings can be grown at home in 5 to 6 weeks and will give a gardener a head start on the autumn growing season. Growing transplants in the summer is easy, and once seeds have emerged, they can be grown outdoors in partial shade to full sun. However, due to the small amount of root growing area, transplants must be continuously watered, at least two or three times per day, especially if they are placed in full-sun conditions. At this time of year, there are no worries of freezing weather and no need to use grow lights for these transplants, and blowing winds and bright sunlight will help produce stocky, vigorous plants. Also, the gardener must keep an eye on cole crop seedlings grown outdoors to prevent them from being skeletonized by cabbage loopers or other foliage-feeding caterpillars (see Fig. 13.2).

FROST PROTECTION

The first frost in the autumn is often followed by 2 to 3 weeks of warm weather. If frost-tender warm-season plants can survive this initial frost, then they will produce fresh vegetables during these few weeks of warm weather until they are killed by repeated nights of freezing weather. As discussed in Chapter 12, lightweight fabric blankets, hotcaps, etc. can be used to protect tender plants. The first frost in the autumn generally develops on a clear, cloudless night, and covering plants will maintain some heat around the plants so that they do not freeze.

Although root crops such as beets, carrots, parsnips, rutabagas, and turnips can be dug in the late fall and stored in a cool, moist location such as a basement, they can be left in the ground through the fall and into the winter months. Before several hard freezes in the autumn occur, cover these plants with a heavy layer of straw mulch, and these root crops can be dug

Figure 13.2 Caterpillar feeding damage to young cabbage seedlings grown to be used for a late summer planting.

as needed throughout the winter by temporarily removing the mulch cover and replacing to protect other roots in the soil that will be dug at a later time.

AUTUMN/WINTER VEGETABLE GUIDE

A quick reference guide is provided for several important vegetable crops that are often grown during the autumn months. The information provided is for growing these crops in zone 6b, with the minimum winter temperatures about –5° to 0°F. Although the first frost at this locale occurs about October 15, consistent freezing temperatures (less than 20°F) really don't occur until about mid to late-December. If a specific region's first frost is earlier or later, just adjust these dates accordingly. It is also important to realize that adequate moisture must be provided to achieve sufficient germination rates of vegetables when planted during the late summer, as dry conditions tend to persist for several weeks this time of year.

Beets—In warm soils, seed do not germinate at a high rate and plants are slow to establish. Therefore, it is best to plant early to mid-September. Also, remember that adequate soil moisture is required to obtain sufficient seed germination rates, so that sufficient plant stands can be realized in a garden. a good stand in the garden. Roots and leaves can handle temperatures down to about 20° and 16°F, respectively, and once temperatures decrease below these levels, the foliage and roots will not recover.

Broccoli—It is best to either directly seed into well-moistened soil from July 1 to July 15 or have transplants ready to put in the garden by the first week of August. Although leaves can handle temperatures down to about 15°F, the heads are much more tender and will not generally survive repeated freezing exposures below 20°F.

Brussels sprouts—This vegetable crop requires a much longer growing season compared to broccoli and many other cole crops. It is best to direct-seed into a well-moistened soil by July 1 or transplant by August 1 for a dependable late autumn or early winter crop. Brussels sprouts are a very cold-hardy plant and in protected areas of a garden, harvest can continue into mid-winter.

Cabbage—Similar to broccoli, either directly seed into a well-moistened soil from July 1 to July 15 or have transplants ready to put in the garden by the first week of August. Cabbage leaves can survive temperatures only down to about 20° to 25°F, and if leaves are damaged by cold temperatures, remove any damaged leaves prior to storage.

Carrots—Plant in a well-moistened soil by late August. Although carrots can survive temperatures only down to about 12°F, covering with straw mulch will provide protection through most of the winter months. This is a great vegetable crop that overwinters very well in the soil with just a little bit of protection. Successful gardeners often have fresh carrots out of the garden all winter and into the spring months from a late-summer planting.

Cauliflower—Similar to broccoli and cabbage, either directly seed in a well-moistened soil from July 1 to July 15 or have transplants ready to put in the garden by the first week of August. Although leaves can survive temperatures down to as low as 15°F, heads are easily damaged by cold temperatures. Cauliflower is much more sensitive to cold weather conditions compared to either broccoli or cabbage.

Chinese cabbage—It is best to directly seed this vegetable in a well-moistened soil by the last week of July, or have transplants ready to put in the garden by the first week of August. Leaves and heads will survive temperatures down to about 25°F with no protection.

Garlic—Cloves can be planted in soil during the first week of October until the first week of November. It is best to get some shoot growth prior to the onset of cold temperatures. Garlic will survive down to about 5°F, although the use of straw mulch is a good insulator and can protect the plant during the winter from excessively low temperatures.

Greens:

Collard—Seeds of this green can be planted into the soil starting early August until mid-September. Although this crop will survive temperatures down to about 12°F, they can be covered with a fabric rowcover for overwintering to provide spring greens.

Kale—This plant has the same temperature requirements as collard (see previous comments on collard).

Mustard—Plant mustard seed into the soil starting early August until mid-September. Mustard leaves are more susceptible to cold conditions than are collards or kale and will generally survive temperatures down to about 25°F. Even if they are covered and protected in some manner during the winter, plants will generally bolt in early spring as days lengthen.

Spinach—Since spinach seeds do not germinate well in warm soils, the seed of this vegetable should be planted after the soil starts to cool in the late summer/early autumn. Spinach plants can withstand temperatures as low as 15° to 20°F.

Swiss chard—Swiss chard should be planted by 1 September. Although leaves can survive temperatures down to about 25°F, the smaller-leaf types are definitely more cold tolerant.

Rutabagas—Seeds of rutabagas should be planted by about 15 August. Rutabaga roots and foliage will consistently survive temperatures down to about 20° and 16°F, respectively.

Turnips—Similar to rutabagas, turnip roots and foliage will consistently survive temperatures down to about 20° and 16°F, respectively. Turnip seed should be planted at the same time as mustard, which is in the late summer/early autumn. Although turnip seed can be planted until about 30 September, the last plantings that are targeted for root production should be planted a few weeks earlier.

Herbs:

Cilantro—Cilantro seed can be planted from about late August until about late September. It is best to plant at the earlier date for fall harvests and later to overwinter for spring harvest. Cilantro seeds often have difficulty germinating in warm soils.

Dill—Dill seed should be planted by mid-September. The foliage will survive temperatures only down to about 25°F, so it will be killed after only a few light frosts.

Parsley—Parsley seeds should be planted from mid- to late July in the garden. Foliage will survive temperatures down to about 15° to 20°F. Curled-leaf types of parsley are slightly more cold tolerant compared to the normal-leaf types.

Kohlrabi—For fall crops of kohlrabi, plant seed from late July to mid-August. The plant will survive temperatures down to about 20°F.

Lettuce—The heading types of lettuce (iceburg and romaine) will survive temperatures only down to about 25°F, so after only a few light frosts, these plants will be killed unless protected in some manner. It is best to seed the heading types of lettuce from early August until late August. The seed of small-leaf types of lettuce can be planted about the same time as the large-leaf types if they are to be harvested in the fall. However, the small-leaf types of lettuce can be planted up until about mid-October if they are going to be protected and overwintered in a garden. These plants should have about 4 to 10 leaves before the onset of winter. The growth of these small-leaf types of lettuce will begin again in February/March. It is important to note that lettuce generally requires cold protection during the winter if they are going to successfully overwinter.

Leeks—Although this plant is generally planted in the spring, a last seeding can be done about mid-July. Leeks can be hilled or mulched for use during the winter, as they can effectively survive temperatures down to about 0° to 10°F.

Green Onions—The seed of green onions should be planted in mid-July to early August for winter to spring harvests. Overwintered onions should be mulched in early to mid-December to minimize winter damage.

Radishes—The roots of radish can survive temperatures down to about 20°F, while leaves can survive slightly lower temperatures down to about 15°F. There are primarily two types of radishes that are grown: the small round or cylindrical spring radishes and the Asian (or Daikon) winter radishes that develop larger, carrot-shaped roots. Although Daikon radishes should be planted before early September (since they require about 60 days to mature) the smaller radishes can be planted up until early November.

Chapter Fourteen

CONTAINER VEGETABLE GARDENING

LIMITED-SPACE GARDENS

Containers are an easy way to incorporate vegetables into home garden landscapes that have limited space. Many different types of containers are available that can be easily used in landscapes. Containers are semi-permanent structures that can be easily moved around landscapes for various reasons. Thus, if a container has vegetables that require limited amounts of sunlight (e.g., lettuce) and is in a site that has too much sunlight, it can easily be moved to a shadier spot in the landscape. A gardener would not be able to do this with plants that are permanently planted into the landscape. Thus, the proper site selection is essential for success when growing vegetables in containers.

Today, many live in apartments or other dwellings that do not have adequate space in which to grow a large in-ground garden. Thus, container gardening is gaining popularity due to the space limitations of many gardeners. Although most do not grow vegetables due to the space limitations, this should not discourage those who would like to have a small garden in which to grow a few vegetable plants. Container vegetable growing is becoming an important part of home gardening, since many people derive personal satisfaction from maintaining an active garden. Many apartments or condominiums have a deck or patio and this area is often not utilized effectively for growing plants. Oftentimes, this space could be utilized to produce a wide array of fresh vegetables with just a little effort.

The lack of adequate space is really not an excuse for not having a garden of some sort, since many kinds of vegetables can be readily grown in containers. Container gardening allows those with limited space or unsuitable gardening conditions to raise a variety of fresh vegetables. Even in small apartments, sunny spaces such as patios, porches, terraces, and window sills can be used to create productive container gardens (Fig. 14.1). While some thought must be given to plant variety selection, the majority of vegetables can be adapted to container culture.

There are several advantages of growing vegetables in containers. Container growing allows gardeners to extend the growing season by being able to move their gardens to frost-proof surroundings when there is a threat of frost. Containers also provide gardeners who have poor soil conditions in the garden the chance to successfully grow vegetables. In addition, weeds are easier to control in containers, and containers can be moved if necessary to capture as much sunlight as possible.

Figure 14.1 Fresh herbs and vegetables grown in compact containers are well suited for backyard or patio gardening.

FACTORS TO CONSIDER WHEN USING CONTAINERS

There are a few important factors to consider when deciding to grow vegetables in containers. The gardener must have a clean usable container, a good soil mix, access to water, adequate sunlight available, as well as having knowledge about planting and spacing requirements, fertilizing, watering, and variety selection.

Sunlight. It is important to remember that the site where the container is placed must have five or more hours of full sun on a daily basis, but it really depends on the vegetable being grown. So, before one can garden successfully in containers, adequate sunlight must be available to allow the plant to grow and develop properly. For most home gardeners who try container gardening, **light** is the most limiting factor and most plants grown in gardens, whether vegetables, fruits, or herbs, require full sun to grow and fruit properly. Most vegetables prefer 6 to 8 hours of direct sunlight each day. Therefore, each container should be filled with plants that are identically suited to specific locations in the landscape and can be moved to take advantage of suitable environmental conditions. There are only a few plants that grow well in limited sunlight conditions and these tend to be cool-season vegetables such as lettuce (Fig. 14.2).

Site selection is an important part of container gardening. In colder climates, pots can be placed near a south-facing wall to get a head start on the growing season. In warmer areas, be cautious of placing containers on concrete, as the excess heat may hinder plant growth. Again, most vegetables prefer to be grown under full-sun conditions. Containers with developing vegetables need to be placed where they will receive at least **5 or 6 hours** of full sunlight every day. If this is not possible, then the gardener should concentrate on growing plants that will be tolerant of shade conditions. Many herbs can be effectively grown in shady conditions, including lemon balm, mints, parsley, and thyme, as well as lettuce, a commonly grown vegetable in many gardens. Although most edible gardens require full-sun conditions, gardeners should not site containers in full mid-day sunlight. Plants may require full-sun conditions, but container gardens heat up much more quickly and intensely than do plants in the ground. Most plants will definitely prefer some relief from the mid-day sun. On the other hand, if a container is in the shade, consider placing it against a wall that can reflect some light onto plants. The plants will generally not suffer from the extra amount of heat, but will benefit from indirect light.

***Figure 14.2** Various vegetables and herbs grown effectively together with a few ornamental plants in a container.*

Containers. There are many types of containers that one could possibly use, depending upon the gardener. Many gardeners have been known to use bathtubs, old whiskey barrels, old tires (Fig. 14.3), and metal buckets as containers. Clay (or terra cotta) pots are probably the most widely used containers in home gardening. They allow a significant amount of air movement in the root zone due to their porous nature, which is highly beneficial for plant growth, but it is important to remember that the soil in a clay pot will typically dry out faster than other nonporous-type containers. In general, plants in terra cotta (clay) need more attention to watering compared to other types of pots, due to their porous nature. Many gardeners prefer other containers such as wooden boxes, wooden tubs, hanging baskets, and window boxes to grow plants. Containers for vegetable gardens can be almost anything from flower pots, pails, buckets, wire baskets, bushel baskets, wooden boxes, nursery flats, window planters, strawberry pots, plastic bags, large food cans, or any number of other things that can be used to hold soil.

Whatever type of container is used, it must have adequate drainage in the base of the container. If there is not adequate drainage, plants will become waterlogged and will eventually die due to the lack of oxygen provided to the roots. To prevent holes from being clogged from

Figure 14.3 An old used tire being used as a container for cilantro.

soil or soil-less mixes placed in containers, it is best to place a layer of rocks in the bottom of the container (if it is deep enough) to allow better water drainage. The bottom of a pot can also be lined with screen or landscape cloth to prevent soil from spilling out and clogging the holes. Drainage is reduced when the container is set on a solid surface such as a cement or patio floor, and raising the container 1 or 2 inches off the floor by setting on bricks or blocks of concrete will partially solve the drainage problem.

The size of the container in which vegetables or herbs are grown is important. A gardener should make sure there is enough room in the container for both the plants and soil. It is also important to understand the mature size of the plants and their growing habits, whether upright or vining. As plants grow, the root systems will fill the pot and the soil will dry more quickly. At planting, a gardener should try to make sure there is plenty of room in the pot for the roots to move downward into soil as they grow. Most suggest that containers at least 10 inches wide and 12 inches deep be used, but it really just depends on the vegetable being grown. Large flowerpots, half barrels, plastic-lined bushel baskets, window boxes, and large containers (e.g., 5-gallon buckets) work well for growing vegetables.

The container depth is an important part of growing vegetables in containers. There must be enough rooting area for the plant to grow, and there are differences among the various vegetables grown with respect to the amount of rooting that is needed for optimum growth. Minimum depths required for some vegetables to grow adequately: 4 inches—lettuce, onions, low-growing herbs; 6 inches—turnips, radishes, beets, chard, kohlrabi, herbs; 8 inches—bush beans, cabbage, eggplant, peppers, cucumbers, spinach; 10 inches—cauliflower, broccoli, Brussels sprouts; and 12 inches—parsnips, carrots, tomatoes, potatoes.

A container can sustain only a certain number of plants, so it is important to limit the number of plants based on the container size and the size of the plant at maturity. A gardener should always plant a few more seeds than the number of plants desired in the container, since 100% germination and emergence rarely occurs even with new seed. Once seeds have germinated and developed about 2 or 3 leaves, the plants can be thinned to the desired number per container.

Some vegetables need particularly large containers. For larger vegetables like tomatoes and eggplants, a large container is most desirable (e.g., a five-gallon container) for growing plants. Although these plants can be grown in smaller containers, they will definitely require more frequent waterings. Standard-size tomatoes and vining crops, such as cucumbers, will do best in containers 20 inches or more across, while peppers prefer pots at least 16 inches in diameter. Generally, when it comes to container size, bigger is better, especially for beginning gardeners, since large pots hold more soil and do not require as frequent watering.

The color of the container can also contribute to the success or failure of a vegetable crop. Dark-colored pots absorb more heat and tend to increase soil temperatures more in the summer months, which can oftentimes damage plant roots, especially in hot-summer areas. Cool-season crops prefer cooler soil temperatures and should be grown in lighter-colored containers to maximize their productivity (Fig. 14.4).

Plants grown in containers that produce tall plants (e.g., tomatoes) or that produce vines (e.g., cucumbers) will be more productive if allowed to grow up some type of support. Many gardeners prefer to just insert a wire cage of some sort into the soil to provide the support required for the plant to maximize its productivity. It is also important to note that larger, heavier containers should be used for trellised plants to minimize the risk of the plant and container from tipping over.

Figure 14.4 A white container used to grow Swiss chard during the summer months. The white color does not provide as much heating of the soil compared to darker pots, which is beneficial for a cool-season vegetable like Swiss chard.

Soil medium. Synthetic soils (or soil-less based potting mediums) are best suited for vegetable container gardening. These mixes may be composed of sawdust, wood chips, peat moss, perlite, vermiculite, or almost any other type of media. However, the mixture components should be sterilized to be free of disease pathogens, insects, and weed seeds. However, **do not** use soil taken from the garden. No matter the composition of the soil, it could contain disease-causing organisms (such as *Rhizoctonia solani*), which can kill young developing seedlings.

The ideal planting medium for containers should provide rapid drainage with sufficient water retention to keep the root zone moist. A sterilized soilless potting mix often works best, since they drain quickly, are lightweight, and are free of soil-borne disease-causing organisms and weed seeds. Peat-based mixes, containing peat, vermiculite, and perlite are excellent choices, although some gardeners avoid mixes that have high amounts of peat, since they tend to dry out quickly. Soil-less based potting mediums can be purchased from garden centers in various sizes under many different names, or they can be prepared by mixing components such as horticultural grade vermiculite, peat moss, perlite, limestone, superphosphate, and garden fertilizer together.

***Figure 14.5** Water is essential for container gardens and the lack of water only for a short period of time can easily kill plants. Watering is such an important part of container gardening and must often be done on a daily basis, especially if containers are sited in full sunlight.*

Water and fertilizer. Watering is an essential container garden activity that often must be completed on a daily basis by a gardener (Fig. 14.5). Unlike plants grown in the ground, container plant roots can't move down into the soil profile in search of subsurface moisture. The most common method used to water plants in containers is with a watering can or sprayer attachment on a garden hose. It is important to make sure that the water used has a relatively cool temperature at application, particularly if the water hose sits in the sunlight for extended periods.

A gardener should check containers every few days during the early part of the growing season to determine when water should be applied, but when the plant is larger and using high amounts of water, the moisture condition of the soil should be checked daily. As the season progresses and as plants mature, root systems will expand and plants will require more water. It is also important to note that shallower containers will require more frequent watering compared to those that have a deeper rooting volume. Containers will require water when the top few inches of soil has dried out, and this is more likely to happen to a plant placed in full sunlight or under windy conditions. It may be easier to install a drip-irrigation system that will automatically irrigate vegetables in containers (Fig. 14.6).

Figure 14.6 Drip irrigation can be placed in containers to make watering much easier for a home gardener.

Soil-less based potting mixes contain few if any nutrients, so nutrients must be added to the soil to maximize plant productivity. It is recommended to feed vegetables in containers about once every two weeks with a complete water-soluble fertilizer. Other fertilization strategies include the addition of a slow-release complete fertilizer to the surface at planting to keep vegetables fed for most of the growing season and/or to add one part compost to two parts planting mix to improve the fertility of the mix. Many gardeners prefer to add a slow-release fertilizer to keep plants fed with nutrients for most of the growing season without having to continuously fertilize.

Wind. Wind can be a real hazard for any container vegetable garden, especially for plants like tomatoes and cucumbers that can become top-heavy with fruit. If possible, gardeners should try to place containers so that they are not in an overly windy location. Although a light breeze will provide air circulation to evaporate excess moisture and aid in the prevention of fungal diseases, a strong wind can topple containers, potentially damaging plants and dislodging fruits. When gardening on a raised deck or a roof top, it may be necessary to provide some type of wind block to prevent wind damage or toppling of containers.

Pest control. A gardener must always be on the lookout for plant pests, even when growing vegetables in containers. While plants grown in containers are not usually as susceptible to disease, insect, and weed pests as those vegetables grown in the ground, a gardener still needs to be aware of the potential problems of these pests. Also, depending on the severity of the insect infestation or disease infection, a plant growing in a container may require disposal or may need to be treated with pesticide to minimize the disease or insect damage.

Harvest. The harvesting of vegetables is also an important part of maintaining a productive container garden. As vegetables ripen, they should be harvested to maintain a productive garden. Most vegetables are more productive if they are harvested often, and if plants go to seed, this will often cause a reduction in fruit set and overall yield. However, we must remember that some gardeners will want to leave the red or multi-colored vegetable fruits on the plant to provide that added splash of color in an edible garden.

Many home gardeners will get only one major harvest of vegetables from their garden, but with proper management, vegetables can be continuously harvested from containers. For a continuous harvest, this requires the planting of containers throughout the growing season. So, some container garden space should be reserved for later plantings. Also, it is important to remember that certain cool-season crops such as cabbage, broccoli, and cauliflower should be grown in early spring to early summer or in late summer to late autumn, and not during the hot conditions of mid-summer.

Sanitation. Fall care and cleanup of a container garden is something that gardeners must think about after the season is finished. At the end of the season, dispose of the soil or add to your compost pile. It is important to remember that reusing soil from year to year can spread disease pathogens and insect infestations. Also, make sure to thoroughly scrub the container to remove all soil and roots attached to the inside of the container. It is best to rinse containers in a solution of 10% bleach, and then rinse with clean water and store in a dry spot until used again the following year.

Varieties. The needs of small space or limited space gardeners have been seen by plant breeders, and many different miniature, dwarf, or bush vegetable varieties are available for container gardening. Many vegetables lend themselves well to container gardening, and almost any vegetable can be adapted to growing in a pot, although those that work best are bush or dwarf varieties. Vegetables that take up only a little space, such as carrots, radishes and lettuce, or crops that bear fruits over a long period of time, such as tomatoes and peppers, are perfect for container vegetable gardens. A container vegetable garden is limited only by the size of the container and the gardener's imagination. Containers of mixed vegetables and herbs often work well together. A tomato planted in a container with parsley or chives grow well together since they have the same water and sun requirements, and will keep producing well into the autumn months. Several compact vegetable varieties will be discussed that have great potential to be grown in containers by the limited space gardener. Vegetables that are great choices for container gardening include beans/peas, beets, carrots, cucumber, eggplant, lettuce, onions, peppers, radish, spinach, and tomato.

EASY VEGETABLES FOR CONTAINERS

Beans and Peas

Green beans and peas are popular plants in home gardens, although they are normally grown in rows. However, these plants can also be grown successfully in limited space container gardens. Both of these legumes will grow well in containers, given the right care and conditions, including good soil drainage and **full-sun growing conditions.**

Green beans make a wonderful addition to a container garden. Graceful vines of a pole-bean plant can extend six feet high or more on a trellis, pole, or fence, while a stocky bush-bean plant will usually reach its peak height at about two feet tall. Either can fit well in any container garden that has a full sun exposure. Some green bean varieties just grow better in containers, such as 'Bush Romano', 'Bush Blue Lake', and 'Bush Kentucky Wonder'. Large pole-bean varieties (such as 'Kentucky Wonder') may get too large to grow easily in containers unless some type of trellising is used.

Peas are also a great vegetable to grow in container gardens. They grow quickly and do not generally require lots of attention. However, they do require full sunlight and moist soils. Peas prefer **cool growing conditions**, so they are generally planted early in the spring. It is important to remember that when it gets too warm, peas will stop producing flowers and pods. Once they become non-productive, they can be pulled out of the container with a warm-season vegetable then started in the same container. Some favorite peas for container gardens include the English peas, 'Little Marvel,' 'Tom Thumb', and 'Early Frosty', the snow pea, 'Oregon Sugar Pod', and the sugar snap pea, 'Sugar Snap'.

The choice of the right container for green beans and peas will increase the chance to maximize productivity of these crops. Any large container with bottom drainage holes is suitable for peas and beans, including traditional planters or plastic tubs. Although these crops require at least a 1-gallon container for successful growth, those at least 5 gallons in size are best for peas and green beans. Larger containers provide the beans and peas with more space for their roots, but containers that are too large can be difficult to move around in a landscape. It is also important to note that an inoculant should be added to the soil—unless seeds that have already been inoculated with *Rhizobium*—as this will improve their productivity.

Unlike some other vegetable plants, several green bean and pea plants can be grown in a single container. It is generally recommended that two or three plants be grown in each container with bush and pole types spaced 2 to 4 inches and 4 to 6 inches apart, respectively. These vegetables can also be planted into a hill in the center of the pot, and later thinned to two or three plants. For pole types, place a trellis above the container so that plants will have something to climb when they begin to grow. To encourage more production, it is best to continually remove pods when they reach the appropriate size for kitchen use and remove any shriveled or underdeveloped pods.

Beets

Beets are not as popular as some other vegetables, but they can easily be grown in containers (Fig. 14.7). The container for beets should be at least 12 inches in diameter and 8 to 12 inches deep. **Beet seeds** should be directly planted about 1 inch deep into a sterilized potting mixture. To obtain proper germination rates, it is important to provide consistent adequate moisture until the beet seeds sprout. It is best to place containers in a sunny location, although the container should be moved into shadier areas on hot, sunny afternoons. A gardener should thin beets when seedlings are about 2 inches tall by cutting the smallest ones with a small pair of scissors. Maintain proper moisture in the pots throughout the growing season, and beets may need to be watered twice per day during hot, dry weather. The soil should be damp, but not over-wet, as the plants will oftentimes rot in soggy soils. Fertilize beets once weekly, using a water-soluble fertilizer for vegetables.

Although there are many beet varieties available, recommended beet varieties for containers include 'Baby Canning', 'Bull's Blood', 'Burpee Golden', 'Detroit Red', 'Ruby Queen', 'Red Ace Hybrid', and 'Early Red Ball'. 'Detroit Red' beets are tailor-made for container growing, since they are compact, bushy plants that require little growing space, but will produce plenty of sweet, tender, deep-red beets.

A gardener should begin harvesting beets when they reach about 1½ inches in diameter, which is normally about 60 days after seeding. If larger beets are preferred, harvest before the roots are more than 3 inches in diameter, as beets allowed to grow for too long will become woody, fibrous, and tough.

Carrots

Carrots can easily be grown in containers that have a sufficient rooting depth, although smaller, immature roots can be grown in more shallow containers and harvested early for tasty

Figure 14.7 Beets grown in a home landscape in a container.

baby carrots. Most gardeners have had trouble at some point in producing carrots in a garden, as carrots are somewhat finicky and require specific soil and weather conditions to grow well. Soil is an important consideration when growing carrot plants. This crop will actually grow better and be more productive in containers when produced in a controlled soil environment. Containers have the advantage of being filled with soil that is free of large rocks and other impediments that can prevent carrot roots from growing straight. The soil is also loose, allowing carrot roots to move through the soil more easily and develop straight roots. Thus, the best and highest quality potting soil should be used when growing carrots. The soil mix should be loose, but not sandy; and, compost mixed with a purchased sterilized soil-less potting mix that is loose should work well. The containers do need to be fairly deep (at least 12 inches) in order to allow the carrots to grow down, although there are a few varieties that do not require the 12-inch depth.

Most carrot varieties will grow well in containers, and although most do not grow longer than 1 foot, try to grow those varieties listed as producing roots in the 6- to 9-inch range. In shorter, more shallow containers, try 'Parmex', which grows as a roundish ball. Suggested varieties include 'Little Finger', 'Danvers Half Long', 'Short 'n' Sweet', and 'Thumbelina'.

Planting of carrots in containers can be done at any time the weather will provide conditions warm enough for the carrots to grow. Carrot seeds should be pushed into the ground about a quarter of an inch or so and then watered carefully soon after they are direct-seeded into the soil. Carrot seed can also be broadcast across the soil surface and covered with a thin layer of soil or vermiculite, and then watered. Moisture is critical to obtain optimum germination rates, so it is best to not allow the soil or media to dry out. Once the seedlings are sprouted and about an inch high, they should be thinned to the proper spacing, which is about one or two inches apart. When carrots reach the young "baby carrot" stage, that is, where the plants are about six or seven inches high and the root is about two or three inches long, they should be thinned again and spaced about six inches apart. These baby carrots are excellent to eat as they are young and tender.

Carrots will require soluble complete fertilizer applications every couple of weeks. They will also need to be watered regularly, which is about every two or three days. Full-sized carrots can be harvested at about 70 to 80 days after seeding, depending on the climate.

Cucumbers

Cucurbit vegetables generally require lots of space, since they are vining-type plants. However, there are many bush-type varieties of cucurbits that don't require as much space. The cucumber is one of these plants in which varieties have been developed for limited space gardens. Anyway, cucumbers are easy vegetables to grow in containers, and they will produce plenty of fresh fruits with very little effort. However, growing cucumbers in pots can sometimes be difficult as their vines can become quite elongated and hard to manage.

Remember that cucumbers are a warm-season vegetable and do not seed into soil in containers until the outside air temperatures are consistently above 70°F. Seeds are typically direct-sown, although some garden centers will sell cucumber seedlings in the springtime and these can be transplanted into containers. Normally, about 3 to 5 seeds are planted in a cluster about a half inch deep into the potting soil. Once the plants reach the two-leaf stage, choose one or two of the seedlings that look strong, and remove the others (Fig. 14.8). A gardener should not pull the cucumber seedlings out of the soil, as you may damage the roots of the remaining ones. So, it is just best to cut or pinch the seedlings off at the soil surface.

Figure 14.8 At the two-leaf stage, cucumber seedlings should be thinned to one in a pot this size.

A good quality potting soil or soil-less mix should be used as the growing medium. Some gardeners will mix in about 25% of the potting volume with a high-quality compost to improve fertility and soil structure. Cucumbers will grow very quickly if grown in full sun, and fertilized and watered properly. Cucumber plants should be given a complete soluble fertilizer about once per week, and it is important to not let the soil dry out completely. Dry soil will stunt cucumber growth, but more importantly, the lack of moisture will make cucumber fruits bitter to the taste from the build-up of cucurbitacins.

About two or three cucumber plants can be grown in a single large container (5-gallon pot) and more plants will definitely improve pollination and fruit set. If indeterminate types are used, it is best to place at least a four-foot-tall trellis behind the container, as this will give the cucumber plants adequate space to grow, keep them off the ground, and maximize growing space.

It is important to harvest your cucumbers on a regular basis. Cucumber fruit should be harvested at a fairly small size, since they will become bitter and seedy as fruit enlarge and mature. The more often a gardener can harvest fruit, the harder the cucumber plant will work to produce more fruit, which allows maximum harvest from plants.

Small or miniature type cucumbers are perfect for the container garden, but larger varieties may also be grown using a trellis. Provide support for these larger vines the same as you would for pole beans. The determinate-plant types of cucumber don't grow as freely as indeterminate types, and they terminate in vine growth, making an ideal container plant. There are several determinate varieties of cucumbers that are available to home gardeners, including the varieties 'Salad Bush', 'Spacemaster', 'Bush Champion', and 'Bush Pickle'.

Cucumbers are generally either **gynoecious** (mostly female flowers) or **monoecious** (male and female flowers), and gynoecious cultivars must be planted with monoecious cultivars to ensure fruit set. Monoecious types can self-pollinate with sufficient bee activity, and have the additional advantage of a longer harvest period. Thus, home gardeners often prefer dwarf, monoecious types for containers.

Other Cucurbits

Although most cucurbit varieties need lots of space to grow, summer squash such as zucchini is often grown in containers, since they have a bush-type growth habit and typically do not form a vine. There are also bush types of pumpkins and watermelons that can be grown in containers, but these are rarely grown in containers due to the significant vine size and the large fruit that are typically produced.

Eggplant

The eggplant is a great choice for a container since it is such an easy vegetable to grow. It is generally a very prolific fruiter, plus the fruit are oftentimes quite showy and can provide additional colors into the landscape as well. Remember that eggplants are warm-season plants that do not tolerate frost. Eggplants are transplanted into containers, with plants either purchased from garden centers as seedlings, or seeds started indoors.

Large containers work best, because they allow ample space for the roots to grow large enough to support the plants. Large containers that are at least 18 inches wide work best. These containers should have a rooting depth of at least 12 inches to allow eggplant roots to grow. Since each plant takes up lots of space, only one plant should be grown in each container. Eggplant grows best in high organic matter and well-drained soils. Therefore, many gardeners will combine compost, peat, and a purchased sterilized potting soil mix in equal parts to use as the medium for growing eggplants in a container.

Eggplant varieties such as 'Orlando', 'Oriental Fingerling', and 'Patio Mohican' are often recommended for containers. **'Orlando'** sets baby fruit that can be picked as small as 2 inches long and are borne on compact space-saving plants. This variety is perfect for containers and the fruit are tender, flavorful, and dark purple to black in color. **'Oriental Fingerling'** produces black-skinned, elongated eggplants that are perfect for container gardening. **'Patio Mohican'** is a unique eggplant that has a dwarf habit, which makes it particularly attractive when grown in patio containers. The long, oval, white-skinned fruits hang like bunches of grapes, and are best picked at a small size to increase yield over a long growing season.

Light, moisture, and high fertility are all essential to maximize eggplant growth in containers. Eggplants will need six to eight hours of direct sunlight each day, so containers should be placed in a sunny area. Gardeners should water the soil when the top 1 to 2 inches feel dry and this can be done by inserting a finger into the top few inches of soil. Treat the eggplant with a water-soluble fertilizer every two weeks, as this plant is a heavy nutrient feeder. It is important to avoid damaging the stem of the plant from fertilizer injury, as this can kill the plant. It is important to note that vegetables grown in containers require more frequent fertilization applications since they lose a significant amount of nutrients through drainage holes when watered.

In most cases, once eggplant fruit reaches 4 to 6 inches in length, and has a dark sheen and glossy shine, they are ripe and ready for harvest. Although some gardeners will leave fruit on the plant to provide an added touch of color to a landscape, the fruit will have to be picked at some point to allow new fruits to develop on the plant.

Lettuce

Lettuce is an easy vegetable to grow in containers and there are numerous varieties to choose from that do well in pots. A benefit of lettuce is that several different varieties can be planted into one container that can display a multitude of colors. Growing lettuce in containers is a common practice for small-space gardeners. Lettuce is a cool-season crop and the foliage develops best in cool temperatures. The use of containers allows an earlier start for lettuce in a garden since pots can be brought indoors during light freezes and left outdoors during early spring or late autumn days. Similar to other vegetables, growing lettuce in containers requires the right type of pot and proper planting medium. It is important to note that lettuce plants require pots large enough—6 to 12 inch diameter and at least 4 inch depths—so that roots have space to grow. Although lettuce needs a consistent supply of moisture, they cannot tolerate wet roots. Thus, a clay container is great for lettuce, as it will provide a permeable surface that can evaporate any excess water and prevent soggy roots.

Planting and maintenance are also critical aspects of growing lettuce grown in containers. Planting lettuce in container gardens is normally accomplished by direct sowing, although transplants can be used. Seeds should be sown when soils are not frozen, about ½ inch deep and 4 to 12 inches apart. Transplants should be buried about ¼ inch deeper than they would be in garden soil and set 6 to 12 inches apart. Remember that leaf lettuces can be planted closer together than heading types.

Professional sterile soil mixes are commonly used for lettuce grown in containers, as they are formulated to hold water and provide nutrients, although compost can be added to improve soil structure and fertility. Loose-leaf lettuces are better suited for container growing than are head types of lettuce. Although some small-heading lettuces may be grown in containers, the leaf-type lettuces are the easiest to grow in pots. Leaf lettuce is well known as a cut-and-come-again vegetable. At harvest, cut the outside leaves of loose-leaf lettuce when they are young, and the leaves will grow back again and then cut the entire plant. Always cut lettuce when it is young and tender, as they are quick to bolt and become bitter. Remember that once the head of heading types are cut, it will not regrow.

There are numerous varieties of lettuce that are well suited for container gardening, although iceberg (or crisphead) is generally not recommended. Lettuce varieties that do well in containers include 'Red Oak' (a looseleaf variety), 'Tom Thumb' and 'Little Gem' (butterhead types), and 'Rosalita' and 'Freckles' (romaine types). Red-colored lettuce varieties are well-suited to be grown in pots with various herbs or ornamental flowers (Fig. 14.9).

Figure 14.9 A red romaine type lettuce growing during the early spring in a container in a home landscape along with an ornamental flower.

Onions

Onions are another vegetable that is well suited for container gardening. Growing onions in containers is a perfect solution for those space-challenged gardeners. The advantage of growing onions is that only in a few weeks, fresh, delicious onions are ready for harvest. However, when large, bulbing types of onions are desired, there are several factors that can influence the overall productivity of these vegetables when grown in containers.

Basic Requirements for Onions Grown in Containers

Correct Onion Type. Large bulb onions are broken down into three types: long day, short day, and day-neutral. Short-day onions are best for more southern latitudes in northern temperate climates, while long-day types grow best in more northern latitudes. Day-neutral types do not depend on day length to bulb and bulbing is influenced primarily by temperature. So, it is important to find out which varieties grow best in your gardening region. 'Candy' is a day-neutral variety that works well in many different climates as a bulbing type onion.

Loose Fertile Soil. Although onions have small and shallow root systems, they are extremely heavy nutrient feeders. To maximize onion production, compost should be mixed at a 1:1 ratio with the potting mix, as this will greatly boost available nutrition and water retention, while maintaining good soil drainage.

Available Sunlight. Onions require full-sun conditions, as this has a direct effect on the critical photoperiod. If the critical photoperiod falls short or exceeds the requirements of the variety grown, onions may fail to produce bulbs. Short-day varieties typically require a 10- to 12-hour day length to start the bulbing process, while long-day varieties need about 14 to 16 hours of daylight for bulbing to begin.

Onion Propagation. The first step for growing onions in containers is to have an understanding of how to propagate them effectively. Onions are generally propagated by seeds, sets, or transplants. **Onion seeds are the most inexpensive option** for the gardener, but do require much more patience and time. For container gardeners, onion seeds must be started a few months before the average last frost to obtain good bulb production. However, gardeners will be able to choose the exact variety they want to grow. When growing in containers, **onion sets are the best way to start gardening. Sets are small, immature onion bulbs that were started the previous year from seed, then dug up and stored.** These small onion bulbs generally set quicker and produce larger bulbs, which are desired by gardeners looking to maximize productivity. **Onion transplants** are normally grown from seed, and although onion sets or transplants can easily be found at a nursery or garden center or even online gardening sites in the spring, you can also start onions from seed and grow your own transplants, but **sets are the easiest and most effective way of planting onions in containers.** Sets are usually planted in March or April depending on the area.

Container selection is very important, and for onions, **container depth is not necessarily as important as the surface area of the soil.** Since onion roots remain fairly shallow, a pot with a minimum depth of 6 to 10 inches of soil works well. Wider, shallower containers offer more room to grow onions compared to containers that are narrow and deep. At about 4 to 6 weeks before the average last frost in your area, fill the containers with potting soil/compost mixture and place outdoors in an area that receives full sun. Space onion plants 3 to 4 inches apart in rows 10 to 12 inches apart. Alternately, if a large container is used, an all-direction style of planting may be used. In this process, the onions are planted in all directions spaced 6 to 8 inches apart from each other. The planted onions should be watered immediately after planting. It is also important to ensure that onion plants are properly watered throughout the growing season. On average, onions planted in containers will need to be watered several times a week to maintain consistent and adequate moisture conditions, with more waterings required during hot and dry weather conditions. It is best to water onion plants only when the top 2 inches of soil has become dry. A gardener should also fertilize onions at 2- to 3-week intervals during the growing season with a water-soluble complete fertilizer, or apply a time-release granular fertilizer to the soil once the onions are actively growing, which will provide fertility throughout most of the growing season.

Green onions can be harvested for the kitchen when the tops are at least 6 inches in height. Green onions are best harvested when they are small and have the best flavor at this stage, as onions tops will develop a stronger flavor as they get older and larger. Harvest large bulbing-type onions during summer when the tops of the plants begin to fall over and begin to dry out. It is best to harvest in the morning and lay onions in the full sunlight to dry out until late afternoon. Onions should then be stored in a cool, dry place. Although fresh onions will have the best taste, they can oftentimes last for a year if stored properly.

Peppers

Peppers make a great addition to a container garden due to the vast diversity of fruit colors, shapes and sizes. Most pepper varieties can easily be grown in containers, so the choice of peppers that can be grown in containers is unlimited. Gardeners will often use several containers just to grow sweet, mild, and hot pepper varieties in the garden. Commonly grown peppers include sweet bell peppers, as well as various hot peppers, like jalapeno, serrano, cayenne, and habanero. Anyway, gardeners just love to grow peppers in containers, since they can add an attractive burst of color to a patio or deck while providing the family with fresh vegetables that can be used to spice up almost any dish (Fig. 14.10).

Pepper seed should be started about eight weeks before they are to be transplanted into a pot to be placed outdoors. Unlike most vegetables, pepper seeds take about 10 days or longer to germinate, depending on soil temperatures. Pepper seeds require soil temperatures of at least 80°F to germinate. After germination, it takes about 6 to 8 weeks to reach the proper size for transplanting into containers outdoors. Pepper plants can be transplanted into outdoor containers when nighttime temperatures are consistently above 60°F, and daytime temperatures are at least 70°F. Many different varieties of pepper transplants can oftentimes be purchased at local garden centers during the spring months, which may be the better option since pepper transplant production can be somewhat difficult unless a gardener has the proper indoor facilities to germinate and grow plants.

Peppers require a container that is at least 8 inches deep and 12 inches in diameter. Most varieties will grow about 12 to 18 inches high, so it is best to have at least a medium-sized pot. A commercial sterilized potting soil medium mixed with about 25% compost by volume works well for peppers. However, it is important to note that peppers are very susceptible

Figure 14.10 Ornamental peppers make magnificent container plants!

to soil-borne diseases and if problems occur, a gardener may have to resort to using only a sterilized potting mix. Peppers require full sun and very warm weather to grow properly. A gardener may have to move containers around in the landscape to make sure they are always in full-sun conditions.

Peppers can be harvested at almost any time after fruit pods develop on a plant. Green peppers can be picked from the plant before fruit are mature, but if colors are desired in a garden, fruit must be left on the plant until they ripen; most varieties ripen to a red color. Peppers are one of the easiest vegetables to grow in containers and almost any type of pepper will grow adequately in containers if maintained properly. However, ornamental edible peppers are well suited for containers and there are several ornamental peppers available (e.g., 'Prairie Fire') that were specifically developed for container use.

Radish

Radishes are one of the fastest-growing vegetables and are a great choice to use in containers. The radish is a cool-season vegetable that is grown for its small, sweet, and spicy roots. Radishes may be grown indoors or outdoors in pots, but must receive at least 6 to 8 hours of full sunlight each day to grow properly. Radish seeds are usually inexpensive and are fast maturing from seeding until harvest. A gardener can often have fresh radishes within one month after seeding. For this reason, radishes are often used to introduce vegetable gardening to young children.

There are numerous radish varieties and most perform well in containers. Most radishes will take about 23 to 30 days from seeding to harvest. It is important to remember that radishes grown in summer or during the hottest times of the growing season will develop a more pungent flavor.

Containers provide an easy way to grow radishes. A shallow container should be used, as radishes do not require a great soil depth to grow roots adequately. It is best to fill a container to be seeded with radishes with a sterilized vegetable potting mix. Since radish seeds are tiny, most gardeners will scatter seeds over the prepared soil and cover with about ¼ inch of the soil mix. It is important that seeds not be planted too deep, since they may not emerge. After germination, seedlings can be thinned to 1 to 2 inches apart, depending on the variety, which is determined primarily by root size at maturity. The soil must be consistently moist, but not overly wet.

The roots are the edible part of the radish plant, and they begin to swell and form soon after seed germination. It is important that the tops of the roots are covered with soil to prevent splitting and drying. Radishes can be harvested as soon as they are an edible size. However, radish roots grow quickly and should be pulled from the soil as soon as they are ready to prevent the roots from getting pithy.

Spinach

Spinach is a **cool-season vegetable** and is most productive in spring and autumn months. The summer heat will cause spinach to bolt and to produce seed, which slows and eventually stops further leaf development on the plant. Growing spinach in containers allows a gardener to extend the growing season into the summer months since the container can be moved to a cooler location (e.g., under a deck or into a cool garage during the hottest time of the day). Spinach is a perfect plant for containers and is also a cut-and-come-again vegetable, so a few plants can produce a lot of edible foliage over a short period of time.

There are numerous varieties of spinach that are well suited for growing in containers. In fact, several spinach varieties have been developed for small spaces. Disease-resistant varieties such as 'Olympia', 'Melody', and 'Tyee' are suitable for growing in containers, although miniature types of spinach such as 'Baby's Leaf Hybrid' and 'Melody' are perfect for container gardening. 'Bloomsdale Longstanding' is an older open-pollinated savoy-type leaf spinach variety that can be used in containers as well, and this variety is often favored by home gardeners since it is slow to bolt (go to seed). Spinach is a cool-season vegetable and when the temperatures are consistently above 75°F, most spinach plants will develop bitter tasting leaves when the plant bolts.

The choice of a container for growing spinach is also very important. Small containers result in spinach plants that develop poor vigor as the season progresses due to root constriction within the pot and plant-to-plant competition that results, as several plants are typically grown together in each container. Thus, containers must be at least 8 inches deep and 10 to 12 inches wide to maximize the growth of one plant. If multiple plants are grown in containers, each additional plant should be spaced 10 inches apart. Terra cotta pots, plastic garden containers, or wooden boxes all work as containers for spinach. Lighter-colored containers are preferred by gardeners since they reflect heat and tend to provide cooler soil conditions. Remember that spinach is a cool-season plant and does not grow well in hot weather conditions. Spinach prefers to be grown in a sunny location with at least 6 hours of sun a day, although spinach plants will also thrive in lightly shaded locations in the landscape.

The potting soil mix should be a sterilized mix of perlite, vermiculite, mature compost, and possibly sand. Compost is recommended, as spinach plants require lots of nutrients to maximize plant growth. Spinach seeds are typically directly planted into the soil. Place seeds on top of the soil and cover loosely with about ¼ to ½ inch of soil or compost. Keep the soil moist, but not overly wet and seeds should germinate within about a week. A gardener should space seedlings about 3 to 5 inches apart in the container once plants reach the 2- or 3-leaf stage. Plant spinach seeds about 6 weeks before the first anticipated hard freeze in the autumn, although transplants may work best in the spring for containers. It is important to remember that temperatures below 20°F will kill seedlings and they may need to be protected in some manner in both the early spring and late autumn.

Spinach plants are heavy feeders, so they require an application of a complete fertilizer about every 2 weeks, or the use of a slow-release fertilizer, as discussed previously for other container-grown vegetables. In addition, ensure that soil is kept moist but not overly wet. The normal period until harvest, depending on variety, is anywhere from 45 to 60 days. It is best to cut individually leaves, starting from the outer parts of the plant going toward the center, making sure to leave some younger leaves on the plant for a continuous harvest. Plants can generally be harvested over about a 4-week period. A once-over harvest can be done by cutting the leaves an inch from the crown, but this will terminate future plant growth.

Tomato

Tomatoes are the most commonly grown vegetable in gardens, and they are also the vegetable most widely grown in containers. Most people with gardens will grow tomatoes and many varieties are well suited to be grown in containers. So, if there is only room for a few plants to grow in containers, tomatoes would be a great choice, since tomatoes are used in so many different ways in the kitchen and are normally very easy to grow. Although tomatoes can easily be grown in containers, there can oftentimes be a few challenges for their production. A few pointers are provided next to ensure successful tomato production in containers.

Container size. The tomato plant is quite large at maturity and at least a 12-inch-diameter container is required to maximize productivity. A large container is necessary so that the plant does not dry out between each watering, as this will increase the likelihood of blossom-end rot appearing on fruit. Many gardeners will use containers such as 5-gallon buckets, half-whiskey barrels, small hanging planters, or even a bushel basket.

Soil Conditions. Tomatoes are heavy nutrient feeders and require high amounts of nutrients, particularly nitrogen, to reach their maximum potential. So, most expert gardeners recommend using a 1:1 mixture of a sterilized soil-less potting soil and compost. Lighter-weight soil-less growing mixes are recommended, as these retain moisture well, which is important for tomatoes. It is necessary to fertilize on a regular basis, or to apply a slow-release fertilizer to ensure sufficient soil nutrients for plant growth. Most recommend that a slow-release fertilizer be applied about a month after planting with supplemental soluble feedings made weekly. Tomatoes prefer small regular fertilizer feedings rather than infrequent large doses.

Variety selection. Dwarf (determinate) varieties are best for containers, although both indeterminate and determinate tomato types can both be grown in containers. Many grow indeterminate types in containers, although determinate types are better suited for containers. Most of the larger-type tomatoes (such as 'Beefmaster,' which can produce 1-lb tomatoes) are indeterminate, and this is what some gardeners prefer. Determinate tomato varieties are best for containers since they terminate in growth, often reaching less than 2 or 3 feet in height. If tomatoes are planted in a container, it is important to choose a variety that was specifically bred for this purpose. Recommended determinate or dwarfing tomato varieties for container gardening include:

'Better Bush' (68 days)—Mature plants reach about 3 feet in height and typically produce 8-inch-diameter tomato fruit.

'Container Choice' (69 days)—Produces 8 oz. bright red fruits on 2- to 4-ft tall plants.

'Early Girl' (60 to 65 days)—This variety will produce numerous bright red, meaty tomatoes with incredible flavor and aroma throughout the growing season. The 5 oz globe-shaped fruit are much earlier than other similar-sized tomato varieties. This variety starts bearing fruit earlier and produces for longer than other medium-sized tomatoes. It is indeterminate, will require staking, and will probably have to be pruned at some point in the growing season. Although indeterminate, it is widely grown in containers.

'Gold Nugget' (55 days)—Produces early golden-colored fruits that reach 1 inch in diameter on 2 ft compact plants.

'Elfin' (60 days)—Mature plants are dwarf, highly productive, and produce red fruits that are approximately 1½ inches in diameter.

'Patio' (70 days)—This small, 2-ft tall tomato plant works great in containers and is highly productive, as it regularly produces numerous 3 to 4 oz red fruits.

'Patio Princess' (65 to 68 days)—Many consider this variety the ideal container tomato. It is a heavy yielder and produces 2½-inch-diameter deep red fruit that weighs 4 to 5 oz apiece.

'Red Robin' (55 days)—This variety only needs an 8-inch pot, as it only grows to a height of about 1 foot, but will produce a bounty of red fruits that are 1¼ inches in diameter.

'Tiny Tim' (45 days)—Produces beautiful cherry tomatoes on a plant that is about 1½ feet in height.

'Tumbling Tom Hybrid' (70 days)—This variety is great choice for hanging baskets, as it will cascade down the containers up to 2 feet. Plants can be grown closely together so that several can be planted in a single 12-inch container. There are both red and yellow fruit varieties, with fruits reaching about 1½ to 2 inches in diameter.

'Yellow Canary' (55 days)—Dwarf variety can be grown in 7-inch-diameter containers and has yellow fruit that are about 1 inch in diameter.

Seeds and Transplants. For tomatoes, transplants are used in the garden. So, this means that either tomato plants will have to be purchased from a local garden center or started from seed indoors at home. Transplants should be dark green, stocky, and healthy, with no visible flower or fruit development on the plant, and not be root-bound in the container. The tomato transplant can be planted into the soil up to the bottom set of leaves.

Sunlight. Tomatoes require direct sun for at least 8 hours each day. This means that if tomatoes are grown in a shaded landscape, they will not be productive. Gardeners will often move tomatoes a few times a day in shadier landscapes to maximize their sunlight exposure. This is easily accomplished by placing containers on a wheeled cart so that they can be moved daily to maximize sunlight exposure.

Water. Water is not only essential to keep the tomato plant growing properly, but consistent soil moisture is also critical to obtain the highest fruit quality possible, otherwise fruits crack or develop symptoms of blossom-end rot. Container tomatoes need more frequent watering because their roots can't search in the soil for extra moisture like tomatoes grown in a garden soil. In the heat of summer, a gardener will most likely need to water daily due to the high soil evaporation and plant transpiration rates.

Blossom-End Rot. Inconsistent soil moisture leads to blossom-end rot, which is the development of a flat, black or leathery spot at the blossom end. This is a very common tomato problem that is exacerbated in container culture. To avoid the development of blossom-end rot, a regular watering schedule should be followed and some type of mulch should be placed on the soil surface to prevent moisture loss between each watering. It is also important to make sure that there is adequate calcium in the soil, as this disorder is really caused by inadequate or inconsistent calcium uptake through the water-conducting tissues of the plant.

Staking and Pruning. Depending on the tomato variety, staking may or may not be required. A dwarf tomato plant that can support itself with a fruit load or a trailing variety will not require support. However, if the plant for a tomato variety gets quite large (e.g., the indeterminate 'Early Girl'), staking will be required. For large dwarfs or indeterminate varieties, the tomato plant will have to be secured to a stake, wire, trellis, or something else to prevent the plant from falling over due to the top foliage and fruit weight. This can really have a negative impact on the tomato plant and reduce its productivity. Besides staking, a larger-sized tomato plant in a container will also require pruning to reduce its top weight, but this will also reduce disease incidence and allow the development of the largest fruit size possible.

Pollination. Once tomato plants have flowered, the pollen must be released from the anthers and transferred onto the exerted stigmas to allow fruit development. In outdoor environments, this is most often accomplished by wind or various bee species shaking the pollen loose from anthers. However, if flowering occurs indoors in a container, most suggest giving the plant a little shake during the middle of the day to aid in pollination. It is important to note that the tomato is not pollinated by insects, although buzzing insects such as bumblebees can aid in releasing pollen from the anthers.

Climatic Conditions. Some growing climates are shorter than others, and in these regions, tomato transplants should be planted as early as possible in early spring once the risk of frost has passed. In addition, climates that have lots of cloud cover will also require more days to obtain mature fruit compared to a climate that has mostly sunny days. So, a gardener should not be discouraged if it seems that the fruit will never ripen, as it often just takes more time in cooler/cloudy climates.

Insect and disease control. A gardener should check tomatoes every few days for insect or disease presence, as these pests can be devastating to container-grown tomatoes. It is oftentimes necessary to apply appropriate insect and disease control measures when needed.

SUMMARY

Many home gardeners do not have the space that is necessary to develop a vegetable garden in their landscape. However, many would like to produce vegetables and herbs to some extent in their outdoor landscapes, and the use of containers is a viable option for many gardeners. It is possible to grow a wide range of vegetables and herbs successfully in container culture in a landscape. This means that anyone with a patio, balcony, or even just a sunny windowsill can harvest their own fresh produce for the kitchen or develop a more beautiful and productive landscape using containers. There is just so much potential using containers to grow edible plants, such as vegetables and herbs, in a home garden. Many vegetables and herbs have some ornamental value, and most gardeners are amazed with the wide array of vegetables and herbs that can add color to a landscape when grown in containers. The use of containers can add to the aesthetic beauty of a landscape as well as provide the gardener with fresh produce.

In a small garden, it is of the utmost importance that gardeners get the most out of all the available space, so many will choose to use large containers and plant fast-maturing crops (such as lettuce) with slower-growing vegetables. One example is the use of red-leaf lettuces, planted around a zucchini plant in a container. Lettuce is a fast grower and will be ready for harvest much sooner than zucchini, while the zucchini plant will produce fruit that can be harvested long after the lettuce is gone.

Other gardeners will also try to be imaginative and use containers of all shapes, sizes, and colors, and arrange them in groups. Many gardeners cluster various-sized containers to enclose a garden space or frame a particular view. This can create walkways through a landscape space by placing pots with different shapes and sizes to provide a sense of enclosure. Some containers can be placed on grilled risers to make plants look taller. Many will also place fragrant plants in containers near walkways to provide a sensory experience. It is almost impossible to walk through a well-designed container garden without detecting the smell of a fragrant garden plant, such as pineapple sage, rosemary, or lavender.

Containers can also be thought of as a movable form of garden design and if certain pots and plants do not work well together, they can just be moved around until some combination works for a garden. There is just so much potential with container vegetable gardening, and the overall aesthetic value and success obtained really depends on the skills and imagination of a gardener.

Chapter Fifteen

HEIRLOOM VEGETABLE GARDENING

WHAT IS AN HEIRLOOM VEGETABLE?

Heirloom gardening is the practice of using plants and seeds that are open-pollinated instead of plants that are hybrids. Open-pollinated plants produce seeds true to type if self-pollinated (see Chapter 4 – Vegetable Propagation Methods for the Home Garden). Heirloom gardeners preserve historic plants that may not be popular or cost effective enough to be mass produced and sold by private seed companies. Thus, a gardener can use their own seeds when saved from heirloom plants, and this can break dependence on always purchasing new seed each year from commercial seed companies.

The term "heirloom" as it relates to vegetable seed can be quite confusing. Although a particular vegetable variety may have been grown for hundreds of years, it can't be grown from seed that your great-grandfather collected many years ago. Vegetable seeds typically remain viable for about 5 to 6 years with a maximum of 20 years. Furthermore, at about 20 years or so, less than 1% germination of those seeds will occur. So, a particular variety is perpetuated only as long as humans continue to grow the variety and save seed to replant each year or every few years. Although older varieties are called heirlooms, the seed must have been propagated and collected recently.

Today, it is difficult to find a catalog or gardening magazine that doesn't list or mention heirloom vegetables. Qualities that make modern hybrid vegetables important to large-scale commercial vegetable growers such as high yields, uniformity of fruit size and shape with few defects, the ability to ripen all at once for mechanical harvesting, ability of fruit to withstand long-distance shipping and handling, and an extended shelf life, are not always wanted or needed by home gardeners, who prize taste, freshness, and an extended harvest over everything else. Most gardeners do not want to grow a tasteless vegetable, like many of the ones we find in supermarkets. For example, many gardeners do not want to grow a thick-skinned, tasteless tomato fruit that was bred for shipping purposes in their home garden but instead prefer a variety that is unique and has great taste (Fig. 15.1). There are some vegetable varieties that can be grown in home gardens that have outstanding flavors and many of those are heirlooms.

An heirloom variety must meet two basic criteria:

The variety must be able to reproduce itself from seed or through asexual propagation methods.

Although most heirlooms are open-pollinated, some are propagated asexually by vegetative plant parts, such as potatoes that are propagated by tubers and not by seed. "Open-pollinated" means that they will grow true to type from seed, as opposed to F_1 hybrids, whose seeds when planted typically segregate toward one of the two parental inbreds. The term "F_1 hybrid" refers to plants that are the product of crossing two separate, genetically uniform parents, referred to as inbreds, and these inbreds can be increased through open pollination.

Figure 15.1 Heirloom tomato variety growing in a garden. Notice the greening of the shoulders and the concentric cracking near the stem-end, which are dead giveaways that this is an heirloom tomato.

The variety must have been introduced more than 50 years ago.

This really makes the variety an heirloom. By having a variety on the market for at least fifty years, this ensures that it has passed through at least one generation. With the recent interest in heirloom vegetables, many old vegetable varieties have been reintroduced by various seed companies. These open-pollinated cultivars have been preserved over the years by families, or by ethnic or religious groups, who handed these seeds down from generation to generation.

Sometimes an heirloom variety has a well-established history and clear date of when it was introduced. However, in most instances, older cultivars have only a probable date of introduction, and can be dated only by their documentation of existence by garden writers or in a commercial seed list. Some heirlooms played important roles as a staple in diets for specific ethnic groups or as a ceremonial crop, while others became well adapted to the climate and growing conditions of a particular region. Many gardeners find that part of the fun in growing heirlooms is in uncovering their storied history.

There are hundreds of heirloom varieties available to the modern gardener. Some of the most beloved heirloom vegetables include 'Kentucky Wonder' pole beans, 'Detroit Dark Red' beet, 'Danvers 126 Half Long' carrot, 'Iceberg' lettuce, 'Brandywine' tomato, and 'Crimson Sweet' watermelon. Several heirloom vegetable varieties are suggested next.

SUGGESTED VEGETABLE HEIRLOOM VARIETIES

Beans—'Blue Lake Bush 274' (standard dark green bean), 'Kentucky Wonder' (standard pole-type bean with 6- to 8-inch-long pods)

Beets—'Detroit Dark Red' (popular, old standard red beets, see Fig. 15.2), 'Bull's Blood' (deep reddish-purplish leaves with pinkish-colored rings inside root)

Broccoli—'Calabrese Green Sprouting' (Italian heirloom with 5- to 8-inch-diameter heads with many side shoots), 'Waltham 29' (compact plants with 4- to 8-inch-diameter heads)

Brussels Sprouts—'Long Island Improved' (standard open-pollinated variety since the late 1800s)

Carrot—'Danvers 126 Half Long' (although it dates back to the late 1800s, the 126 strain was released in the 1940s, produces thick, 7-inch-long roots), 'Chantenay Red Core' (released in 1929, large carrot that tapers from top to bottom of root and has a deep orange color)

Cabbage—'Early Jersey Wakefield' (small, pale-green pointed heads), 'Late Flat Dutch' (large flattened heads that are 10 to 15 lb each), 'Mammoth Red Rock' (5 to 8 lb, reddish-purple colored heads)

Cauliflower—'Snowball Self-Blanching' (standard open-pollinated favorite for over 100 years with white curd)

Figure 15.2 'Detroit Dark Red' beets.

Cucumber—'Boston Pickling' (popular open-pollinated variety in early 1900s, produces large yields of green pickle-type fruit), 'Lemon' (has the shape, size, and color of a lemon, introduced in late 1800s), 'Straight Eight' (slicer-type that has 8-inch-long straight green fruits)

Lettuce—Looseleaf: 'Black Seeded Simpson' (bolting-tolerant variety that has large, loose, green leaves that are crisp, tender, and sweet); **Crisphead (Head Lettuce)**: 'Iceberg' (old standard open-pollinated iceberg type that has compact, medium-sized heads); **Butterhead**: 'Buttercrunch' (standard butterhead type for many years, heat tolerant and slow to bolt), 'Tom Thumb' (dates back to the 1850s and produces small, 3- to 4-inch diameter heads); **Cos (Romaine)**: 'Paris White Cos' (pale-green uniform heads inside and dark green outside, developed in late 1940s)

Melons—'Banana' (yellow-skinned elongated melon 16 to 24 inches in length, weighing 5 to 8 lb with salmon-colored flesh), 'Hale's Best' (a standard open-pollinated muskmelon variety for many years, has tasty orange flesh and the typical netted skin of muskmelon), 'Honey Rock' (introduced in 1920, 3 to 4 lb with netted ribbed skin and sweet, firm, deep salmon flesh), 'Piel de Sapo' (oval, 8-inch-long fruit with sweet white flesh and a green mottled skin that resembles a toad's skin, see Fig. 15.3)

Okra—'Cajun Jewel' (favorite in Cajun country since the 1950s, dwarf and spineless plants are 3 to 4 ft tall, and produce tender 1-inch-diameter pods that reach up to 7 inches in length),

'Clemson Spineless' (1939 All-America Selections winner is still one of the most popular varieties on the market, vigorous 4 ft tall plants produce an abundance of dark green, grooved pods without spines), 'Emerald' (developed by the Campbell Soup Company and was released in 1950, semi-dwarf plants produce pods that are thick-walled, round, slender, long, spineless, and a deep green color)

Onion—'Red Creole' (developed by the Desert Seed Company in California and released in 1962, spicy red short-day type onion has a somewhat flattened globe shape with pale purple red skin), 'Texas Early Grano 502 PRR' (released by the Texas Agricultural Experimental Station in 1944, forms large globe-shaped Vidalia-type onions that have a uniform white flesh and nice flavor that is not pungent, and although it is a short-day variety, it also works well as an day-neutral type), 'White Sweet Spanish' (produces sweet-flavored large, white globe-shaped bulbs with white flesh that can reach up to 1 lb in size), 'Yellow Sweet Spanish' (long-day variety, extra-large globe-shaped onion, skin is straw-colored with white flesh, popular due to mild and sweet flavor).

Peas—Garden (shelling peas): 'Alaska' (very early and is good for short-season regions), 'Little Marvel' (from early 1900s, high yields on a bushy plant); **Snow (edible flat pods)**: 'Dwarf Grey Sugar' (pods 2 to 3 inches long and slightly curved with plants about 3 feet in height), 'Mammoth Melting Sugar' (large, sweet pods produced on 4-ft tall plants); **Sugar snap types** were first developed in 1979, thus no varieties have been on market for at least 50 years

Peppers—Hot: 'Tabasco' (introduced into Louisiana in the mid- 19th century, plants grow to 4 feet in height and are covered with small, thin hot peppers, see Fig. 15.3) 'Long Red Cayenne' (hot, pungent, thin pepper that is 6 inches long, often dried and used in various spicy dishes). 'Pico de Gallo' (fruit are 2 inches long by ¼ inch wide, peppers turn from green to red when mature); 'Anaheim' (introduced into California in the early 1900s, prolific bearers of long fruits about 6 inches long on 2- to 3-ft tall plants, only mildly hot, as they are used primarily for roasting and stuffing); **Sweet:** 'California Wonder' (standard open-pollinated sweet bell pepper since the 1920s); 'Marconi' (Italian heirloom that is very large at 7 to 10 inches in length and sweet, perfect for eating fresh, frying, grilling, or stuffing), 'Sweet Banana' (released in the early 1940s and even now this is still an extremely popular pepper, has large, pointed fruits that measure 6 to 7 inches in length with mild yellow peppers turning a brilliant red at maturity)

Potato—'French Fingerling' (medium-sized with red skin, and yellow flesh that is mottled with red, thought to have arrived in the U.S. in the 1800s in a horse's feedbag), 'Garnet Chile' (developed from Chilean seedstock and released in 1853, small, round, pink- to rose-colored potato that became a direct ancestor of most varieties we know today), 'Kennebec' (tubers are large with white skin and flesh, grows well even in difficult environments), 'Irish Cobbler' (well-respected heirloom and still popular despite its release in 1876, has round to oblong-shaped tubers with white skin, flesh, and deep eyes), 'Red Pontiac' (widely adaptable red skinned potato with deep eyes and white flesh, released in 1945 and does well in both southern and northern temperate climates), 'Russet Burbank' (released in 1914 but still the standard potato due to excellent baking and processing quality, produces large, long and cylindrical or slightly flattened tubers with russeted skin, shallow eyes, and white flesh, see Fig. 15.3)

Radish—'Cherry Belle' (AAS winner in 1949, standard radish for many years, roots are small, round, crunchy, crisp and have a bright red color with an outstanding flavor). 'French Breakfast' (roots oblong with white top and a white bottom, pre-1885 French heirloom), 'Long Scarlet' (roots are long and slender with a bright red color, pre-1870 release date),

'White Icicle' (slender 6-inch long, ice-white roots with crisp flesh, pre-1865 release date but still popular with home gardeners)

Spinach—'Bloomsdale Long Standing' (released in 1926, standard spinach variety for many years and really the best heirloom spinach variety, slow to bolt and has heavy, continuous yield of thick-textured, crinkled, glossy, dark green leaves)

Squash and Pumpkins—*Cucurbita maxima*: 'Amish Pie' (slightly pale orange flesh that can be several inches thick, with fruits weighing up to 75 lb, firm moist flesh is excellent for making pies and freezing, see Fig. 15.3), 'Blue Hubbard' (tough-skinned oblong bluish-green fruits weighing 10 to 15 lb, with golden yellow flesh), 'Boston Marrow' (unusual New England winter squash heirloom, fruits are about 12 inches in diameter by 16 inches in length, have reddish-orange skin, weigh about 20 lb each, and have fine-grained, yellow-orange flesh), 'Buttercup' (fruit weigh about 3 to 5 lb each and have green skin with deep orange flesh), 'North Georgia Candy Roaster' (pink, elongated banana-shaped fruit have a blue tip on the blossom end, weigh up to 15 lb each, and have smooth orange flesh), 'Turk's Turban' (pre-1800s heirloom, edible but great for decoration because of its multi-colored turban-shaped small fruit that weigh about 5 lb); ***Cucurbita moschata***: 'Long Island Cheese Pumpkin' (fruit resemble a 10 lb wheel of cheese and are flattened, ribbed, and buff in color with orange flesh), 'Waltham Butternut' (often marketed as an heirloom, although its release date was 1970, has an elongated fruit with a bulbous end, small seed cavity, and smooth great-tasting orange flesh); ***Cucurbita pepo***: 'Black Beauty Zucchini' (traditional, dark green zucchini that was introduced into American markets in the 1920s), 'Cocozelle' (Italian heirloom with long dark green cylindrical zucchini fruit that are striped in lighter green), 'Connecticut Field' (large 15 to 20 lb orange pumpkin pre-1700s heirloom, many call this the original traditional American pumpkin, the fruit rind is hard but thin and slightly ribbed), 'Early Prolific Straightneck Squash' (AAS winner from 1938, uniform lemon-yellow club-shaped fruit), 'Early Yellow Summer Crookneck' (an old favorite heirloom squash, fruits have small crooknecks and bright yellow, bumpy skin), 'Small Sugar' (also called 'New England Pie', first described in mid-1800s, small orange fruit weigh 4 to 5 lb each and have a fine flesh suitable for eating), 'Table Queen Acorn' (six-inch-diameter fruit have a hard green shell and resemble an acorn, often used by cutting in half and baking, introduced by the Iowa Seed Company in 1913), 'Vegetable Spaghetti' (fruits are shaped like a cylinder, eight inches long, weighing about four pounds, and rind is hard, smooth and yellow in color), 'White Bush Scallop' (originally known as "pattypan squash" due to the resemblance to antique scalloped pie plates, fruit are pale green turning white as scalloped fruits mature); ***Cucurbita argyrosperma***: 'Green-Striped Cushaw' (a true American heirloom as it was grown by Native Americans pre-dating European arrival, fruit are large, white with green striping, oblong-shaped with crooked necks and bulbous ends, and weigh 10 to 15 lbs each).

Sweet corn—'Golden Bantam' (old standard golden yellow sweet corn that has been a favorite of home gardeners since the beginning of the 20th century), 'Country Gentleman' (produces tender milky white kernels on 8-inch-long ears)

Swiss chard—'Fordhook Giant Chard' (introduced in 1924 by W. Atlee Burpee, has large green leaves and white stems)

Tomato—The interest in **heirloom tomatoes**, more so than any other vegetable, has spurred demand for these older or unusual varieties. Heirloom tomatoes have a real old-fashioned tomato taste, and many say that the newer F_1 varieties being developed today are having the taste bred out of them. The taste is the most important characteristic of these tomato heirloom varieties. However, the yields of these older open-pollinated varieties are much lower compared to modern hybrid cultivars and they are softer and more perishable when ripe. These varieties also have little to no disease resistance, and are more prone to cracking compared to the newer hybrid varieties. When growing heirloom tomatoes, mulch heavily with an organic amendment, such as straw, to maintain constant moisture to the plant, which will reduce the amount of fruit cracking. There are many heirloom varieties that can be easily found in catalogs and at local garden centers, and include varieties such as 'Amish Paste' (Roma-type tomato heirloom that dates back to the late 1800s, giant, blocky, bright red, plum-shaped fruit that weigh about 8 oz), 'Brandywine' (widely popular and probably the most known heirloom vegetable, potato-leaf variety from about 1885, gourmet quality fruit are boat-shaped, dark pink, and average about 16 to 24 oz, not a heavy yielder), 'Cherokee Purple' (a Cherokee Indian heirloom that pre-dates the 1890s, fruit are a deep pink-purple color with darker purple shoulders and weigh about 12 to 16 oz), 'Costoluto Genovese' (Italian heirloom has been grown since the early 1800s, red fruit are somewhat flattened with deep ribbing, see Fig. 15.3), 'Mortgage Lifter' (introduced in the 1930s, fruits are large, smooth, pink, and weigh about 1 lb each), 'Roma VF' (introduced in 1963, although the original 'Roma' was released in 1956 from the USDA Agricultural Research Center in Beltsville, Maryland, plum-shaped, bright red, paste-type tomato and fruits average about 2 oz and have a high solids-to-water ratio making them ideal for paste, sauces, or puree), 'Rutgers' (developed at the New Jersey Agricultural Experimental Station and introduced in 1934, bright red fruits are globular, slightly flattened with smooth, thick walls that are crack-resistant and weigh about 6 oz), 'Traveler' (or 'Arkansas Traveler', released from the University of Arkansas in 1970, so cannot really be classified as an heirloom yet, but a great open-pollinated tomato that has pink crack-resistant fruit that weigh about 6 to 7 oz), 'Yellow Pear' (a pre-1800 variety that has small, yellow pear-shaped fruit with mild flavor due to low acidity, see Fig. 15.3)

Turnip—'Purple Top White Globe' (a pre-1880s variety, purple color on top of root with white color on bottom, foliage can also be eaten)

Watermelon—'Charleston Gray' (released from USDA in 1954, elongated light green fruit with red flesh that weigh 20 to 25 lb each), 'Crimson Sweet' (released in 1963 from Kansas State University and has resistance to anthracnose and fusarium wilt, fruits are light green with dark-green stripes, average about 25 lb, and have a dark red flesh with a fine texture that is very sweet), 'Desert King' (extremely drought tolerant and resistant to sunburn, fruits are round to slightly oblong with a light-green rind, has deep-yellow flesh, and weigh about 20 lb), 'Georgia Rattlesnake' (pre mid-1800s southern variety, fruit are long, light green with irregular, dark-green stripes, weigh up to 30 lb and have a red, crisp, and sweet flesh), 'Kleckley's Sweet' (developed in the late 1880s, but was introduced by W. Atlee Burpee in 1897, fruit is oblong, dark green in color, weigh up to 40 lb, has a very sweet, dark red flesh with a string-less heart and large, white seeds), 'Moon & Stars' (released in 1924, the dark green skin of the fruit is speckled with bright yellow specks and splotches up to silver dollar size and sometimes larger, has pink flesh that is very sweet and melons can weigh up to 25 lb), 'Sugar Baby' (released in 1955, produces small red-fleshed melons, with tough, thin solid dark green rinds, and weigh about 8 to 10 lb each).

Figure 15.3 Some heirloom vegetables: 'Piel de Sapo' melon (top left), 'Tabasco' Peppers (top right), 'Amish Pie' pumpkins (middle left), 'Russet Burbank' potatoes (middle right), 'Yellow Pear' tomatoes (bottom left) and 'Costoluto Genovese' tomatoes (bottom right).

SUMMARY

There are no special rules or requirements for heirloom gardening, though most gardeners who cherish heirloom plants would likely recommend the use of strict organic gardening methods. The preservation of old vegetable varieties is not the only reason many develop an interest in heirloom gardening. Growing heirloom varieties also gives some gardeners a deeper feeling of connection to the earth and with history. Many of these varieties have storied histories that can be interesting to those who grow them. In some cases, gardeners are planting seeds that have been coveted in their family for generations, perhaps brought from around the world when their ancestors first migrated to a new home in America. Heirloom varieties are easy to find in local garden centers, vegetable seed catalogs, or through online seed sources. Since most heirloom seeds are less expensive compared to hybrids, and the fruit that develop from them are oftentimes better tasting (especially tomatoes), many gardeners will place at least a few of these varieties in their garden.

Chapter Sixteen

CULINARY HERBS FOR THE HOME GARDEN

Although herbs are not generally thought of as vegetables per se, they are, however, grown in many home gardens similar to vegetable plants. It is important that gardeners are aware of various culinary herbs that can be grown in home gardens. Some culinary herbs are annuals, while others are either biennials or perennials. Many prefer warm growing conditions, while others prefer cooler climatic conditions to maximize their growth.

Herbs are annual, biennial, or perennial plants that have culinary or other practical uses. Herbs are horticultural crops utilized as secondary ingredients for flavoring and seasoning in food preparation, for their ornamental or decorative value, and medicinally in the preparation of pharmaceuticals. Most herbs are found in the mint (Labiatae), parsley (Apiaceae), or bulb (Amarylladiace) family.

The use of herbs in home gardens has become popular in recent years. Now various culinary herbs can easily be found in most garden center stores for purchase in either seed or plant form during the early spring. Many home gardeners have become more aware of the practical value of fresh herbs, and today, few people would consider a garden complete without at least a few herb plants. Most herbs are grown for culinary or decorative use, although some are grown for medicinal uses.

An introduction to home vegetable gardening would not be complete without some discussion of herbs. Herbs are widely grown in many vegetable gardens, and a gardener should have some idea of the most common herbs that are grown. A few herbs are in the Apiaceae family, and these will be discussed first, followed by those in the mint family (Lamiaceae or Labiatae) and, lastly, those in the Amarylladiace.

Culinary herbs in the Apiaceae family are widely grown in gardens and include important herb plants such as coriander (cilantro), dill, and parsley.

CORIANDER/CILANTRO: *CORIANDRUM SATIVIUM*

Coriander is a herb in which the seed are used in all types of culinary dishes. The fresh leaves of coriander are called cilantro, and leaves are widely utilized in Mexican and Asian dishes. The leaves have a pungent odor that is characteristic of this herb. The seeds are called coriander and are an important spice, having been used for centuries. Cilantro is often described as having a citrus flavor by some, while others dislike the flavor and describe it as having a soapy taste.

Although coriander is primarily used in a culinary sense, it does have medicinal uses, and it is often used to treat digestive ailments and colic.

This plant is best propagated by seed and grows best in full sun conditions. This plant produces the greatest and highest quality leaf growth when placed in a sunny well-drained location and grown under cooler weather conditions. This is an easy to grow annual (although it is technically a biennial), and should be planted as early as possible in the spring after all chances of frost have past, since it will bolt in midsummer as temperatures increase (Fig. 16.1). Seeds or transplants should be placed into the garden soil soon after the danger of frost has passed. Plant cilantro seeds ½ inch deep and space them 2 inches apart if the plant will be used for its leaves. To produce coriander seed, it is best to space plants about 8 to 10 inches apart. Rows in either case should be at least 15 inches apart. Successive plantings may be done until late summer if a continued harvest is desired. It is also recommended to mulch around the seedlings as soon as they emerge or right after transplants are placed into the garden. This plant will bolt easily, and it will not do well when the weather gets excessively hot.

It is also recommended to fertilize this herb once or twice during the growing season by applying about ¼ cup of a nitrogen-based fertilizer (e.g., 10-10-10) per 25 ft^2 of growing area. Be careful to avoid over-fertilization since too much nitrogen can reduce the flavor of the leaves. The plant grows to a height of about 2 feet with a spread of 9 inches. The bright green leaves are somewhat fan shaped (Fig. 16.2) and develop a more feathery appearance towards the terminal portion of the plant. The small, white flowers develop from mid- to late-summer in umbel-like clusters (characteristic of the Apiaceae).

Figure 16.1 Cilantro in flower.

Cilantro leaves can be harvested from plants when they are about 6 inches in length (Fig. 16.2). For coriander seeds, the gardener must wait until the plant has flowered and the inflorescence has turned brown. The inflorescence should then be placed into a container of some sort until the seeds fall off.

Although most soils are somewhat suited to grow this plant, it is recommended to incorporate 2 to 4 inches of well-composted organic matter into the top 4–6 inches of soil before planting, as this provides a loosened soil that allows the seeds to emerge easily from the soil. It also

Figure 16.2 Cilantro growing in a home garden.

does well as a container plant on a sunny porch or balcony, although stems are weak and the plant may require staking.

DILL: *ANETHUM GRAVEOLENS*

Dill is a very aromatic and easy to grow herb. It is often suggested as a herb for those who have little experience with growing plants. This herb is best propagated by seed. Although dill transplants can be purchased at most local garden centers, this plant is easily grown from seeds. Most gardeners will just plant dill seeds directly into the garden soil, where they will thrive with minimal care. Many gardeners say that dill should be directly seeded into the garden soil in the early spring, as seedlings oftentimes do not transplant well. Since seeds ripen on plant and often fall to the ground, self-sown seedlings can be problematic in the garden.

Dill is a multi-branched, upright plant with finely dissected leaves, and some varieties are more highly dissected than others, having almost a fern-like appearance. The wide, flat flowers often make the plant top heavy and cause it to bend over towards the ground. However, it is a gorgeous plant when it is in flower, displaying its large, yellow, umbel-shaped flowers (Fig. 16.3). The rounded chartreuse flowers of dill are compound umbels characteristic of the Apiaceae and are about 3 to 4 inches across. Dill plants can reach anywhere from 18 to 40 inches in height. Most will space plants about 12 to 18 inches apart.

Figure 16.3 Close-up of the dill flower.

The entire plant of this herb is extremely fragrant and can be used in various ways as a culinary herb. Dill as a culinary herb has played a major role in seasoning pickled foods to be canned and stored for winter. Both leaves and seeds can be used in cooking. A sprig of dill can be used to provide flavor to almost any soup, salad, or main dish. The taste of dill leaves resembles that of caraway, while the seeds are pungent and aromatic. Freshly cut, chopped dill leaves enhance the flavor of dips, herb butter, soups, salads, fish dishes, and salads. Although dill seed are utilized in many dishes, many associate their use primarily with the preparation of pickling cucumbers. Seeds can also improve the taste of roasts, stews, and vegetables. Additionally, the flowering heads and seeds of dill are used in flavored vinegars and oils. Many herbalists recommend combining dill and fennel to ease colic in infants.

Dill plants should be grown under full sun conditions. Although the dill plant is somewhat delicate in appearance, it is actually a fairly cold hardy plant. However, under hot weather conditions, dill plants will bolt and produce seed quickly. Dill is a cool-season plant and prefers to be planted and grown during cool weather. In those warm winter areas that don't experience a hard killing frost, dill can be planted in autumn or winter, but in cooler areas that have severe cold winters, plant dill several weeks before the last hard frost in the autumn. For a continuous crop of dill, sow seed repeatedly about every 2 weeks from mid spring to early summer. Dill, if not kept cut, will go to seed, so it is important to cut the plant often until the gardener is ready to switch to seed production.

There are a few varieties of dill available for homeowners to purchase including 'Long Island Mammoth' and 'Fernleaf.' 'Long Island Mammoth' is a dill variety grown both for its leaves and seeds. It will reach a height of about 30 inches. This variety can be grown in a container, but make sure the container is at least 12 inches deep. 'Fernleaf' is a dwarf variety of dill that grows to a height of about 18 inches and is a good option if a gardener is looking for a shorter dill plant to grow in the garden. However, most dill plants reach 3 feet or more in height in a home garden, and the stems of this plant are somewhat weak. Dill grown outdoors should be planted in a protected place or staked in some manner to prevent plants from falling over. High winds will often flatten dill plants grown in the garden, so oftentimes, strings will be strung around stakes that surround the dill planting to keep the plants upright (Fig. 16.4).

***Figure 16.4** Dill flowering in a home vegetable garden. Notice the stakes with string around the dill planting to prevent plants from falling to the ground.*

PARSLEY: *PETROSELINUM CRISPUM*

Parsley is widely grown in many vegetable and herb gardens. It is a hardy cool-season biennial in the family Apiaceae, but it is usually grown as an annual. Since parsley is a cool-season crop, it will tolerate a light frost. Although there are three types of parsley, two are most commonly used as herbs; these are curly leaf (*Petroselinum crispum* var. *crispum*) and Italian, or flat leaf types (*P. crispum* var. *neapolitanum*) (Fig. 16.5). The Hamburg or turnip-rooted parsley (*P. crispum* var. *tuberosum*) is a lesser known type of parsley grown for its enlarged, edible root. The flat leaf type more closely resembles the natural wild species. Flat-leaved parsley is preferred by some gardeners, as it is easier to cultivate, more tolerant to both moist and dry conditions, and is typically thought to have a stronger flavor. 'Italian flat leaf' is a commonly grown flat leaf variety of parsley. The curly leaf parsley is preferred by some because it has a more decorative appearance when used in garnishing foods. 'Moss curled' is a common curly leaf variety that is grown by home gardeners. A different texture can be added to a garden with the curly leaf type of parsley. The moss-like foliage of the curly leaf type will form a plant 9 to 12 inches in height and will eventually produce tiny clusters of greenish yellow flowers in the second year, since it is a true biennial.

Parsley does best in fertile, moist, sandy loam soils having good drainage. This herb will do well in full sun or partial shade conditions, although in very hot climates, it is best to grow plants in partial shade.

Figure 16.5 Flat leaf type parsley (left) and curly leaf parsley (right) growing in the home garden.

Parsley is best propagated by seed, and plants can be started indoors or outside in the garden. It is suggested that seeds be planted about ¼ inch deep. Most gardeners suggest that the tiny seed of parsley be started indoors and allowed to germinate and grow for a few weeks prior to planting outdoors. Parsley plants usually transplant easily into the garden and can be transplanted from about a month before the last expected frost in the spring until about a month before the first expected frost in the autumn. Parsley seed have a reputation for being slow or difficult to germinate, but seed will usually germinate if the soil in which they are planted is warm and has sufficient moisture. If seed do not germinate well, soak the seed in lukewarm water for a couple of hours (or overnight) before sowing, and make sure that soil doesn't dry out before germination occurs. Most gardeners will sow seed at various times throughout the growing season to have fresh parsley throughout most of the year. Transplants can also be purchased from local garden centers. Some gardeners will also sow seed in a cold frame during the autumn months to have parsley for winter use. Parsley can usually be harvested about 2 to 3 months after seeding. Plants are usually spaced about 6 to 8 inches apart in the row with about 2 feet between rows. The leaves of parsley are often utilized as a garnish or are mixed into vegetable dishes to add additional flavor.

Parsley plants grow from the center of the plant outward. Once parsley plants are established, the older stems can be cut from the outside part of the plant as desired. This method will keep the plant producing for an entire season and possibly into the next spring, and it also helps shape the plant by keeping it smaller and more compact in its growth. To extend parsley harvest, an organic mulch can be placed several inches deep around the base of plants and then later covered by fabric rowcovers; it can also be placed in some type of a winter protection structure which will also allow some growth to occur over the winter months. During the winter, a few leaves can be harvested at each picking, but do not disturb a plant's dormancy by picking all foliage.

Most culinary herbs are in the mint family (Lamiaceae or Labiatae). Many of these will be discussed, including basil, lemon balm, mints, oregano, rosemary, sage, sweet marjoram, and thyme.

BASIL: *OCIMUM BASILICUM*

Basil is a member of the mint family (Lamiaceae or Labiatae) and is one of the most important culinary herbs. There are many types of basil, each with a unique taste, aroma, color, size, and potential culinary use. Basil is most closely identified with Italian foods and cooking. It is often used as a companion plant for tomatoes in the garden.

This herb is grown for its foliage, which has a pungent and distinctive odor. There are green- and purple-leaved types, and many grow them in combination to get a good color contrast in a garden. There are also ruffled-leaved types, which can give a garden a much different appearance compared to the normal-leaved types. The extremely aromatic leaves also have a delightful variety of flavors, from the slightly lemony-mint of sweet basil to cinnamon and licorice. The flowers are somewhat insignificant but very popular with bees. Although the leaves are most commonly used in food preparation, the flower buds are also edible.

Basil is an easy-to-grow tender annual herb that grows best in full sun. It definitely prefers to be grown in warm, sunny weather conditions. The basil plant is a heat lover, and gardeners should not bother planting this herb until the daytime temperatures consistently remain about 70° F and night temperatures are above 50° F. Basil is very sensitive to cold weather conditions; it should be planted at a time when there is no chance of frost. Additionally, a sudden chill in the air or very cold wind may also severely injure or even kill basil without a frost.

Basil is propagated by seed, and seed can easily be obtained from garden centers in spring, mail-order garden catalogs, or internet gardening sites. From seed, basil is ready to start harvesting in about 60–90 days. Basil seeds are tiny and should not be planted too deeply. It is best to just sprinkle them on the surface and cover with a thin layer of soil. Seeds are usually started indoors and allowed to grow for about 6 weeks before they are transplanted outside into a garden. Once basil plants are outside, they should be fertilized every 2 to 3 weeks. Basil plants prefer fertile, well-drained soils and should be watered frequently. This herb is very sensitive to dry soils and requires consistent water, although excessive amounts of moisture on foliage should be avoided to prevent the development of fungus diseases. Many gardeners will provide some type of mulch around the plant to keep roots cool and the soil slightly moist.

Basil does well in containers (Fig. 16.6), but remember it will have the most growth under full sun conditions. Since most basil plants can get quite large, containers should be at least 18 inches in diameter with only one plant grown in each container.

There are many varieties of basil from which to choose. Most varieties of basil grow between 1 and 2 feet in height. Suggested varieties include 'Lemon Sweet Dani' (has a strong lemon scent), 'Genovese' (giant leaf basil that grows up to 3 feet high, with leaves up to 3 inches in length), 'Italian Large Leaf' (sweet type of basil, large plant with medium-dark green leaves up to 4 inches in length), 'Minette' (compact plant with tiny leaves that are very fragrant), 'Osmin' (has dark purple foliage and pink flowers), 'Monstruoso' (giant leaf basil with leaves that grow up to 6 inches long), 'Purple Ruffles' (purple-leaved variety that also makes a nice ornamental), 'Red Rubin' (keeps purple color throughout the season with great flavor and is wonderful to use in vinegars), and 'Thai Siam Queen' (a large-leaf basil with spicy flavor and aroma).

Figure 16.6. Basil is another culinary herb that grows well in containers.

Basil plants should be spaced about 10 to 12 inches apart in the garden. It is important to begin pinching the terminal portion of the plant off once the plants reach about 6 inches in height. If a gardener doesn't pinch or harvest the top portion of the basil plant, they will tend to grow tall and spindly, have few leaves, and rapidly bolt and produce seed. Most gardeners want to prevent basil from producing flowers for as long as possible by harvesting or pinching off the top sets of leaves as soon as the plant reaches about 6 inches in height. Once a basil plant goes to seed, remember that the leaves are not as flavorful.

Basil plants shouldn't be relegated just to vegetable or herb gardens. They can make a beautiful edging plant for any garden, and certain flowering varieties make an attractive addition to flower gardens. Basil plants are relatively easy to grow and are very useful for a variety of culinary purposes. For the most flavor, leaves should be torn, not chopped. Basil can be dried or frozen for later use, or used fresh. This herb is usually added to dishes late in the cooking process to preserve its taste and color. It is a common addition to tomato sauce, vinegar, and pesto. Although basil is known primarily for its uses in Italian cooking, it can also be served fresh in salads or on sandwiches.

Basil is a widely utilized herb that is used in many dishes, and it should be grown in every garden since it is so easy to grow. Basil can be used in cooking and is generally added near the end of dish preparation to keep its fresh taste and color. Basil can also be used fresh in salads, on sandwiches, or even as a wrap—for instance, around cheese cubes.

LEMON BALM: *MELISSA OFFICINALIS*

Lemon balm, also known as lemon mint, is a low maintenance hardy herbaceous perennial herb found in the mint family (Lamiaceae or Labiatae) (Fig. 16.7). It can grow to a height of around 2 feet and produces ovular to heart-shaped leaves about 2 to 3 inches in diameter. This herb is hardy from USDA zones 4 to 9, but in the colder zone 4, it will require a winter organic mulch and a well-drained soil to survive. In some cold temperate zones, the stems of this plant will die at the onset of winter, but new growth will develop again in the spring. Lemon balm is relatively easy to cultivate in most gardens and has become widely naturalized in most temperate zone areas in America. In late spring to summer, the lemon balm plant will produce small, light blue flowers, but it is most often grown for its lemon scented and flavored leaves. Lemon balm is oftentimes placed next to a path or at the edge of a raised bed in a garden, where the strong lemon scent of the foliage will be released as the leaves are brushed. Lemon balm will also attract bees, butterflies, and birds, and it has fragrant blossoms.

Lemon balm prefers to be grown in full sun or partial shade conditions. This herb is moderately shade-tolerant, much more so than most other culinary herbs. Lemon balm prefers a fertile,

Figure 16.7 Lemon balm waking up after a long winter nap.

well-drained clay or sandy loam soil and will grow in a relatively wide pH range between 6 and 9, although it prefers a range of 6.0 to 7.5. Lemon balm plants should be spaced between 12 and 15 inches apart in a garden. This herb does require consistently moist soil, so it is best to water on a regular schedule and to not let the soil dry out between water applications. However, it is also important to not overwater lemon balm plants. Lemon balm is also susceptible to a variety of pests, including whitefly, spider mites, thrips, and powdery mildew.

Lemon balm tends to grow in clumps and can spread vegetatively, as well as by self-seeding. Lemon balm plants can be easily propagated by division, cuttings, or seed. However, seeds are often slow to germinate, but some self-sown seedlings can usually be found around an existing plant. The easiest way to propagate lemon balm is by dividing the clump in spring, or by stem cuttings in the late spring. Lemon balm may be considered a noxious weed or invasive plant in some areas. Lemon balm self-sows freely, and it is best to remove flowers to prevent seed development, which will reduce volunteer seedling numbers the following season. It is best to replace plants or rejuvenate plantings every 3 to 4 years, since older plants often become large, lose their vigor, and become unsightly in a garden.

Similar to many other herbs, the golden or variegated types of lemon balm are often preferred by gardeners due to the extra color they will add to a garden. Suggested varieties of lemon balm that are readily available to home gardeners include: 'All Gold,' 'Aurea,' 'Compacta,' 'Gold leaf,' 'Lime,' 'Lime Balm,' 'Quedlinburger Niederliegende,' and 'Variegata.'

Although lemon balm is grown in gardens primarily for its culinary uses, it also has medicinal properties. The leaves of lemon balm are utilized in many dishes and are often used to replace lemon in seafood dishes. In the culinary world, lemon balm is prized for its light, refreshing aroma and taste, which is often described as lightly lemon with a hint of mint. The fresh leaves provide the best flavors for cooking, although dried leaves and stems are often used in making teas. The leaves produce a mild, lemony tea, and are often added to black or green teas to provide lemon flavor. Whole or chopped fresh lemon balm leaves also add flavor to green salads, fruit salads, vegetable dishes, poultry stuffing, and marinades for fish, and they provide a nice accompaniment to corn, beans, broccoli, asparagus, shell fish, and lamb dishes.

Lemon balm has also been used medicinally for thousands of years, and even today the herb is used to provide relief for numerous ailments. Lemon balm is a calming, slightly sedative and mildly antibacterial herb, and it is believed to prevent and fight off microbial infections in the human body. Herbal teas made from the leaves of lemon balm are used to treat insomnia, anxiety, herpes, fever, menstrual cramps, muscle spasms, colic, vomiting, and digestive problems. Crushed lemon balm leaves rubbed on the skin may help prevent insect bites or heal existing wounds, and a lemon balm infusion added to warm bath water may promote menstruation.

MINTS: *MENTHA SPP.*

Mints are hardy perennial herbs that are easy to grow in most situations. They have squared, four-sided stems with opposite leaves (indicating they are a member of the mint family—Lamiaceae or Labiatae) and small-lipped flowers. Mint plants produce 2- to 3-foot upright stems, and all parts of the plants are pungent. Mints produce spikes of small bluish-mauve or purple flowers in late summer. These flowers are generally not attractive and can be removed to make the foliage more visible in a garden.

Most mints are rampant spreaders, forming a thick mat of spreading rhizomes (creeping underground stems) just under the surface of the ground. So mints are highly aggressive in their growth and will often take over areas of a garden in which they are planted, since their rhizomes spread quickly. Their excessive overgrowth must be controlled in some manner, and many gardeners do not grow mints due to their aggressive nature. Home gardeners often combat their aggressiveness by planting them in containers such as plastic or clay pots, or in barrels (Fig. 16.8), or into garden areas that are separated from other portions of the garden. However, one of the best ways to grow mint in the garden without the threat of rampant spreading is by using some type of bottomless container sunk into the soil to prevent it from spreading, which prevents the underground rhizomes from spreading out of the confined area. This is done by sinking containers into the soil and leaving only the top portion protruding an inch or so above the soil line.

Mint can be grown from seed or purchased as seedlings. However, mint is oftentimes extremely variable when grown from seed, especially when collected from plants in the garden; thus, it is best to purchase plants from a reputable source, or to visit a nursery/garden center to find plants with appealing flavor and aroma. Once plants have been established in a garden, they can be propagated by division of plants or by rhizomes. Division can be accomplished by dividing established clumps of mint into two or more pieces using a shovel or spade during either autumn or spring and replanting the separate sections. Also, pieces of rhizomes taken from mint plants can be used for propagation purposes. The divisions or rhizomes should placed into the soil a few inches deep and placed at least 1 foot apart. It is also important to note that after plants have grown for several years in a garden, perennial herb plants, such as mint, tend to have low vigor. Pruning the plant to a just few inches above the ground will stimulate a flush of new growth and improve plant vigor. To help prevent mint diseases, the stems and foliage can be cut down to the soil surface in the fall and then burned or composted.

Figure 16.8 Growing mint in a container will help control it from spreading all over a garden.

There are several different mints that are widely available to be grown by home gardeners. The genus *Mentha* has many species and cultivars. The various types include varieties of applemint, chocolatemint, pineapplemint, orangemint, peppermint, spearmint, gingermint, and numerous variegated forms of mints. The most commonly grown mint is probably spearmint. Most mints are hardy in zones 3 to 8. All mints prefer a cool, moist spot in partial shade but will also grow adequately in full sun. Although most varieties of mint are easy to grow in a wide range of soils, they thrive best when located in a moist, high organic matter, well-drained soil.

Mint leaves can be harvested for use in the kitchen once plants have begun to flower. To harvest mint, snip leaves or sprigs as needed. To harvest in quantity, cut stems to within an inch or so above the ground. A gardener can make several harvests during the growing season, depending on the length of the season. Mint can be hung in loose bunches to air dry, while individual leaves can be dried or frozen for later use.

Mint is an herb with many different culinary and medicinal uses. Aromatic mint teas can be enjoyed in hot or iced form. Peppermint tea, a centuries-old remedy, can calm an upset stomach. Whole or chopped fresh mint leaves are often added to improve the flavor of lamb, rice, salads, or various cooked vegetables.

OREGANO: *ORIGANUM VULGARE*

Oregano is a widely utilized culinary herb in the mint family (Lamiaceae or Labiatae), and it is easy to grow in the home garden. The leaves of this herb are used in the preparation of many different dishes. Another name for oregano is wild marjoram, and it prefers to be grown under full sun conditions. Unlike sweet marjoram (which will be discussed next), oregano is a hardy perennial. Oregano will die down for the winter, but in the spring it will quickly establish itself and grow to about 2 feet in height with a spread of about 1 ½ feet. Oregano starts out as a ground-hugging rosette of leaves, but it can easily grow to about 2 feet tall. Oregano is hardy from zones 5 to 10. Oregano plants should be cut back to the ground and covered with a layer of mulch to over-winter outdoors. Container-grown oregano plants can be brought inside for growing indoors during the winter months and then placed back outdoors in the spring.

Oregano does best under full sun conditions, although the golden-colored types do best in partial shade conditions, as their leaves tend to scorch in the full sun. Oregano plants tend to make an attractive mound of aromatic foliage in a garden, with small pink flowers appearing from mid- to late-summer (Fig. 16.9). Oregano can add a lot to a garden since it is a highly decorative and attractive plant. Although it is grown predominately as a culinary herb, oregano can make an excellent bedding plant for perennial garden beds or foundation landscape areas and can develop into a beautiful, low-maintenance, scented groundcover. Oregano makes a nice edging plant and ground cover, requiring little maintenance, and the smaller, dwarfing varieties also do well in rock and alpine gardens. Since oregano requires very little attention, many gardeners grow oregano in containers, which makes it a great plant for those with limited garden space. It is also important to note that oregano flowers are attractive to many insects, including butterflies, bees, and wasps.

This herb can be propagated by stem cuttings, division of plants, or seed. Since different species of oregano will cross pollinate, seed collected in a garden may produce off-types. However, seed can be purchased from reputable sources or local garden centers that will produce consistent high quality plants. It is also important to note that oregano tends to become woody and less attractive after several years of growth in a garden. At this time, gardeners should consider dividing plants and replanting (as described for mint), or raising new plants from cuttings or

Figure 16.9 Freshly picked oregano flowers. The flowers of this herb make a beautiful fresh bouquet for the home.

seed. The simplest way to propagate this herb is by dividing a mature plant in the spring or autumn. This is done by simply digging up the entire plant and dividing it into two or more pieces by slicing with a spade or shovel. The divided oregano plants should be planted immediately. Many gardeners divide oregano plants in the autumn with some divisions placed into containers and brought indoors for the winter months.

Growing oregano is easy, but choosing the right variety is sometimes confusing. Plants in the genus *Origanum* can be perennial ground covers, tender perennials, or even small perennial sub-shrubs. There are many types of oregano, and those that are most commonly grown are briefly described next. *Origanum vulgare* is known by several names including 'Common Oregano,' 'Wild Marjoram,' and 'Pot Marjoram.' Greek oregano (*O. heracleoticum)* is the type most often used in Mediterranean cooking, and is the one that is usually associated with oregano flavor. Thus, Greek oregano is the type most commonly used for cooking purposes.

There are several popular oregano varieties, including 'Aureum' (also often known as Golden Oregano or Golden Marjoram; has bright-gold leaves in the spring that darken as they age and become light green by late summer or autumn), 'Compact Pink Flowered' (has pungent, dark green leaves and dark pink flowers), 'Aureum Crispum' (a selection of 'Aureum' that has, crinkly leaves), 'Compactum' (compact low-growing variety with leaves that turn purple in winter), 'White Anniversary' (has attractive, bright green foliage with white margins, and produces white flowers that are barely noticeable; great variety for gardeners that want to use oregano in a container or as a groundcover or edging plant), and 'Greek oregano' or 'Italian oregano' (most popular for those interested in growing oregano predominately for kitchen uses).

Oregano is a culinary herb that prefers well-drained soil and full sun conditions. Climate, soil, and moisture can definitely cause variation in the flavor of oregano. The genus *Origanum* is native to the Mediterranean area, but *Origanum vulgare* has become naturalized in many areas, including the eastern United States.

As with most herbs, oregano leaves taste best before the plant flowers (Fig. 16.10). Once the plant has reached 4 to 5 inches in height, stems can be cut for culinary uses. Harvesting before the plant flowers will produce the most flavorful leaves, as levels of essential oils diminish as the flowers begin to develop on the plant. Cutting stems all the way back to the ground will encourage more growth and a more vigorous plant. The flowers should be pinched and removed to maintain bushy or compact plant growth which prevents bolting and the plant from producing seed.

Established oregano plants do not require much attention. In fact, these drought-tolerant herbs require water only during excessively dry periods. Oregano doesn't require much fertilization either, and only about a tablespoon of a complete fertilizer per plant in the spring is needed before new growth starts.

The leaves of oregano are the plant part used for flavoring foods, although the flowers are also edible. They retain their flavor better in dishes if added toward the end of food preparation. Dried oregano has a stronger taste than fresh. Oregano leaves can be stored whole, placed in freezer bags, and frozen. They can also be dried in a dark, well-ventilated area and stored in airtight containers until ready to use.

Figure 16.10 Picking fresh oregano from a garden.

ROSEMARY: ROSMARINUS OFFICINALIS

Rosemary is an outstanding culinary herb and should be included in most every vegetable garden. It is a beautiful plant in the mint family (Lamiaceae or Labiatae), and it is worth planting purely as an ornamental herbaceous plant. This herb is often the centerpiece of formal herb gardens. Due to the ease with which rosemary can be pruned and trained, it is often used as an herbal bonsai plant or topiary. The plant size of this herb can vary, depending on the variety, with some developing a bushy or compact habit of 3 feet or less ('Irene'), while others, like 'Tuscan Blue,' are somewhat tall, often reaching 6 feet or more in height. However, rosemary will tolerate a great amount of pruning, so it can be kept small and compact if necessary.

Rosemary is classified as a tender perennial and will persist year to year in gardens in frost-free climates. Although classified as a perennial, it is not very hardy in areas with cold winters. In climates where winters are somewhat severe with sustained temperatures well below 30°F for extended periods, rosemary plants will have to be brought indoors during the coldest months. Rosemary is perennial in zones 7 to 10 and will also grow as a perennial in some areas of zone 6, especially if grown in an area protected from cold weather and high winds. Many gardeners say that when grown in a sheltered area with a southern exposure, rosemary will survive short periods of time at temperatures between 10°F and 20°F. When grown indoors, water rosemary only as needed and do not overwater. Rosemary doesn't need a lot of water whether indoors or out, but it does need to be put in front of a sunny south-facing window, and if this is not possible, use some type of artificial light. Rosemary plants can be moved back outdoors once the last frost date has passed in the spring. It is important to note that rosemary does well as a container plant and makes an elegant plant for a deck or patio; besides, growing this herb in a container makes it easier to protect from the cold weather during the winter months.

Rosemary has small gray-green colored needle-like leaves that are evergreen in frost-free climates. The distinctive rosemary fragrance is released just by simply touching or brushing against the leaves. They give off a very distinct and aromatic fragrance when crushed or rubbed between the fingers. The flowers of rosemary are small, and they vary from dark blue to pale blue or white, appearing in late spring or early summer (Fig. 16.11). The flowers are edible and are oftentimes used along with the leaves in food preparation.

Rosemary plants prefer full sun for their growth, although they will also grow in light shade conditions. So, this herb should be planted in a landscape where it will receive either full or partial sun. Rosemary also prefers a moderately to well-drained soil, and if a soil is heavy and not well-drained, some sand or mulch can be added into the soil to provide better water drainage. This herb also prefers to be grown under dry or drought conditions. It is also important to note that rosemary is very susceptible to root rot, so make sure the soil is dry before the plant is watered again. The resiliency of the plant makes it a great addition to any garden, regardless of the soil or weather conditions.

Rosemary does not require any special care, especially when planted in the ground within a garden. This herb is drought hardy and really not particular about soil type or its fertilization requirements. There is little need to continually fertilize the plants, and a basic complete fertilizer applied in the spring will keep the plants supplied with enough nutrients for the entire growing season.

Although stem cuttings of existing established plants are usually used to propagate this herb, most gardeners will purchase a young plant from a local nursery or garden center. There are numerous varieties of rosemary available to home gardeners. Some display an upright growth habit and are capable of growing more than 6 feet tall, while others are trailing and spread their branches closer to the ground. The upright rosemary plants are better suited to planting in the garden landscape and will produce a larger quantity of stems and leaves for culinary use than those grown in containers. The trailing varieties are very ornamental and great for growing in hanging baskets or containers. Popular varieties of rosemary include: 'Arp Rosemary' (upright), 'Blue Boy' (dwarf/trailing), 'Creeping Rosemary' (trailing), 'Collingwood' (semi-upright), 'Huntington Carpet' (trailing), 'Irene' (trailing), 'Mozart' (semi-upright), 'Pink

Figure 16.11 Rosemary in flower. Also, notice the needle-like leaves of this plant.

Majorca' (trailing), 'Prostrata' (trailing), 'Santa Barbara Blue' (semi-upright), 'Tuscan Blue' (upright), and 'White Rosemary' (upright).

Rosemary is oftentimes picked and then used directly from the garden. Stems are typically cut from the plant with shears, pruners, or a knife. It is important to not remove more than one-fourth of all stems on a plant, as this ensures that the plant will not be stunted in its growth and will have the ability to produce more growth in the future. Rosemary may also be preserved for later use either by drying or freezing the stems with attached foliage.

Rosemary is best known for its culinary uses, and it is a common ingredient in a variety of foods. The leaves of rosemary are also used to add a distinctive flavor to many culinary dishes. Rosemary is a popular component of herbal vinegars and infused oils, as this herb will impart its aroma, flavor, and a tinge of color to various oils or vinegars. Rosemary is also popular in the cosmetic industry, and the herb is included in many hair care products, soaps, lotions, oils, and even toothpaste.

There are numerous health benefits that have been associated with Rosemary. Rosemary exhibits various medicinal properties and is oftentimes placed in ointments, teas, and essential oils. This herb may improve memory, relax muscles, and stimulate the nervous system; it has also been used in alternative medicine to aid in digestive problems as well as for depression.

SAGE: *SALVIA OFFICINALIS*

Sage is a hardy perennial, semi-evergreen plant in the mint family (Lamiaceae or Labiatae) that usually grows no more than about 2 feet in height. Sage plants will grow in USDA hardiness zones 4 to 9. It will often suffer damage in very cold winters, but even in colder climates at the limits of its range (zones 4 to 5), it can survive with protection. It may look rough at the onset of growing season in the spring with many damaged leaves, but by early summer, plants will regain their vigor with fresh new leaves. In warmer areas, sage will survive the winter with little damage, while in areas with severely cold winters, it may be killed. Sage will grow in most climates, but will most likely be damaged in some manner when grown in areas with cold winters. Gardeners may consider bringing sage indoors for the winter months when it is being grown in these climates. Additionally, some gardeners will apply a heavy layer of mulch in the late autumn around and on top of sage plants in a garden to protect them from the cold winter weather.

Sage is often thought of as a shrubby, sprawling gray-green perennial herb. This herb will produce spikes of white, pink, or pale blue flowers that appear in late spring and early summer, but many gardeners will remove the flowers to focus more of the plant's growth into leaf production. Sage will provide not only an attractive plant that lends color and texture to a garden, but one that is also useful in attracting important pollinators (e.g., various bees). However, similar to many other perennial herbs, sage plants will tend to lose vigor after 3 to 4 years in a garden and should be replaced with new plants. Pruning sage plants a few times during the growing season will keep them attractive and prevent this herb from becoming woody and long-stemmed (Fig. 16.12).

Figure 16.12 Common sage planting that has lost much of its vigor. Notice the long woody stems with most growth occurring at the top portion of plants. This planting needs to be rejuvenated in some manner or be replanted with new plants.

Sage grows best in full sun conditions and prefers lighter well-drained soils with a pH of 5.5 to 6.5, although they will also grow in more heavy soils. For optimal growth, it is best to use a general purpose fertilizer (e.g., 12-12-12) once or twice during the growing season. Sage is also a drought tolerant herb, and most suggest to water plants only during dry periods or at the most, once or twice per week. If sage is grown in a wet climate, a gardener should consider using a raised bed garden or containers for this plant. Anyway, sage is an excellent choice of a culinary herb for container gardening. Few pests tend to bother sage plants, and it is more frequently killed by excess water, lack of sufficient sunlight, and poor pruning methods than by insects or disease.

Sage is usually propagated by seed, but for those varieties that are variegated or have leaf colors other than the typical gray-green, stem cuttings are oftentimes used by home gardeners to get true-to-leaf types. Seed are generally germinated indoors and transplanted into the garden during mid spring about 18 inches apart. To root stem cuttings, trim 6 inch shoots from plants in early summer and remove bottom leaves from the stem. It is best to then dip the bottom 1 inch of the stem in a root-inducing hormone powder and plant in a moist, light medium consisting of perlite, vermiculite, and peat moss. The foliage should be kept moist by misting a few times a day until adventitious roots form, and this can be easily accomplished by using a spray bottle. Once the cuttings have established roots, they can then be transplanted into containers. Once plants have developed some growth and established themselves somewhat, they can be transplanted outdoors. Fortunately, reasonably priced, small sage plants can be found in most garden centers in the spring, and this is really the easiest way to get an established sage plant quickly into a garden.

The true wild species, *Salvia officinalis*, has leaves that are pale green to gray, but many varieties are variegated or purple-leaved, which will add more colors to a garden. Suggested varieties include: 'Common,' 'Tricolor,' 'Purpurescens' (also known as purple sage), and 'Aurea' (also known as golden sage). 'Common' sage tends to be more cold hardy than the ornamental varieties. 'Tricolor' sage has variegation of green, white, and pink/purple in its leaves, which makes it a great ornamental plant for the home garden. *Salvia officinalis* 'Purpurescens' has deep purple young leaves that mature to a burgundy (Fig. 16.13), and as a culinary herb, it is one of the most flavorful sages and can be used interchangeably in dishes with regular garden sage. *Salvia officinalis* 'Aurea' has compact growth with yellow leaves and purple flowers.

Figure 16.13 Purple sage growing in a home garden. This herb makes a great addition to the home landscape since it has both ornamental and culinary value.

The young, tender leaves should be harvested from the sage plant for the best flavor, and this should be done before the plant develops flowers. Many gardeners pick leaves during the morning, since flavors and oils in leaves tend to be at their highest levels in the early morning. Sage leaves are used as an herb for culinary flavorings, aromatic scents, and health-promoting properties. However, although sage leaves are utilized in many dishes, it is mainly known by Americans as a primary ingredient in holiday poultry stuffings. Sage can be used both in the fresh and dry forms for poultry and meat seasoning. Besides seasoning, sage leaves and branches are often featured in crafts, like wreaths, where they add color, texture, and fragrance.

Sage oil is known to have antiseptic and antibiotic properties, which have led to its reputation for helping fight infections in the human body. It is commonly used in treatment of inflammations of the mouth, throat, and tonsils, as its volatile oils are believed to be soothing to mucous membranes. A gargle of a sage tea/infusion is thought to be helpful for a variety of throat/mouth infections.

A sage infusion is sometimes used for treating night sweats and hot flashes of menopause, because it is believed to reduce sweating. Other suspected benefits of sage include regulation of insulin in those with diabetes and providing a calming effect for those suffering from stress and nervous headaches. A sage compress can help ward off infections and speed the healing of wounds. Sage is also thought to stimulate the muscles of the uterus, so it should be avoided during pregnancy.

SWEET MARJORAM: *ORIGANUM MAJORANA*

Sweet Marjoram (also known as annual marjoram or knotted marjoram) is gaining popularity as an herb for home gardeners (Fig. 16.14). It is a low-growing tender perennial native to the Mediterranean region that will overwinter outside where frosts do not occur, such as in USDA hardiness zones 9 and 10. Sweet marjoram grown in zones 7 and 8 must be mulched for winter protection, and even then, there is no guarantee that this herb will survive. This plant should be grown only as a summer annual in zone 6 and below.

Sweet marjoram is a delicately flavored cousin of oregano. It is also a member of the mint family (or Lamiaceae or Labiatae). Similar to oregano, leaves are the main part of the plant utilized, and they are used in many culinary dishes. Sweet marjoram is slow-growing and prefers to be grown in full sunlight. Plant this herb 12 inches apart in full sun in a well-drained soil with a pH between 6.5 and 7.0. It is best to add a slow-release fertilizer to the soil at or before planting. Water the plants during extended dry periods, but be sure not to over-water, as sweet marjoram prefers a slightly dry climate. This herb will grow to about 10 inches in height (although some types grow upwards to 2 feet in height) with tough trailing stems that are covered in summer with clusters of small dainty flowers (typically mauve-pink in color but sometimes white or purple). The flowers are produced from pea-like buds that resemble knots, which is why the herb became commonly known as knotted marjoram. It makes an attractive container plant, and it can be brought indoors during the winter months, but it should be placed where it gets at least 5 to 6 hours of sunlight on a daily basis. The plant also makes a pretty summer groundcover or edging plant and works well for rock gardens and borders.

This herb can be propagated by cuttings of stolons (above ground stems that run along the ground) or by division of plants, but it is usually easier to raise new plants from seed each year. Sweet marjoram can be divided in either the spring or autumn to start new plants. However, seeds are most commonly used by gardeners to establish new plants in a garden. Seeds are quite small and are best started indoors about 6 weeks before the last frost date. Plants can be placed about 6 to 8 inches apart or in several clumps of two or three plants set 12 to 14 inches apart in the garden. It is also important to note that Sweet marjoram can also be cultivated

Figure 16.14 Sweet marjoram growing in a garden.

in containers, both indoors and outdoors. It is best to remove old, woody stems of sweet marjoram and allow the newer stems to produce new leaves which definitely have more flavor.

Sweet marjoram can provide multiple cuttings in each growing season. Typically, leaves and stems are harvested as they are needed for kitchen use. When flowers appear, the plants should be cut down to only a few inches above the soil line. They can be cut again later as more flower heads form. The leaves can be used fresh or they can be dried for later use. In order to dry marjoram, pick the leaves just after flower buds appear but before they open, removing no more than a third of the plant's leaves in a single harvest. Once the leaves have dried, strip them from the stem. Leaves can be harvested again when flower buds reappear later in the season.

Sweet marjoram is used in many different types of foods, and some say it is a necessary ingredient for many different Mediterranean dishes. Sweet marjoram also works well in spaghetti sauces, and it is suggested to add at the end of food preparation to achieve the greatest amounts of flavor. As a medicinal tea, sweet marjoram will soothe an upset stomach. It also has anti-microbial properties and can be used as a skin wash. Sweet marjoram, used lightly at the end of the cooking process, adds a nice, mellow flavor to vegetables such as spinach, beans, peas, and carrots. It is good in salads and herbal butters, as well as in vinaigrettes.

It is important to note that sweet marjoram provides more flavor than a similar amount of oregano.

THYME: *THYMUS SPP.*

Thymes are versatile and beautiful plants in the mint family (Lamiaceae or Labiatae), and they definitely have a place in every garden. Although there are more than 300 different species of thyme (*Thymus* spp.), common thyme (*T. vulgaris*) is most frequently used for cooking purposes. The creeping types, such as mother-of-thyme (*Thymus serpyllum*), will cover bricks and stones or low walls and can tolerate a small amount of foot traffic. Golden lemon thyme (*T. x citriodorus* 'Aureus'), another widely grown species of *Thymus*, has yellow-edged leaves and a strong lemon odor.

All thymes are low-growing hardy perennial herbs with very small leaves and tiny flowers ranging in color from white through pink to deep rose-magenta. Thymes are hardy from USDA zones 5 to 9 and typically require little attention once plants become established. All thymes are low-growing (6 to 10 inches in height at maturity), and most have a spreading or creeping habit. Some varieties will form an almost flat carpet of growth. Thymes are widely grown since they are extremely fragrant and attractive, grow well in tough, dry conditions, and can be used to add flavoring to various types of foods.

Thyme prefers to be grown in full sun and under dry growing conditions. This herb is ideal for container gardening, and since they are small in stature at maturity, many different varieties can be grown in one container. Trailing varieties are attractive when grown in pots, especially the golden and variegated varieties. Furthermore, a garden can oftentimes be made more appealing by growing various types of thymes amidst stones, as edging plants, or in containers. A collection of thymes having different colors provides an attractive feature for almost any garden. However, over a period of years thyme will tend to lose its vigor, and it is usually best to propagate a few new plants every year to use as replacements for the older plants.

Thyme can be propagated by layering, division, stem cuttings, or seed. This herb tends to grow slowly from seed and should be grown at least a few months before cutting. If a gardener is trying to form a thyme ground cover, plants should be spaced about 6 inches apart, so that it will quickly form a covering over the soil.

Established thyme plants can be harvested at any time just by simply snipping off a few stems. The cut stems are often bundled together and then dried in some manner (Fig. 16.15). The flowers are also edible and have the best flavor when they are first opening. As plants begin to flower, cut off the top portion of the plant and hang to dry in a shady place, such as a basement. Once the leaves are completely dry, strip them from the stems and store in a dark, cool place until ready to use. Although thyme can be frozen, it dries and keeps easily without freezing. Gardeners can harvest stems from thyme plants all summer, but it is important to reduce pruning as autumn approaches, as this can reduce their ability to survive the winter.

Most gardeners will grow a few different varieties of thyme in their gardens. Some are grown for the ornamental characters they add to a garden (e.g., variegated and golden types), while others, like the common thyme varieties are primarily grown for culinary uses. There are several varieties of thyme, and many have different variations in flower and foliage color. Suggested varieties include: *Thymus x citriodorus* 'Aureus' (golden lemon thyme; lemon-scented thyme with a strong lemon scent and golden yellow-edged variegated leaves), *Thymus pseudolanuginosus* 'Woolly Thyme' (very soft texture and spreads flat over the soil like a carpet; has no scent and is limited to USDA zones 5 to 6), *Thymus herba-barona* 'Caraway Thyme' (low growing, with pale pink flowers and the scent of caraway), and *Thymus serpyllum* 'Coccineus' (red creeping thyme; mat-forming with showy reddish purple flowers and bronze fall foliage; tolerates moderate foot traffic but requires a well-drained soil).

Figure 16.15 Stems of thyme bundled together for either fresh use or for drying.

The leaves of thyme are utilized in many dishes. Thyme is a Mediterranean herb. It holds its flavor in cooking and blends well with other flavors of this region, like garlic, olive oil, and tomatoes. Thyme is flavorful as a fresh and dried product. It makes a nice complement to tomato-based sauces, cheeses, pasta, salad dressings, soups and stews, stuffings, meat loafs, and vegetables. Thyme is an especially flavorful seasoning with poultry, fish, and eggs. It can also be used to flavor jellies, breads, vinegars, marinades, and other sauces. Thyme also has antiseptic and preservative properties. This herb has a long history of medicinal uses and has also been used in the preservation of meats. One of the essential oils in thyme is thymol, which is still used medicinally, especially in cough medications.

ALLIUMS AS CULINARY HERBS

There are several *Alliums* that are versatile herb plants which can be utilized throughout the garden. Although many *Alliums* are classified as vegetables, several are often grown as herbs for their culinary, ornamental, and medicinal uses. Several *Alliums* were covered previously in the Onion, Garlic, and other *Alliums* section of Chapter 6—Cool-Season Vegetables for the Home Garden. Specifically, garlic, leek, shallot, chives, and garlic chives are *Alliums* often grown in herb gardens, and these warrant mentioning again. The most popular culinary *Allium* herb is probably chives (*Allium schoenoprasum*), since it is easy to grow and has various culinary uses in the kitchen (Fig. 16.16). So, refer back to the Onion, Garlic, and other *Alliums* section of Chapter 6 for a discussion of these herbs.

***Figure 16.16** Chives are an* Allium *widely grown in containers, since they can be quickly clipped and then used for culinary purposes.*

CREDITS

Cover: Winter Vegetable Garden / Copyright © 2011 Depositphotos Inc./Steve Cukrov. Yellow potato root harvest / Copyright © 2013 Depositphotos Inc./Анна Курзаева. Carrot Growing Closeup / Copyright © 2012 Depositphotos Inc./Anna Subbotina. Urban Garden / Copyright © 2013 Depositphotos Inc./Daniel Korzeniewski. Vegetable garden / Copyright © 2013 Depositphotos Inc./Arina Habich.

Design image: Organic Carrots / Copyright © 2013 Depositphotos Inc./Mark Hammon.

interior images:

Map / Adapted from: http://commons.wikimedia.org/wiki/File:2012_USDA_Plant_Hardiness_Zone_Map_%28USA%29.jpg.

Seeding Lettuce / Copyright © Dwight Sipler (CC by 2.0) at http://commons.wikimedia.org/wiki/File:Seeding_lettuce_%28464342804%29.jpg.

Plugs ready for transplant / Copyright © Dwight Sipler (CC by 2.0) at http://commons.wikimedia.org/wiki/File:Plugs_ready_for_transplant_%287027255351%29.jpg.

Real Compost / Copyright © Kessner Photography (CC by 2.5) at http://commons.wikimedia.org/wiki/File:Real_Compost.jpg.

Worms just added / Mathijs Romans, http://commons.wikimedia.org/wiki/File:Worms_just_added.jpg. Copyright in the Public Domain.

Worm Power Raised Beds / Copyright © WormPower (CC BY-SA 3.0) at http://commons.wikimedia.org/wiki/File:WormPowerRaisedBeds.jpg.

Chinese cabbage seedling / Copyright © 2011 Depositphotos Inc./Denis Pogostin.

Chocolate mint / Copyright © 2011 Depositphotos Inc./Stephanie Frey.

Garlic cloves / Jon Sullivan, commons.wikimedia.org/wiki/File:Garlic_cloves_garlic.jpg. Copyright in the Public Domain.

Allium sativum / Copyright © H. Zell (CC BY-SA 3.0) at http://commons.wikimedia.org/wiki/File:Allium_sativum_003.JPG.

Potato-sprouts / Copyright © Sanjay Acharya (CC BY-SA 3.0) at http://commons.wikimedia.org/wiki/File:Potato-sprouts-base1.jpg.

Wortel bloeiwijze Daucus / Copyright © Rasbak (CC BY-SA 3.0) at http://commons.wikimedia.org/wiki/File:Wortel_bloeiwijze_Daucus_carota.jpg.

Artichoke Flower / Copyright © Little Mountain 5 (CC BY-SA 3.0) at http://commons.wikimedia.org/wiki/File:Artichoke_Flower.jpg.

Rapsdolde / Copyright © böhringer friedrich (CC BY-SA 2.5) at http://commons.wikimedia.org/wiki/File:Rapsdolde.JPG.

Ipomoea batatas / Copyright © H. Zell (CC BY-SA 3.0) at http://commons.wikimedia.org/wiki/File:Ipomoea_batatas_002.JPG.

Asparagus at Kroger / Copyright © Quadell (CC BY-SA 3.0) at http://commons.wikimedia.org/wiki/File:Asparagus_produce-1.jpg.

Asparagus in Birlingham / Copyright © Philip Halling (CC BY-SA 2.0) at http://commons.wikimedia.org/wiki/File:Asparagus_in_Birlingham_-_geograph.org.uk_-_965852.jpg.

Asparagus beetle / Copyright © Beentree (CC BY-SA 3.0) at http://commons.wikimedia.org/wiki/File:Crioceris_asparagi_1_beentree.jpg.

Asparagus beetle eggs / Copyright © Vmenkov (CC BY-SA 3.0) at http://commons.wikimedia.org/wiki/File:Asparagus_beetle_-_eggs_-_P1000034.JPG.

7carrots / Jonathunder, http://commons.wikimedia.org/wiki/File:7carrots.jpg. Cleared via the GNU Free Documentation License (http://commons.wikimedia.org/wiki/Commons:GNU_Free_Documentation_License_1.2).

Celeriac growing at RHS / Copyright © Mrmatiko (CC BY-SA 3.0) at http://commons.wikimedia.org/wiki/File:Celeriac_growing_at_RHS_Wisley.jpg.

Fennel / Copyright © Jamain (CC BY-SA 3.0) at http://commons.wikimedia.org/wiki/File:Fennel_J1.jpg.

Red cabbage / Copyright © Rick Heath (CC by 2.0) at http://commons.wikimedia.org/wiki/File:Red_cabbage.jpg.

فوفلم / Copyright © مغلبلا سراف (CC BY-SA 3.0) at http://commons.wikimedia.org/wiki/File:%D9%85%D9%84%D9%81%D9%88%D9%81.JPG.

Au marché / Dinkum, http://commons.wikimedia.org/wiki/File:Au_march%C3%A9_-_chou.JPG. Copyright in the Public Domain.

Broccoli / Copyright © Ixitixel (CC BY-SA 3.0) at http://commons.wikimedia.org/wiki/File:Broccoli2.jpg.

Broccoli bloem / Copyright © Rasbak (CC BY-SA 3.0) at http://commons.wikimedia.org/wiki/File:Broccoli_bloem.jpg.

Cauliflower-on-Plant-view / Copyright © Navaneeth Krishnan S (CC BY-SA 3.0) at http://commons.wikimedia.org/wiki/File:Cauliflower-on-Plant-view3.JPG.

Brussels-sprouts-on-stalk / Copyright © Phr (CC BY-SA 3.0) at http://commons.wikimedia.org/wiki/File:Brussels-sprouts-on-stalk.jpg.

Collard-Greens-Bundle / Evan-Amos, http://commons.wikimedia.org/wiki/File:Collard-Greens-Bundle.jpg. Copyright in the Public Domain.

Turnips / Copyright © thebittenword.com (CC by 2.0) at http://commons.wikimedia.org/wiki/File:Turnip_2622027.jpg.
A root of rutabaga / Garitzko, http://commons.wikimedia.org/wiki/File:Steckr%C3%BCbe_%28Brassica_napus_subsp_rapifera%29.jpg. Copyright in the Public Domain.
White and red radishes / Copyright © Visitor7 (CC BY-SA 3.0) at http://commons.wikimedia.org/wiki/File:Raphanus_sativus-1.jpg.
Picture of a pile of Daikon / Copyright © Chris 73 (CC BY-SA 3.0) at http://commons.wikimedia.org/wiki/File:Daikon.Japan.jpg.
Fresh Ground Horseradish / Copyright © Horseradish29 (CC BY-SA 3.0) at http://commons.wikimedia.org/wiki/File:Fresh_Ground_Horseradish.jpg.
Artichoke Cynara cardunculus / Copyright © Derek Ramsey (CC BY-SA 3.0) at http://commons.wikimedia.org/wiki/File:Artichoke_Cynara_cardunculus_Plants_2000px.jpg.
Iceberg lettuce / Copyright © Rasbak (CC BY-SA 3.0) at http://commons.wikimedia.org/wiki/File:Iceberg_lettuce_%28IJssla_krop%29.jpg.
Frisée / Derby, http://commons.wikimedia.org/wiki/File:Fris%C3%A9e.jpg. Copyright in the Public Domain.
Cichorium endivia / Copyright © Jamain (CC BY-SA 3.0) at http://commons.wikimedia.org/wiki/File:Cichorium_endivia_J1.JPG.
Garlic from a recent harvest / Copyright © Steve46814 (CC BY-SA 3.0) at http://commons.wikimedia.org/wiki/File:Korea-Goheunggun-Garlic_harvest_4196-06.JPG.
Scapes / Copyright © Dwight Sipler (CC by 2.0) at http://commons.wikimedia.org/wiki/File:Scapes_%28583373543%29.jpg.
Garlic cloves / Leon Brooks, http://commons.wikimedia.org/wiki/File:Garlic_cloves.jpg. Copyright in the Public Domain.
Leeks at Kroger / Copyright © Quadell (CC BY-SA 3.0) at http://commons.wikimedia.org/wiki/File:Leeks_produce-1.jpg.
Onions / Hans Braxmeier, http://commons.wikimedia.org/wiki/File:Shallot-5774_-_Hans_Braxmeier.jpg. Copyright in the Public Domain.
Different potato varieties / Scott Bauer, http://commons.wikimedia.org/wiki/File:Patates.jpg. Copyright in the Public Domain.
Eggs of the Colorado potato / Copyright © Björn Appel (CC BY-SA 3.0) at http://commons.wikimedia.org/wiki/File:Leptinotarsa-decemlineata-eggs.jpg.
Leptinotarsa decemlineata / Copyright © Stephan Czuratis (CC BY-SA 2.5) at http://commons.wikimedia.org/wiki/File:2006-07-16_larve2.jpg.
A Colorado potato beetle / Copyright © Björn Appel (CC BY-SA 3.0) at http://commons.wikimedia.org/wiki/File:Leptinotarsa-decemlineata-beatle.jpg.
Rabarber stelen / Copyright © Rasbak (CC BY-SA 3.0) at http://commons.wikimedia.org/wiki/File:Rabarber_stelen.jpg.
Rheum rhabarbarum / Copyright © Dieter Weber (CC BY-SA 3.0) at http://commons.wikimedia.org/wiki/File:Rheum_rhabarbarum.2006-04-27.uellue.jpg.
Tetragonia tetragonioides / Copyright © Forest Starr & Kim Starr (CC by 3.0) at http://commons.wikimedia.org/wiki/File:Starr_050222-4149_Tetragonia_tetragonioides.jpg.
Chard in the Victory Garden / Copyright © mercedesfromtheeighties (CC BY-SA 2.0) at http://en.wikipedia.org/wiki/File:Chard_in_the_Victory_Garden.jpg.
Cucumis sativus Melita / Copyright © Rasbak (CC BY-SA 3.0) at http://commons.wikimedia.org/wiki/File:Komkommer_Cucumis_sativus_%27Melita%27.jpg.
Cucumis sativus / Copyright © Jerzy Opioła (CC BY-SA 3.0) at http://commons.wikimedia.org/wiki/File:Cucumis_sativus.jpg.
Gurkenproduktion / Copyright © Karin Weigel (CC BY-SA 3.0) at http://commons.wikimedia.org/wiki/File:Gurkenproduktion.JPG.
Lemon cucumber / Copyright © Jamain (CC BY-SA 3.0) at http://commons.wikimedia.org/wiki/File:Lemon_cucumber_J1.JPG.
Armenian cucumbers / Copyright © _e.t (CC BY-SA 2.0) at http://commons.wikimedia.org/wiki/File:Armenian_cucumbers.jpeg.
Chinese Cucumber / Copyright © 2011 Depositphotos Inc./Digifuture.
Melon plant / Copyright © Lionel Rich (CC BY-SA 2.0 FR) at http://commons.wikimedia.org/wiki/File:Melon_plant.jpg.
Cantaloupes / Scott Bauer, http://commons.wikimedia.org/wiki/File:Cantaloupes.jpg. Copyright in the Public Domain.
Honeydew / Yotoen, http://commons.wikimedia.org/wiki/File:Honeydew.jpg. Copyright in the Public Domain.
Honeydew.Melon / Copyright © Church of emacs (CC BY-SA 3.0) at http://commons.wikimedia.org/wiki/File:Honeydew.Melon.2.jpg.
Galia melon / Tomwsulcer, http://commons.wikimedia.org/wiki/File:Galia_melon_in_a_supermarket.jpg. Copyright in the Public Domain.
Futuro Melone / Copyright © Frank C. Müller (CC BY-SA 3.0) at http://commons.wikimedia.org/wiki/File:Futuro_Melone_1_%28fcm%29.jpg.
yellow squash / Copyright © 2012 Depositphotos Inc./John Trax.
Squash / Copyright © 2013 Depositphotos Inc./Digidream.
USDA petit pan squash / http://commons.wikimedia.org/wiki/File:USDA_petit_pan_squash.jpg. Copyright in the Public Domain.
Fresh marrow fruits with green leaves / Copyright © 2011 Depositphotos Inc./Yasonya.
Vegetable marrows / Copyright © 2012 Depositphotos Inc./Galina Afanasyeva.
Zucchini / Copyright © 2011 Depositphotos Inc./Newlight.
Yellow squash / Downtowngal, http://commons.wikimedia.org/wiki/File:Yellow_squash_with_green_spots_on_the_skin.jpg. Copyright in the Public Domain.
Acorn Squash / Copyright © 2011 Depositphotos Inc./Cornelia Pithart.
Spaghetti squash / Copyright © 2013 Depositphotos Inc./John Trax.
Delicata squash / Copyright © 2013 Depositphotos Inc./John Trax.
Butternut squash / Copyright © 2011 Depositphotos Inc./Cornelia Pithart.
Red kuri squash / Copyright © 2012 Depositphotos Inc./Cynoclub.
Pink banana squash / Copyright © 2011 Depositphotos Inc./Chris Brignell.
Buttercup winter squash / Copyright © 2012 Depositphotos Inc./John Trax.
Blue hubbard squash / Copyright © 2012 Depositphotos Inc./John Trax.
Squash / Copyright © 2012 Depositphotos Inc./Masaki Nagura.
Jarrahdale squash / Copyright © 2012 Depositphotos Inc./John Trax.

Red Turban Squash / Copyright © 2011 Depositphotos Inc./ Cornelia Pithart.
Green striped cushaw squash / Copyright © 2012 Depositphotos Inc./Larry Masseth.
Pumpkin Patch / Copyright © 2013 Depositphotos Inc./Mary Sligar.
Long Island Cheese Pumpkin / Copyright © 2012 Depositphotos Inc./Fabrizio Troiani.
Squash / Copyright © Evelyn Simak (CC BY-SA 2.0) at http:// commons.wikimedia.org/wiki/File:Squash_%28Cucurbita_ spec.%29_-_geograph.org.uk_-_935471.jpg.
Gourds / Copyright © 2013 Depositphotos Inc./John Anderson.
Green gourd / Copyright © 2013 Depositphotos Inc./Tatiana Belova.
Smooth loofah / Copyright © 2012 Depositphotos Inc./Leung Cho Pan.
Watermelon / Copyright © 2012 Depositphotos Inc./Barbara Helgason.
Sunroot growing / Copyright © Paul Fenwick (CC BY-SA 3.0) at http://commons.wikimedia.org/wiki/File:Sunroot_growing.jpg.
Tobinambur / Hagen Graebner, http://commons.wikimedia.org/ wiki/File:Tobinambur1.jpg.
yellow okra blossom / Copyright © 2012 Depositphotos Inc./ Bennyartist.
Bucket of raw okra pods / Bill Tarpenning, http://commons.wikimedia.org/wiki/File:Bucket_of_raw_okra_pods.jpg. Copyright in the Public Domain.
Scarlet Runner Beans / Copyright © 2012 Depositphotos Inc./ Verena Matthew.
Cranberry Beans / Copyright © 2013 Depositphotos Inc./Bryan Mullennix.
soybean / Copyright © 2013 Depositphotos Inc./HONGYAN JIANG.
Yardlong bean plantation / Copyright © 2013 Depositphotos Inc./ Boonsom.
Tomatoes / Copyright © Jamain (CC BY-SA 3.0) at http://commons.wikimedia.org/wiki/File:Tomatoes_J1.JPG.
Tomato garden / Copyright © 2011 Depositphotos Inc./Tanya Tsoncheva.
HornWORM / Khanm83, http://commons.wikimedia.org/wiki/ File:HornWORM_%283%29.JPG. Copyright in the Public Domain.
Peppers / Daderot, http://commons.wikimedia.org/wiki/ File:Peppers,_Cambridge_MA_-_DSC05394.jpg. Copyright in the Public Domain.
Peppers / Jon Sullivan, http://commons.wikimedia.org/wiki/ File:Chilies_chillies_peppers_seeds_pods.jpg. Copyright in the Public Domain.
Dried Thai Dragon Peppers / Copyright © Adam (CC BY-SA 2.0) at http://commons.wikimedia.org/wiki/File:Dried_Thai_ Dragon_Peppers.jpg.
Sweet corn / Copyright © 2012 Depositphotos Inc./John Trax.
Corn / Copyright © Carl Lewis (CC by 2.0) at http://commons.wikimedia.org/wiki/File:Corn_%28Zea_mays%29_-_ Aug._2007.jpg.
Weeding garden / Copyright © 2010 Depositphotos Inc./Jure Porenta.
Green Fertilizer / Copyright © Mark Edwards (CC BY-SA 2.0) at http://commons.wikimedia.org/wiki/File:Green_Fertilizer_-_ geograph.org.uk_-_259856.jpg.
Wheel Bug / Copyright © jeffreyw (CC by 2.0) at http://commons.wikimedia.org/wiki/File:Wheel_Bug_aka_Assassin_ Bug_%285060660429%29.jpg.
Larva of Coccinella / Copyright © Stu Phillips (CC BY-SA 3.0) at http://commons.wikimedia.org/wiki/File:Coccinella_septempunctata,_7_spot_Lady_Beetle,_larva,_UK.jpg.
Broscus cephalotes / Copyright © Aiwok (CC BY-SA 3.0) at http://commons.wikimedia.org/wiki/File:Broscus_cephalotes_1.JPG.
Praying Mantis / Copyright © Amos Oliver Doyle (CC BY-SA 3.0) at http://commons.wikimedia.org/wiki/File:Praying_ Mantis_in_Illinois.jpg.
Green lacewing / Copyright © Bruce Marlin (CC BY-SA 2.5) at http://commons.wikimedia.org/wiki/File:Green_lacewing.jpg.
Geocoris punctipes / Jack Dykinga, http://commons.wikimedia.org/wiki/File:Geocoris_punctipes.jpg. Copyright in the Public Domain.
Tomato Hornworm / Copyright © Max Wahrhaftig (CC BY-SA 3.0) at http://commons.wikimedia.org/wiki/File:Tomato_ Hornworm_5.JPG.
Parasitic wasp Cotesia / Copyright © Beatriz Moisset (CC BY-SA 3.0) at http://commons.wikimedia.org/wiki/ File:Cotesia9061.8.15.07.c.jpg.
Raised bed / Copyright © Srl (CC BY-SA 3.0) at http://commons.wikimedia.org/wiki/File:Raised_bed.jpg.
Container Cardening / Copyright © 2013 Depositphotos Inc./Eric Krouse.
Vegetable growing in old used tire pot / Copyright © 2013 Depositphotos Inc./Praiwun Thungsarn.
Detail of watering can with pouring water / Copyright © 2013 Depositphotos Inc./Magdalena Kucova.
Cultivation of cucumber sprouts in a pot / Copyright © 2012 Depositphotos Inc./MIKHAIL IAKOVLEV.
Bolivian Rainbow fruits / Copyright © Tommi Nummelin (CC BY-SA 3.0) at http://commons.wikimedia.org/wiki/ File:Bolivian_Rainbow_fruits.JPG.
Tomates Auch / Jean-noël Lafargue, http://commons.wikimedia.org/wiki/File:Jardin_potager_7_tomates_buffle.jpg.
Detroit dark red beets / Copyright © BriannaWalther (CC BY-SA 3.0) at http://commons.wikimedia.org/wiki/ File:Detroitdarkredbeets.png.
Piel de Sapo Melon / Copyright © 2013 Depositphotos Inc./ Antonio Ribeiro.
Tabasco peppers / http://commons.wikimedia.org/wiki/ File:Tabasco_peppers.JPG. Copyright in the Public Domain.
Russet potato / Copyright © Steve Caruso (CC BY-SA 2.5) at http://commons.wikimedia.org/wiki/File:Russet_potato_.jpg.
Yellow Pear Tomatoes / Copyright © cygnus921 (CC by 2.0) at http://commons.wikimedia.org/wiki/File:Yellow_Pear_ Tomatoes_012.jpg.
Pomodoro Costoluto / Brücke-Osteuropa, http://commons.wikimedia.org/wiki/File:Pomodoro_Costoluto_Genovese.JPG. Copyright in the Public Domain.

INDEX

C

D

E

F

G

H

P

T

9 781516 552955